Y0-AAA-461

FINITE SIMPLE GROUPS II

Proceedings of a London Mathematical Society Research Symposium in Finite Simple Groups held at the University of Durham in July-August 1978.

FINITE SIMPLE GROUPS II

edited by

Michael J. Collins
University College, Oxford

1980

ACADEMIC PRESS
A Subsidiary of Harcourt Brace Jovanovich, Publishers
London New York Toronto Sydney San Francisco

Academic Press Inc. (London) Ltd
24–28 Oval Road
London NW1

US edition published by
Academic Press Inc.
111 Fifth Avenue
New York, New York 10003

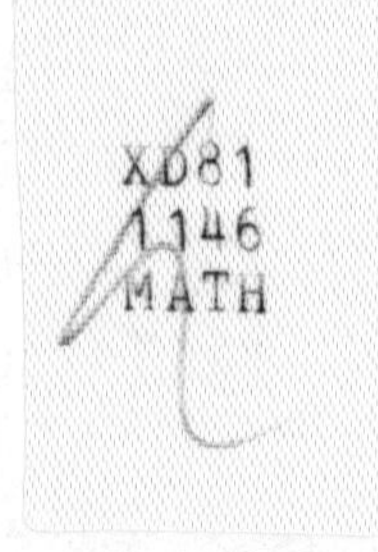

British Library Cataloguing in Publication Data

Research Symposium in Finite Simple Groups,
University of Durham, 1978
Finite simple groups II.
1. Finite groups—Congresses
I. Title II. Collins, M J III. London
Mathematical Society
512′.2 QA171 77-149703

ISBN 0-12-181480-7

Printed in Great Britain by
John Wright & Sons, Ltd., at the Stonebridge Press, Bristol

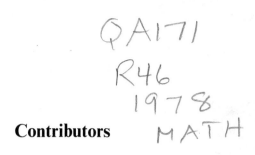

Contributors

M. Aschbacher, *Department of Mathematics, California Institute of Technology, Pasadena, CA 91125, U.S.A.*

R. W. Carter, *Mathematics Institute, University of Warwick, Coventry, CV4 7AL, U.K.*

M. J. Collins, *University College, Oxford, OX1 4BH, U.K.*

G. Glauberman, *Department of Mathematics, University of Chicago, Chicago, Illinois 60637, U.S.A.*

D. M. Goldschmidt, *Department of Mathematics, University of California, Berkeley, CA 94720, U.S.A.*

D. Gorenstein, *Department of Mathematics, Rutgers University, Hill Center, New Brunswick, N.J. 08903, U.S.A.*

R. L. Griess, *Department of Mathematics, University of Michigan, Ann Arbor, Michigan 48104, U.S.A.*

J. E. Humphreys, *Department of Mathematics and Statistics, University of Massachusetts, GRC Tower, Amherst, Mass 01003, U.S.A.*

J. C. Jantzen, *Mathematisches Institut, Universitat Bonn, 5300 Bonn, W. Germany.*

R. Lyons, *Department of Mathematics, Rutgers University, Hill Center, New Brunswick, N.J. 08903, U.S.A.*

G. Mason, *Department of Mathematics, University of California, Santa Cruz, CA 95064, U.S.A.*

G. M. Seitz, *Department of Mathematics, University of Oregon, Eugene, Oregon 97403, U.S.A.*

C. C. Sims, *Department of Mathematics, Rutgers University, Hill Center, New Brunswick, N.J. 08903, U.S.A.*

R. Solomon, *Department of Mathematics, Ohio State University, Colombus, Ohio 43210, U.S.A.*

J. G. Thompson, *Department of Pure Mathematics and Mathematical Statistics, 16 Mill Lane, Cambridge, CB2 1SB, U.K.*

F. Timmesfeld, *Mathematisches Institut, Justus-Liebig Universität, D-6300 Giessen, W. Germany.*

J. Tits, *Collège de France, 11, pl. Marcelin-Berthelot, 75231 Paris Cedex 05, France.*

Preface

This book is the outcome of a Research Symposium in Finite Simple Groups which was held at the University of Durham, July 31–August 10, 1978, under the auspices of the London Mathematical Society. The dominant area of discussion was the classification of simple groups, a programme which is now almost complete. In our original proposal for the symposium, we did not feel that this should be the exclusive topic, and there were also lectures and seminars on the representation theory of simple groups of Lie type and on geometry and sporadic groups. This book is correspondingly divided into three parts.

Part A covers finite simple groups and their classification. While formally this book is a proceedings of the symposium, this material has been presented in a form which will serve as a survey of the eventual complete classification, and as an introduction to the more detailed arguments that appear in the individual research papers in the literature, both now and in the future. This is the purpose of the first eight chapters, and, with this in mind, the opportunity has been taken during the preparation for publication to update the factual information presented as much as possible. Because of their origin, the emphasis is naturally on recent work, and no details are given of much of the necessary work which preceded Gorenstein's formal programme for a classification, though the first chapter does include an historical background and a brief account of the known simple groups. The last two chapters are a little different, and are related more to a re-examination and axiomatization of the ideas and methods employed.

The core of this part of the symposium consisted of series of lectures given by M. Aschbacher, G. Glauberman and D. Gorenstein, and they have contributed their lectures, in Gorenstein's case with R. Lyons on their joint work. There were a number of seminars on various aspects of the classification and related topics; for most of these, their organisers have written unified accounts of the subjects. In addition, Gorenstein gave an opening lecture as an introduction and survey of the entire classification programme. In writing the corresponding sections of Chapter 1 (Sections 4–6), I have leaned heavily on his notes, but have considerably expanded on his account so as to link the remaining chapters, and give the necessary notation and definitions not special to a particular chapter. In addition, because of the unity of the material, I have prepared a common list of references for Part A; this should *not* be considered as a complete bibliography for the subject. Books are separated from articles; in particular, most of the basic group theory which is freely assumed can be found in the texts by Gorenstein [G] and Huppert [H], while a more specialised background can be found in Carter [C] and the Proceedings of the 1969 Oxford L.M.S. Instructional Conference [FSG].

The remainder of the book covers the other topics. R. W. Carter and J. E. Humphreys gave series of lectures on the representation theory of finite simple groups of Lie type as "surveys for group theorists". These appear in Part B, together with an account by J. Jantzen of his own work, given in the seminar on representation theory. We invited J. Tits and J. G. Thompson to give lectures. Tits gave four lectures entitled "The geometry of sporadic groups". In contributing two chapters, he has excluded his lecture on the Mathieu groups. Thompson lectured on an application of representation theory to questions about finite projective planes and his work appears here. These comprise Part C, together with a general account of his computer constructions of sporadic groups which Sims gave to the sporadic groups seminar. (This seminar also included a considerable discussion of Fischer's Monster, but this is not included; in fact, relatively little was known then as compared with now!)

It is a pleasure to acknowledge the financial support given to the Symposium by the Science Research Council, and also to thank the National Science Foundation for their partial support of many of the participants from the United States. Personally, I should like to thank Roger Carter for his advice and assistance with the programme as co-organiser, Tom Willmore and Lyndon Woodward in Durham, and Valerie Willoughby for her assistance with the organisation in Oxford. Not least, I should like to thank all those who participated in the Symposium and contributed to its success.

In the preparation of this volume, I am extremely grateful to all the contributors for their cooperation and to the Academic Press for their patience in waiting for the manuscript. I am also grateful to the California Institute of Technology where I have held a visiting appointment while preparing this volume, to David Wales for his advice and criticism while I was writing Chapter 1, and to Frances Williams and Lillian Chappelle who have typed portions of the manuscript.

Pasadena, California
October 1979

M. J. Collins

Contents

Part B: Representation Theory of Groups of Lie Type

Notation and Terminology

Much of the basic notation which is used in this book is standard, and the following list is not complete, but only intended for clarification. (Definitions required for only one chapter and given there are excluded.)

$\{\cdots \mid \cdots\}$	set of $\cdots$ such that $\cdots$
$A \subseteq B (B \supseteq A)$	A is a subset of B (also used without distinction for subgroups)
$A \subset B (B \supset A)$	A is a proper subset of B
$A - B$	$\{a \in A \mid a \notin B\}$

If G is a group:

$H \lhd G (H \unlhd G)$	H is a normal subgroup of G
$H \lhd\lhd G (H \unlhd\unlhd G)$	H is a subnormal subgroup of G
$o(g)$	order of an element $g \in G$
x^g	$g^{-1}xg \quad (g, x \in G)$
A^g	$g^{-1}Ag \quad (g \in G, A \subseteq G)$
$\langle \cdots \mid \cdots \rangle$	group generated by $\{\cdots \mid \cdots\}$
$[x, y]$	$x^{-1}y^{-1}xy \quad (x, y \in G)$
$[H, K]$	$\langle [h, k] \mid h \in H, h \in K \rangle \quad (H, K \subseteq G)$
$[H, K, L]$	$[[H, K], L]$
$G' = G^{(1)}$	$[G, G]$
$G^{(n)}$	$[G^{(n-1)}, G^{(n-1)}]$ (inductively)
$G^{(\infty)}$	$\bigcap G^{(n)} \quad (n \in \mathbf{N})$
$C_G(H)$	centralizer of H in G
$N_G(H)$	normalizer of H in G
$Z(G)$	centre of G
$F(G)$	Fitting subgroup of G
И$_G(H; \pi)$	set of π-subgroups of G normalized by a subgroup H (π set of primes)
И$^*_G(H; \pi)$	maximal elements of И$_G(H; \pi)$ under inclusion

If p is a prime, p' denotes the complementary set of primes:

$O_p(G)$	largest normal p-subgroup of G
$O_{p'}(G)$	largest normal p'-subgroup of G
$O_{p',p}(G)$	subgroup of G containing $O_{p'}(G)$ such that $O_{p',p}(G)/O_{p'}(G) = O_p(G/O_{p'}(G))$
$O(G) = O_{2'}(G)$	*core* of G
$Z^*(G)$	subgroup of G containing $O(G)$ such that $Z^*(G)/O(G) = Z(G/O(G))$
$O^p(G)$	smallest normal subgroup of G of index a power of p

$O^{p'}(G)$	smallest normal subgroup of G of index prime to p
$Syl_p(G)$	set of Sylow p-subgroups of G
If P is a p-group:	
$\Omega_1(P)$	$\langle x \in P \mid x^p = 1 \rangle$
$\mho^1(P)$	$\langle x^p \mid x \in P \rangle$
$\Phi(P)$	Frattini subgroup of P
Also:	
$F_q = GF(q)$	finite field of order q
$H \circ K$	central product of groups H, K
$H \wr K$	wreathed product of H by K
Z_n	cyclic group of order n
D_{2n}	dihedral group of order $2n$
S_n	symmetric group of degree n
A_n	alternating group of degree n
$G(q)$	group of Lie type defined over $GF(q)$
$Chev(p)$	$\{G(q) \mid p \text{ divides } q\}$ (includes groups of Ree type if $p = 3$)

See Chapter 1, Section 3, for detailed notation for the known simple groups.

The following table contains notation which is defined in Chapter 1.

Notation	Meaning	Page of definition
$m(P)$	rank of a p-group P	26
$r(P)$	sectional rank of P	27
$m_p(G)$	p-rank of G	27
$m_{2,p}(G)$	2-local p-rank of G	36
$e(G)$	odd 2-local rank of G	36
$SCN(P)$		27
$SCN_3(P)$		27
$\Gamma_{P,k}(G)$	k-generated (p-)core of G	28
$\Gamma^0_{P,2}(G)$	weak 2-generated p-core	39
$F^*(G)$	generalized Fitting subgroup of G	32
$E(G) = L(G)$	layer (extended socle) of G	32
$L_{2'}(G)$	2-layer of G	33
$B(G)$		33

The following terminology, additional to that above, is introduced in Chapter 1.

Part A

Finite Simple Groups and Their Classification

1

Introduction: A Survey of the Classification Project

MICHAEL J. COLLINS

1. Background

The current attempt to classify the finite simple groups has its origin, as a systematic programme, in Brauer's address at the International Congress in Amsterdam in 1954 [33]. Brauer had proved that if G was a finite simple group of even order and t was an involution (element of order 2) in G, then

$$|G| \leqslant (|C_G(t)|^2)!$$

Thus, if one specifies a group H having centre of even order, there can be only finitely many simple groups G which contain an involution t for which $C_G(t)$ is isomorphic to H. The bound in question is not a useful one, but its mere existence encouraged Brauer to suggest a systematic characterization of known simple groups in this way and he was able to obtain such results, using character-theoretic methods to obtain enough information to construct G from a knowledge of H in certain specific cases.

What does one mean by "classification of finite simple groups"? Results such as those obtained by Brauer are often termed *specific* characterizations, even though H may have been specified only as a member of a class of groups. Indeed it can happen, as we shall see later, that a judicious choice of a group H may result in the characterization of a previously unknown simple group. By contrast, a *general* characterization will consist of the determination of those simple groups (or suitable information about unknown groups) satisfying some general inductive property; for example, having odd order (yielding the empty set) or having abelian Sylow 2-subgroups (allowing for the then indeterminacy over the Ree groups, for which the term *group of Ree type* was introduced). Because the nature of proofs is usually inductive, it is normally the case that one characterizes *all* groups with a given property and deduces the list of simple groups as a corollary; for example, one often knows nothing about the structure of normal subgroups of odd order, and this is why the term *characterization* is used. However, the ultimate *classification* theorem will be one which says, "If G is a finite simple group, then G is isomorphic to a known group".

In practice, we may settle for a little less. In effect, the current attempt to classify finite simple groups, as will be described in this book, is a proof by induction. One should start with a minimal counterexample. However, the methods are such that one can ignore certain "indeterminacies"; all that is required is that one know those properties of a possible unknown simple group that are needed for the arguments. Then the question of actual existence or uniqueness can be left as a separate problem; at worst, resolution of this question can result in some of the previous arguments becoming superfluous.

The first ingredient needed in this process, therefore, is a list of the known simple groups. (Here, as throughout Part A, we shall suppress the word "finite" as understood.) The search for simple groups has always been a part of the subject, as much as a source of examples as in their own right. By early this century, the finite analogues of the classical groups had been discovered, plus two exceptional families found by Dickson (now known as $G_2(q)$ and $E_6(q)$, but discovered as groups leaving certain forms invariant), in addition to the alternating groups and the five Mathieu groups. The existence of the Mathieu groups was long questioned; they were found as multiply transitive permutation groups and were described by Burnside as "apparently sporadic groups" since they did not fit into any of the infinite families.

Despite the work of Burnside and Frobenius, there was an enormous gulf between the general theory of finite groups that developed around that time and the particular knowledge of the simple groups then known; in particular, and highly relevant to modern study, the conjectures of Frobenius and Burnside on the nilpotence of the kernel of a Frobenius group and the solubility of groups of odd order, respectively, were, as far as we know, completely untouched. In both cases, a minimal counterexample is simple. For Burnside's conjecture, this is trivial; in the case of Frobenius' conjecture, one first needs to know that the conjecture holds for soluble groups, and this fact is usually ascribed to Witt.

While progress in group theory continued, most notably p-groups, soluble groups, permutation groups and representation theory, the specific study of simple groups remained largely dormant for nearly fifty years. Then three major developments regenerated interest in the subject.

The first was the application of character theory. Brauer's project has already been mentioned, but general results were also obtained. The first general characterization of the groups $L_2(q)$ was given by Brauer, Suzuki and Wall [35] using the theory of exceptional characters, and this was followed by two nonexistence theorems. Suzuki [205] showed that there cannot be a simple group of odd order in which the centralizer of every element other than the identity is abelian ((CA)-group). This was the first special case of the Burnside conjecture to be treated and, as is described in Chapter 9, can be viewed as a basic model for its ultimate solution. Then Brauer and Suzuki [34] showed that a finite group having either quaternion or generalized quaternion Sylow 2-subgroups cannot be simple.

Next, in 1955, the first new simple groups since Dickson's time were discovered by Chevalley [42]. These fall into a framework which now includes all the known infinite families with the exception of the alternating groups. Chevalley showed that, associated with each simple complex Lie algebra, one could define a group over an arbitrary field which, except in certain small cases, was simple, and the finite fields yield a *family* of groups for each such Lie algebra—these were classified by Cartan in the last century. With the exception of the isomorphisms already known for classical groups that occurred in this construction, the groups were distinct and, in particular, included three new families. Steinberg and others were then able to construct the *twisted* groups as the fixed point subgroups of certain automorphisms of these groups. These groups are known collectively as the *groups of Lie type*; we shall restrict the term *Chevalley groups* to the obvious subclass. In Section 3 we shall give a brief account of these groups, together with their identification with the classical groups where applicable. For a complete account, the reader is referred to Carter's book [C] or, for an alternative approach, to Steinberg [197], and for fuller surveys to Carter [39] or Curtis' chapter in [FSG].

In the same period, working more geometrically, Tits established his axioms for a BN-pair and a theory of buildings (see Chapter 15). The finite simple groups of Lie type satisfy the axioms for a BN-pair, and Tits was able to associate with each such group a building of spherical type. Since a doubly transitive permutation group possesses a BN-pair trivially, for a brief period it was hoped that this might be the "universal property" required to determine all finite simple groups, but Janko's discovery of the simple group of order 175 560 provided a counterexample. However, Tits [T] has been able to show that a finite simple group which possesses a BN-pair of rank at least three is necessarily a group of Lie type, though he observes that in the course of a group-theoretic characterization, by the time that one can show that a group does have a BN-pair, enough information is available to construct the group directly without appealing to his general result.

The third development was the work of Thompson. Using ideas developed by P. Hall in his work on soluble groups and the techniques introduced in the Hall–Higman paper [114], but to an extent never previously envisaged, Thompson first settled the Frobenius conjecture. This was an immediate consequence of his normal p-complement theorem ([210] and, later, [211]), as also was the solubility of a group having a nilpotent maximal subgroup of odd order. The latter yielded another special case of the Burnside conjecture, as did the extension, by Feit, M. Hall and Thompson [51], of Suzuki's work to the solubility of (CN)-groups of odd order, groups in which every nonidentity element has nilpotent centralizer. While the group-theoretic reductions in Suzuki's case were trivial, here they were not, and required the techniques that Thompson had developed. With this case solved, however, Feit and Thompson [53] were able to go on to the complete solution of the Burnside conjecture. These results led to an upsurge in interest in the classification of finite simple groups, and the Feit–Thompson theorem of course added impetus to the Brauer project.

The influence of Thompson on the subsequent strategy for classifying finite simple groups cannot be overemphasized. In the case of the odd order theorem, a counterexample of minimal order is a *minimal* simple group—namely, a simple group, all of whose proper subgroups are soluble. We shall not go into this further, but refer to Chapter 9 of this book or to Chapter 16 in [G]; the point that we wish to make is that Thompson's next step was to classify the minimal simple groups of *even* order. However, he saw that it would be no more difficult to characterize the larger class of *N-groups* [214]—a group G is an N-group if, whenever H is a subgroup such that $H = N_G(P)$ for some nonidentity p-subgroup P of G, then H is soluble. (Note that G need not be simple.) In an arbitrary group, the normalizer of nonidentity p-subgroup is called a *(p-)local subgroup*. In the characterization of N-groups, only the local subgroups, in particular the 2-local subgroups, play a role, and the basic subdivisions of Thompson's work carry over to the general classification programme.

The 1960's saw the characterization of most of the then known finite simple groups in terms of the structure of the centralizer of an involution. In some cases, as for example in the work of Suzuki [207] culminating in his determination of those simple groups in which the centralizer of every involution has a normal Sylow 2-subgroup, the characterization was in terms of a general property. (It is worth remarking that during this work, while studying certain doubly transitive permutation groups, he discovered the groups that bear his name—the only infinite family found as the result of a characterization theorem!) In other cases, it was by specific characterizations following Brauer's project. Here, Glauberman's celebrated Z^*-theorem [77] proved invaluable as a first step, forcing every involution to be conjugate to another involution in its centralizer, and this was used in conjunction with Alperin's fusion theorem [1]. These specific characterizations also gave rise to the discovery of some sporadic simple groups, the first so found being Janko's group of order 175 560.

There were also general characterization theorems; in particular, we mention the characterizations of groups with dihedral Sylow 2-subgroups by Gorenstein and Walter [103] and of groups with abelian Sylow 2-subgroups by Walter [228]. The former is modelled on the odd order paper and is needed in the N-group classification; it serves as the link with the ideas which are needed for subsequent characterizations of simple groups whose Sylow 2-subgroups are, in a sense which we shall describe in Section 4, not too big. The latter focused attention for the first time on *components*, which we will define in Section 5; indeed, many of the problems which have to be faced in the general classification arise for the first time here.

The 1960's reintroduced Burnside's term *sporadic* groups. Thirteen sporadic simple groups were discovered; they are listed together with others found in Section 3. Some arise out of the Brauer project, while others were found using methods developed by M. Hall to show the existence of the Hall–Janko group as a permutation group of rank 3. By contrast, the groups discovered by Conway and Fischer arose in totally different ways yet, seemingly as a coincidence, Conway's biggest group contains not only his other two, but also the previously discovered rank 3 permutation groups.

At that point, it was not clear what effect these discoveries, with their potentially unlimited continuation, would have on the attempt to determine all finite simple groups, and for a period the emphasis switched to the study of general properties of abstract groups; in particular, it is appropriate to mention the work of Glauberman on the relationship between the local and global properties of finite groups, and the development of signalizer functors and their application to the study of the structure of the centralizer of an involution by Gorenstein and Walter. The reader is referred to the appropriate chapters of [FSG] for an account.

The most important innovations were due to Bender. First he characterized groups which contain a (proper) *strongly embedded subgroup* [29]: by definition, a proper subgroup M of a group G is strongly embedded in G if $|M|$ is even and if

(i) $C_G(t) \subseteq M$ for every involution t of M, and
(ii) $N_G(S) \subseteq M$ for each Sylow 2-subgroup S of M.

An equivalent definition is that

(iii) $|M|$ is even and $|M \cap M^g|$ is odd whenever $g \in G - M$,

and this is the same as saying that in the action of G on the cosets of M, involutions fix precisely one point. It is in this light that Bender attacked the problem and, in particular, he showed that the only simple groups which arise are the groups $L_2(q)$, $U_3(q)$ and $Sz(q)$, for q even. This theorem serves two purposes. First, it has been applied in many characterization theorems for which these groups form the desired conclusion, and for this reason they are often referred to as the *Bender groups*. Secondly, in proving various general theorems by way of contradiction, one can often produce a strongly embedded subgroup in a counterexample, and this (or the weaker condition of a *proper 2-generated core*, which will be defined in Section 4) is a target in applying signalizer functor methods.

The other innovation was what is now called the *Bender method*. In two short papers ([27] and [28]), Bender effected dramatic reductions first in a portion of the odd order paper (see Chapter 9), and then in Walter's characterization of groups with abelian Sylow 2-subgroups. The Bender method contains considerable technical sophistication; however it also introduced a new philosophy into the approach to characterization problems. While the basis was still to start with a minimal counterexample, instead of looking at small subgroups and building up, he started by studying a suitable collection of *maximal* subgroups and investigating their intersections.

Of a rather different nature were the ideas being introduced by Fischer. It was well known that two involutions in a group generate a dihedral group, and this had been heavily exploited by Brauer and others. However, Fischer studied properties related to the generation of a group by an *entire* conjugacy class; in particular, his first groups were discovered as groups generated by 3-transpositions, a conjugacy class of involutions the product of any two of which has order at most 3 [67]. Fischer was able to characterize all groups satisfying this property together with a minor restriction not relevant to the determination of the appropriate *simple* groups. Subsequent generalizations due to Aschbacher and Timmesfeld have provided a key step in identifying groups of Lie type defined over fields of characteristic two.

This is a brief description of some of the major steps prior to 1970, in particular those that have shaped the classification programme as it has developed. The description is far from complete, but should provide a sufficient background for what is to follow. The year 1970 is a convenient dividing line in view of Feit's survey article [50] and this book's predecessor [FSG]. The previous year had seen Thompson's work on quadratic pairs [215] which was to have an influence on almost all subsequent characterizations of groups of Lie type at the stage of their identification; the following two years were to see the important work of Goldschmidt first to improve on the signalizer functor theorems and then to extend Bender's arguments from the abelian Sylow 2-subgroup problem to a characterization of groups containing a strongly closed abelian 2-subgroup [91], and Aschbacher's extension of the strongly embedded subgroup theorem to groups having a proper 2-generated core [9].

With these results available to him and with his own work with Walter, Gorenstein proposed, in a series of lectures at the University of Chicago in 1972, a sixteen point programme for the classification of finite simple groups.

This has been published as an appendix to his recent article [98]. The following year, in lectures in London and Rehovot, he reorganized it into four main parts, and this with a fuller discussion appeared in [97]. Today, that programme has been almost completed, and the only significant deviation (and a rather minor one) from the original proposal is that the prime 3 has played a less distinguished role amongst odd primes than had been anticipated. What has been done, and how the completion should take place, is the subject of the first eight chapters of this book.

The reader will soon see the constant importance of Aschbacher's work. In it he has fused the geometrical thinking of Fischer with the purely group-theoretical approach of Thompson and Bender, and he has combined this with his own tremendous insight and technical ability. It is highly unlikely that the classification would be so near to completion without his contribution.

The attempt to classify finite simple groups started with applications of character theory. Gorenstein's programme and its description in this book do not call for its use. Character theory, however, remains an essential tool when the classification of finite simple groups is viewed in its entirety; as with the N-group classification, it is simply that the necessary applications of character theory occur in prior characterization and general theorems. This is a particular example of the way in which this book does not treat the entire classification problem, as was explained in the Preface.

The classification of finite simple groups, if completed as envisaged, will have been a monumental effort. Because of the length and complexity of the proofs, attempts are already under way to try to unify and simplify them. This is the subject of Revisionism, which is discussed in Chapter 9. Some significant improvements have already been made. Whether reductions of a "logarithmic" proportion can be made with these methods remains to be seen; possibly a radically new approach will ultimately be necessary. Whatever the outcome of such revisions, however, a classification of finite simple groups will answer many of the old questions of finite group theory.

Note. At the beginning of this section, we referred to groups of Ree type. Bombieri [239] has recently announced that he has been able to show that the considerable analysis of Thompson [241] can be completed to prove that a group of Ree type is necessarily a Ree group. This solution of the "Ree group problem" does not affect any of the details given in this book, except in the obvious way, so the term "group of Ree type" has been allowed to remain.

2. Gorenstein's Programme

The attack on the classification of finite simple groups is primarily by an analysis of the *internal* structure of such groups. If G is a simple group, the aim is to show that the internal structure "closely resembles" that of some known simple group G^*. Once this is achieved, one tries to prove that G must be isomorphic to G^*.

This last phase of the analysis is known as *recognition theory* or the *identification problem*, and the methods are very different from the purely local analysis up to this point. One must show that some set of *internal* conditions suffice to characterize each of the known simple groups. If such a characterization already exists, for example in terms of the structure of the centralizer of an involution or as a primitive permutation group on the cosets of some subgroup of specified structure, then one need only establish that the relevant conditions hold in the situation that one is studying, though this should not obscure the fact that someone else has already solved the identification problem. With certain exceptions, however, all the known finite simple groups possess at least one such characterization.

The exceptions are the family $\{^2G_2(3^{2n+1})\}$ of Ree groups of characteristic 3‡ and two possible sporadic groups, Janko's fourth group J_4 and the Fischer monster F_1, whose existence has yet to be determined††. In each of these cases there is a precise internal condition, namely the exact centralizer of an involution, which yields considerable information including a group order and considerable information about characters, but it is not known whether these determine a unique group, if one at all. To allow for this ambiguity, we define groups *of Ree type*, *type* J_4 or *type* F_1. These indeterminacies may remain even after the classification is "completed" and so may be viewed as isolated problems independent of the general classification.

Definition. A group G is a $\mathscr{K}$*-group* if every composition factor of G is one of the following:

(i) cyclic of prime order;
(ii) an alternating group $A_n, n \geqslant 5$;
(iii) a simple group of Lie type;
(iv) one of the 24 currently known sporadic simple groups;
(v) a group of Ree type; or
(vi) a group of type F_1 or J_4.

One should, of course, add to this list whenever a new simple group is discovered; however, this is the list which will be taken in this book. In the next section, we shall discuss the simple $\mathscr{K}$-groups.

Assuming, then, that sufficiently good recognition theorems exist or can be proved, the classification of simple groups is now reduced to the problem of

‡ Now settled: see note at end of Section 1.

†† The existence and uniqueness of J_4 have been established by Norton and others in Cambridge. Griess has announced the explicit construction of a group which he believes to be F_1 (see page 26).

proving that the internal structure of an abstract simple group resembles that of a simple $\mathscr{K}$-group sufficiently closely that such theorems or methods may be applied. The work on this problem divides into four major parts:

(A) nonconnected groups;
(B) groups of component type;
(C) small groups of noncomponent type; and
(D) the general group of noncomponent type.

We shall define these terms in the later sections of this chapter where they are discussed. Work in (A) is complete and already appears in the literature; we shall give the results in Section 4, but otherwise not discuss this case further in this book. That in (B) is now nearly complete; the reduction to two major problem areas was established several years ago, and we discuss this in Section 5. Chapters 2 and 3 then give an account of work in these two areas and, in particular, collate the work of many contributors.

The situation with (C) and (D) is a little different, and we discuss this in Section 6. The recent major work on (D) by Gorenstein and Lyons and by Aschbacher is described in Chapters 4 and 5 respectively. This may be considered as analogous to the reductions in (B) and, when it is complete, will leave (C) and (D) at a similar stage to that of (B), or possibly even slightly ahead. The work in (C) falls into two parts and will be covered in Chapters 6 and 7, while the situation to which the Gorenstein–Lyons work reduces (D) is taken up in Chapter 8.

Much of the notation and terminology of this work has evolved and changed as progress has been made through the programme. In some cases alternative notation is given in this book. Hopefully the current usage will have been adopted, but the reader is warned that this may well differ from the original papers to which reference is made.

3. The Known Simple Groups

In this section, we shall give a brief account of the known simple groups together with their Schur multipliers. Many detailed properties of these groups are required in the course of the classification programme, but it is beyond the scope (or space) of this book to include such material. However, the reader will find references, both direct and indirect, in the individual chapters of this book. Knowledge of the Schur multipliers, however, is necessary for a precise formulation of some of the steps in part (B) of the classification, including those which are incomplete.

An account of the basic properties of the Schur multiplier is given in [H: V. 23]. Given a group G, a group $\hat{G}$ is called a *covering group* of G if $\hat{G}$ has a normal subgroup $A \subseteq Z(\hat{G}) \cap \hat{G}'$ such that $\hat{G}/A \cong G$; if A is cyclic of order n, then $\hat{G}$ is called an *n-fold* cover of G and will often be denoted by $n.G$ (or simply nG). If A is as large as possible, namely the Schur multiplier of G, then $\hat{G}$ is called a *representation group* of G; if G is perfect (in particular, simple) then the representation group is uniquely determined and is called the *full* covering group of G.

3.1. The alternating groups

For $n \geqslant 1$, S_n will denote the *symmetric group* of degree n, the group of all permutations on n letters. The subgroup of even permutations is called the *alternating group* of degree n and will be denoted by A_n.

For $n \geqslant 5$, the group A_n is simple. Schur [170] showed that the Schur multiplier of A_n is Z_2 except for $n = 6$ and $n = 7$; in those cases, it is Z_6.

3.2. The classical groups

The first complete treatment of the finite analogues of the complex matrix groups was given by Dickson [Dc]; a more modern approach, allowing arbitrary fields, has been given by Dieudonné [De] to which the reader is referred for details. Here, since we wish to describe the finite simple groups, we shall always assume that fields are finite. (For infinite fields, some statements may be false.) The classical groups can be identified with certain of the groups of Lie type; see Section 3.3. The exceptional isomorphisms between classical groups will be given there (Table 3) as will all Schur multipliers. (Tables 1 and 2 for the general case; Tables 3 and 4 for exceptions.)

Linear groups. The multiplicative group of nonsingular $n \times n$ matrices over F_q is the *general linear group* $GL_n(q)$ (or $GL(n, q)$). The subgroup consisting of those matrices of determinant 1 is the *special linear group* $SL_n(q)$ and the factor group

$$PSL_n(q) = SL_n(q)/(\text{scalars})$$

is the *projective special linear group*, which will normally be denoted by $L_n(q)$. (Here, as with the other matrix groups discussed, the centre of $SL_n(q)$ consists of the scalar matrices lying in $SL_n(q)$.)

For $n \geqslant 2$, the groups $L_n(q)$ are simple, except for $L_2(2)$ and $L_2(3)$.

Unitary groups. Let V be a vector space over a finite field K. Then a *Hermitian* inner product on V is a nonsingular form (x, y) defined on $V \times V$, linear in the first variable, for which

$$(x, y) = \overline{(y, x)}$$

where $\overline{}$ denotes an automorphism of K of order 2. This forces $|K| = q^2$ and

$\bar{t} = t^q$ for $t \in K$. All (nonsingular) Hermitian products on V are equivalent, and a *unitary group* is a group of nonsingular transformations which preserve this product.

If we define the *general unitary group* $GU_n(q)$ as the subgroup of $GL_n(q^2)$ leaving invariant the product

$$(x, y) = x_1\bar{y}_1 + \cdots + x_n\bar{y}_n,$$

then $GU_n(q)$ consists of those matrices A for which $A = (\bar{A}')^{-1}$. (i.e. If $A = (a_{ij})$, then $A = (\bar{a}_{ji})^{-1}$.) The *special*, and *projective special*, *unitary groups* are defined by

$$SU_n(q) = GU_n(q) \cap SL_n(q^2)$$

and

$$PSU_n(q) = SU_n(q)/(\text{scalars});$$

throughout, $PSU_n(q)$ will be written as $U_n(q)$. Then $U_2(q) \cong L_2(q)$ and, for $n \geqslant 3$, $U_n(q)$ is simple, except for $U_3(2)$.

(*Note.* Writing $GU_n(q)$ instead of $GU_n(q^2)$ is consistent with the convention that will be adopted for labelling the twisted groups.)

Symplectic groups. Let V be a vector space endowed with an alternating bilinear form (i.e. a bilinear form (x, y) for which $(x, x) = 0$ for all x). If the form is nonsingular, then V necessarily has even dimension $2n$ and there is a basis with respect to which the form is given by

$$(x, y) = \sum_{i=1}^{n} (x_{2i-1}y_{2i} - x_{2i}y_{2i-1}).$$

The subgroup of $GL_n(q)$ leaving this form invariant is the *symplectic group* $Sp_{2n}(q)$, and

$$PSp_{2n}(q) = Sp_{2n}(q)/\langle \pm I \rangle.$$

$Sp_2(q) = SL_2(q)$; for $n \geqslant 2$, the groups $PSp_{2n}(q)$ are simple, except for $Sp_4(2)$ which is isomorphic to S_6.

Orthogonal groups: odd characteristic. Let V be a vector space of dimension n over a finite field K of odd characteristic. If (x, y) is a nonsingular symmetric bilinear form on V, then there is an associated *quadratic* form f given by

$$f(x) = (x, x),$$

from which the bilinear form may be recovered. (This requires char $K \neq 2$.) The group of nonsingular linear transformations of V which preserve the bilinear form (and, equivalently, the quadratic form) is called the *orthogonal group associated with* f.

If $n = 2m+1$ is odd, there are two inequivalent nonsingular symmetric bilinear forms on V, which may be taken to correspond to the quadratic forms

$$x_1x_{m+1}+\cdots+x_mx_{2m}+x_{2m+1}^2$$

and

$$x_1x_{m+1}+\cdots+x_mx_{2m}+\alpha x_{2m+1}^2$$

where α is a nonsquare in $K=\boldsymbol{F}_q$; however, they determine isomorphic orthogonal groups, denoted by $O_{2m+1}(q)$.

If $n=2m$ is even, there are two forms corresponding to

$$x_1x_{m+1}+\cdots+x_mx_{2m}$$

and

$$x_1x_m+\cdots+x_{m-1}x_{2m-2}+x_{2m-1}^2-\alpha x_{2m}^2;$$

these have Witt index m and $m-1$ respectively and determine nonisomorphic orthogonal groups $O_{2m}^+(q)$ and $O_{2m}^-(q)$.

We consider only the cases $n \geqslant 3$. Let $O_n^\varepsilon(q)$ be any orthogonal group defined above, where $\varepsilon = +$ or $-$ if n is even, and is suppressed if n is odd. Let

$$SO_n^\varepsilon(q) = O_n^\varepsilon(q) \cap SL_n(q).$$

Then $O_n^\varepsilon(q)' = SO_n^\varepsilon(q)'$; this commutator group has index 2 in $SO_n^\varepsilon(q)$ and is denoted by $\Omega_n^\varepsilon(q)$, and we put

$$P\Omega_n^\varepsilon(q) = \Omega_n^\varepsilon(q)/(\text{scalars}).$$

$\Omega_3(q) \cong L_2(q)$ and, when $n=4$,

$$\Omega_4^+(q) \cong SL_2(q) \circ SL_2(q)$$

and

$$\Omega_4^-(q) \cong L_2(q^2).$$

If $n \geqslant 5$, the groups $P\Omega_n^\varepsilon(q)$ are always simple; for $n=5$ or 6, they are isomorphic to previously described groups, namely

$$P\Omega_5(q) \cong PSp_4(q), \quad P\Omega_6^+(q) \cong L_4(q) \quad \text{and} \quad P\Omega_6^-(q) \cong U_4(q).$$

Corresponding to each orthogonal group, there is the "spin" group $\mathrm{Spin}_n^\varepsilon(q)$ which is a 2-fold cover of $\Omega_n^\varepsilon(q)$. These spin groups are obtained as projective representation groups in the Clifford algebra associated with the corresponding forms. (See [De: II, Sections 7,8].)

Orthogonal groups: even characteristic. Let V be a vector space of dimension n over $\boldsymbol{F}_q$, where $q=2^a$. A form f on V is said to be *quadratic* if there is a bilinear form (x, y) such that

$$f(\lambda x+\mu y) = \lambda^2 f(x)+\mu^2 f(y)+\lambda\mu(x, y)$$

for all $\lambda, \mu \in \boldsymbol{F}_q$. It follows that $(x,x) = 0$ and $(x,y) = (y,x)$, although the bilinear form may be singular. However, $V = V_0 \oplus V_1$ where

$$V_0 = \{x \in V \mid (x,y) = 0 \quad \text{for all} \quad y \in V\}$$

and the bilinear form (x,y) is nonsingular on V_1. Then V_1 has even dimension, and the dimension d of V_0 is called the *defect* of f. If $f(x) \neq 0$ whenever $x \in V_0 - \{0\}$, then f is *nondegenerate*, and we consider only this case. The *orthogonal group associated with* f is defined as the group of nonsingular linear transformations preserving the form f.

First suppose that $n = 2m+1$ is odd. Then, up to equivalence, there is only one nondegenerate quadratic form which may be taken as

$$x_1 x_{m+1} + \cdots + x_m x_{2m} + x_{2m+1}^2;$$

then $d = 1$ and the orthogonal group is just $Sp_{2m}(q)$, so this case need not be considered further.

If $n = 2m$ is even, there are two inequivalent quadratic forms, both with $d = 0$, which may be taken as

$$x_1 x_{m+1} + \cdots + x_m x_{2m}$$

and

$$x_1 x_m + \cdots + x_{m-1} x_{2m-2} + \alpha x_{2m-1}^2 + x_{2m-1} x_{2m} + \alpha x_{2m}^2$$

where $\alpha t^2 + t + \alpha$ is an irreducible polynomial over $\boldsymbol{F}_q$, having Witt indices m and $m-1$ respectively; the respective orthogonal groups, denoted by $O_{2m}^+(q)$ and $O_{2m}^-(q)$, are both subgroups of symplectic groups. Again, define

$$\Omega_{2m}^\varepsilon(q) = O_{2m}^\varepsilon(q)'$$

where $\varepsilon = +$ or $-$.

The groups $O_2^\pm(q)$ are dihedral of order $2(q \mp 1)$, and

$$O_4^+(q) \cong SL_2(q) \wr Z_2;$$

in all remaining cases, $\Omega_{2m}^\varepsilon(q)$ is a simple subgroup of $O_{2m}^\varepsilon(q)$ of index 2 with, for small dimensions, the isomorphisms

$$\Omega_4^-(q) \cong L_2(q^2), \quad \Omega_6^+(q) \cong L_4(q) \quad \text{and} \quad \Omega_6^-(q) \cong U_4(q).$$

3.3. Groups of Lie type

Chevalley groups. Let $\mathfrak{g}$ be a simple complex Lie algebra with Cartan decomposition

$$\mathfrak{g} = \mathfrak{h} \oplus \sum_{r \in \Phi} \mathfrak{g}_r$$

where Φ is the root system. Let Π be a set of fundamental roots and

$\{h_r | r \in \Pi\}$ the corresponding set of coroots. Chevalley showed that it was possible to choose elements e_r in $\mathfrak{g}_r (r \in \Phi)$ for which the set

$$\mathscr{C} = \{h_r, r \in \Pi; e_r, r \in \Phi\}$$

was a basis for $\mathfrak{g}$ such that the multiplication constants with respect to this basis were rational integers. (Such a basis is called a *Chevalley basis*; it is not necessarily unique.) Furthermore, he showed that if $\zeta \in C$, then the map

$$x_r(\zeta) = \exp(ad \zeta e_r)$$

was an automorphism of $\mathfrak{g}$ which transformed each element of $\mathscr{C}$ into a linear combination of elements of $\mathscr{C}$ with coefficients which were polynomials in ζ with rational integer coefficients.

Let $\mathfrak{g}_Z$ be the subset of $\mathfrak{g}$ consisting of all Z-linear combinations of the elements of $\mathscr{C}$ and let K be an arbitrary field. Chevalley defined the Lie algebra

$$\mathfrak{g}_K = K \otimes \mathfrak{g}_Z$$

over K, and showed that the formal map $x_r(t)$ obtained by replacing ζ by an element t of K in its action on $\mathscr{C}$ yielded an automorphism of $\mathfrak{g}_K$. The groups that he constructed,

$$G(K) = \langle x_r(t) \,|\, r \in \Phi, t \in K \rangle,$$

are called the *(adjoint) Chevalley groups.* Except for certain cases when $|K| = 2$ or 3 (see Table 3), the groups $G(K)$ are simple; up to isomorphism they are independent of the particular Chevalley basis chosen, and they are finite if K is finite.

The classification of the Chevalley groups follows from a knowledge of the simple complex Lie algebras. There are four infinite families, A_ℓ, B_ℓ, C_ℓ and D_ℓ (where $\ell = |\Pi|$ denotes the *rank*), and the five exceptional algebras of types E_6, E_7, E_8, F_4 and G_2. The infinite families yield all the finite classical groups except for the unitary groups and the orthogonal groups of $-$type. In Table 1, we give these identifications, and also the Dynkin diagrams for each type. (The groups $\tilde{G}$ will be discussed below.)

Chevalley's construction gives rise to a collection of homomorphisms

$$\phi_r : SL_2(K) \to G \qquad (r \in \Phi)$$

where

$$\phi_r \begin{pmatrix} 1 & t \\ 0 & 1 \end{pmatrix} = x_r(t), \qquad \phi_r \begin{pmatrix} 1 & 0 \\ t & 1 \end{pmatrix} = x_{-r}(t).$$

In particular, this implies that each of the subgroups $\langle x_{\pm r}(t) \,|\, t \in K \rangle$ is a homomorphic image of $SL_2(K)$ and that one has relations

(A) $$x_r(t) x_r(u) = x_r(t+u) \qquad (r \in \Phi; t, u \in K).$$

Additionally, Chevalley showed that the elements $x_r(t)$ satisfied the "commutator formulae"

$$\text{(B)} \qquad [x_s(u), x_r(t)] = \prod_{i,j>0} x_{ir+js}(c_{ijrs}t^i u^j)$$

whenever r, s are independent roots; the $\{c_{ijrs}\}$ are integers depending only on $\mathfrak{g}$ but not t or u, and the product is taken over the finite number of pairs i,j of positive integers for which $ir + js \in \Phi$, in some fixed order. In the case that $\mathfrak{g}$ has rank greater than 1, Steinberg [195] was able to show that the relations (A) and (B) give a presentation for a covering group of $G(K)$; if $|K| > 4$, he showed that it is the *full* covering group. An important refinement of this result has been given by Curtis [46], and this will be discussed later.

By starting with the crystallographic root systems and their Weyl groups, Steinberg has given an alternative approach which generalizes the original concept of a Chevalley group ([197]; see also Curtis [FSG]). His construction yields covering groups of the adjoint groups, including the *universal Chevalley groups* (of which all others are images) which, for $|K| > 4$, are precisely the groups for which he had previously given presentations. The method also gives an explicit construction for the centre of a universal group.

Where appropriate, the universal Chevalley groups $\tilde{G}$ are identified as finite classical groups in Table 1; in any case, their centres are given for the finite groups. The Schur multipliers of the adjoint Chevalley groups $G(q)$ for $q \leqslant 4$ have now all been determined (see [106] and references there, and [107]): those which differ from the generic situation are listed in Tables 3 and 4.

Let $G(K)$ be an adjoint Chevalley group of rank $\ell > 1$. Curtis' refinement of Steinberg's presentations was to observe that some of the relations are redundant. If Φ is the corresponding root system and Π a set of fundamental roots $r_1, \ldots, r_\ell$, let

$$\mathrm{P}_{ij} = \{\lambda r_i + \mu r_j \in \Phi \mid \lambda, \mu, \in \mathbf{Z}\} \qquad (i,j = 1, \ldots \ell)$$

and let $\mathrm{P} = \bigcup \mathrm{P}_{ij}$. Then Curtis showed that it was sufficient to take generators $\{x_r(t) \mid r \in \mathrm{P}, t \in K\}$ subject to relations of type (A) for $r \in \mathrm{P}$ and type (B) for independent pairs r,s belonging to some P_{ij} in order to have a presentation for a covering group of $G(K)$ (which would always be a covering of the associated universal Chevalley group). The effect of this is that, if $K = \boldsymbol{F}_q$, and G is a finite group having subgroups $H_1, \ldots, H_\ell$ in 1–1 correspondence with the nodes of the Dynkin diagram for Φ, each isomorphic to $SL_2(q)$, and such that any pair "canonically generate" a rank 2 Chevalley group (where nonconnected nodes correspond to commuting subgroups), then $\langle H_1, \ldots, H_\ell \rangle$ is isomorphic to a covering group of $G(K)$. This is a useful device in "building up" a Chevalley group in classification theorems, and an example is given in Chapter 2, Section 3.

TABLE 1. *Chevalley groups*

Type	Dynkin diagram of root system	Classical identifications of adjoint group G	Classical identifications of universal group $\tilde{G}$	$\lvert Z(\tilde{G})\rvert$	Structure of $Z(\tilde{G})$
$A_\ell(\ell \geqslant 1)$		$L_{\ell+1}(q)$	$SL_{\ell+1}(q)$	$(\ell+1, q-1)$	cyclic
$B_\ell(\ell \geqslant 2)$		$P\Omega_{2\ell+1}(q)$, q odd	$\mathrm{Spin}_{2\ell+1}(q)$	2	
		$Sp_{2\ell}(q)$, q even	$Sp_{2\ell}(q)$	1	
$C_\ell(\ell \geqslant 3)$		$PSp_{2\ell}(q)$	$Sp_{2\ell}(q)$	$(2, q-1)$	
$D_\ell(\ell \geqslant 4)$		$P\Omega^+_{2\ell}(q)$, q odd	$\mathrm{Spin}_{2\ell}(q)$	$\begin{cases}(4, q-1) \\ 4\end{cases}$	cyclic (ℓ odd) el. abelian (ℓ even)
		$\Omega^+_{2\ell}(q)$, q even	$\Omega^+_{2\ell}(q)$	1	
E_6	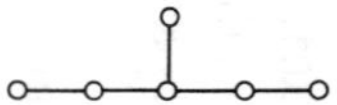			$(3, q-1)$	
E_7	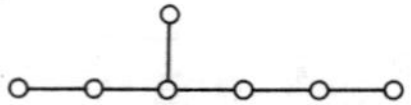			$(2, q-1)$	
E_8	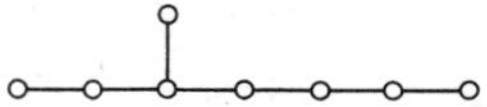			1	
F_4				1	
G_2				1	

Twisted groups. Let $G = G(K)$ be an adjoint Chevalley group over an arbitrary field K with root system Φ, and let $\tilde{G}$ be the corresponding universal group. Assume that G is not of type A_1. If $|K| > 4$, the Steinberg presentation given by (A) and (B) is a presentation for $\tilde{G}$; otherwise certain additional relations implied by the homomorphisms $\{\phi_r\}$ may be necessary, but these will not affect the conclusions of the following discussion.

We consider two particular types of automorphism of $\tilde{G}$. First, suppose that σ is an automorphism of the field K. Then it is clear from the presentation for $\tilde{G}$ that the map

$$x_r(t) \to x_r(t^\sigma) \qquad (r \in \Phi, t \in K)$$

extends to an automorphism of $\tilde{G}$ called a *field automorphism* of $\tilde{G}$, also denoted by σ. Now suppose that the Dynkin diagram of Φ admits a nontrivial symmetry, ρ. Then, in particular, all the roots in Φ have the same length, and ρ can be extended to an isometry of Φ. It can be shown that there exist signs $\gamma_r = \pm 1$ such that if $\rho : r \to \bar{r}$, then the map

$$x_r(t) \to x_{\bar{r}}(\gamma_r t) \qquad (r \in \Phi : t \in K)$$

extends to an automorphism of $\tilde{G}$. This automorphism, also denoted by ρ, is called a *graph automorphism* of $\tilde{G}$ and has the same order as the original symmetry (necessarily 2 or 3). It is clear from their definition that graph and field automorphisms commute; also, since $Z(\tilde{G})$ is a characteristic subgroup of $\tilde{G}$, they induce automorphisms of the corresponding adjoint group G.

Let $\tau = \rho\sigma$ where ρ is a graph automorphism of $\tilde{G}$ and σ is a field automorphism having the same order as ρ. Let $\tilde{G}_1$ be the fixed point subgroup of τ. Then $Z(\tilde{G}_1)$ consists of the fixed points of τ on $Z(\tilde{G})$. If G_1 is the image of $\tilde{G}_1$ in G, then, with only a few exceptions, G_1 is a simple group, but not a Chevalley (or alternating) group. These groups depend only on G and the *order* of ρ; they are known as the *Steinberg variations.* (In his original paper where these groups were defined, Steinberg worked directly with the adjoint groups [194]; the above treatment was given later by Steinberg [196]. See also [198].)

Assume now that K is finite. The existence of the field automorphism σ imposes restrictions on K. In fixing notation, we shall adopt the convention that a Steinberg variation is labelled by the fixed subfield of K and the order of ρ; thus (in "adjoint" form) they are ${}^2A_\ell(q)$ $(\ell \geqslant 2)$, ${}^2D_\ell(q)$ $(\ell \geqslant 4)$, ${}^3D_4(q)$ and ${}^2E_6(q)$, defined as subgroups of $A_\ell(q^2)$, $D_\ell(q^2)$, $D_4(q^3)$ and $E_6(q^2)$ respectively. The first two families turn out to be the unitary groups and the orthogonal groups of $-$type; the last two families are new.

Next we consider finite groups of types B_2, F_4 and G_2. There is an apparent symmetry in the Dynkin diagram, yet because roots of different lengths are involved, there can be no corresponding isometry of the root system. After Suzuki constructed his simple groups $Sz(q)$ as subgroups of

$Sp_4(q)$ [206], Ree noticed that the construction had an interpretation within the Lie framework. For $q = 2^m$, it had been known that $Sp_4(q)$ possessed an "extra" automorphism. Ree showed that if m was *odd*, then there was a field automorphism such that the composite automorphism had $Sz(q)$ as its fixed point subgroup. He also showed that the extra automorphism of $Sp_4(q)$ acted on the root system of type B_2, interchanging roots of different lengths. He was then able to show that for the groups $F_4(2^m)$ and $G_2(3^m)$, as well as $B_2(2^m)$, there was a map which interchanged the roots of different lengths and which extended to an automorphism α of the group. In each case, α^2 was the field automorphism corresponding to the map $t \to t^p$ (where p is the characteristic of the field). For m odd, there is a field automorphism β such that $\alpha^2\beta^2 = (\alpha\beta)^2 = 1$, and the respective families of groups, ${}^2F_4(2^{2n+1})$ and ${}^2G_2(3^{2n+1})$, are the *Ree groups*. With the exception of ${}^2F_4(2)$ and ${}^2G_2(3)$ (and also ${}^2B_2(2)$), these groups are simple.

(In an attempt to characterize the Ree groups of characteristic 3 by the

TABLE 2. *Twisted groups*

Twisted group G_1	Classical or other identification	$\lvert Z(\tilde{G}_1)\rvert$
${}^2A_\ell(q)$ ($\ell \geqslant 2$)	$U_{\ell+1}(q)$	$(\ell+1, q+1)$
${}^2D_\ell(q)$ ($\ell \geqslant 4$)	$P\Omega_{2\ell}^-(q)$, q odd	$\begin{cases}(4,q+1), & \ell \text{ odd}\\ 2, & \ell \text{ even}\end{cases}$
	$\Omega_{2\ell}^-(q)$, q even	1
${}^3D_4(q)$		1
${}^2E_6(q)$		$(3,q+1)$
${}^2B_2(q)$ ($q = 2^{2n+1}$)	$Sz(q)$	1
${}^2F_4(q)$ ($q = 2^{2n+1}$)		1
${}^2G_2(q)$ ($q = 3^{2n+1}$)		1

TABLE 3. *Isomorphisms and nonsimplicity*

G	Exceptional nature	Schur multiplier of G'
$L_2(2)$, $L_2(3)$, $U_3(2)$	soluble	
$L_2(4)$, $L_2(5)$	isomorphic to A_5	Z_2
$L_2(7)$, $L_3(2)$	isomorphic	Z_2
$L_2(9)$	isomorphic to A_6	Z_6
$L_4(2)$	isomorphic to A_8	Z_2
$Sp_4(2)$	isomorphic to S_6	Z_6
$U_4(2)$, $PSp_4(3)$	isomorphic	Z_2
$G_2(2)$	$[G:G'] = 2$, $G' \cong U_3(3)$	1
${}^2G_2(3)$	$[G:G'] = 3$, $G' \cong L_2(8)$	1
${}^2F_4(2)$	$[G:G'] = 2$, G' simple (the Tits group)	1

TABLE 4. *Remaining exceptional Schur multipliers*

G	Schur multiplier	G	Schur multiplier
$L_3(4)$	$Z_4 \times Z_{12}$	$G_2(4)$	Z_2
$Sp_6(2)$	Z_2	$U_4(3)$	$Z_3 \times Z_{12}$
$P\Omega_7(3)$	Z_6	$U_6(2)$	$Z_2 \times Z_6$
$\Omega_8^+(2)$	$Z_2 \times Z_2$	$Sz(8)$	$Z_2 \times Z_2$
$F_4(2)$	Z_2	${}^2E_6(2)$	$Z_2 \times Z_6$
$G_2(3)$	Z_3		

centralizer of an involution, Ward was able to obtain considerable local information [232], but he was unable to identify the groups. A group *of Ree type* is a simple group having abelian Sylow 2-subgroups, in which the centralizer of an involution is isomorphic to $Z_2 \times L_2(3^{2n+1})$.)

The Schur multipliers of all the twisted groups are known. In the generic situation, they are obtained as the groups $Z(\tilde{G}_1)$, all of which are cyclic. These are described in Table 2, which is an analogue of Table 1. The exceptional Schur multipliers are given in Tables 3 and 4. In the generic case, Steinberg obtained these results for the Steinberg variations ([196] and unpublished.) Alperin and Gorenstein computed the Schur multipliers of the Suzuki groups and the Ree groups of characteristic 3 [5]. The remainder are due to Griess ([106], [107]).

3.4. Sporadic groups

It remains to consider the 24 sporadic groups whose existence and uniqueness have been determined, plus the two possible types of group for which considerable evidence exists. Gorenstein has given an extensive account in his survey [98]; here we shall be far more brief.

Table 5 gives a list of these groups. The first column gives the notation used for them in Part A of this book, with alternative common notation in parentheses, and the second column gives the group order. The third column, headed *description*, gives a reference for the original defining property and the information derived from it; the fourth column does likewise for *existence* and *uniqueness*. Question marks indicate that existence has yet to be established. Finally, for each group we give its Schur multiplier together with those responsible for its computation.

We now discuss the origins of each of the sporadic groups.

Mathieu groups. These groups were discovered by Mathieu as multiply transitive permutation groups. The groups M_{12} and M_{24} are 5-fold transitive permutation groups on 12 and 24 letters respectively; the groups M_{11}, M_{22} and M_{23} are the natural one or two point stabilizers. They can be

TABLE 5. *The sporadic groups*

Group	Order	Description
M_{11}	$2^4.3^2.5.11$	
M_{12}	$2^6.3^3.5.11$	
M_{22}	$2^7.3^2.5.7.11$	
M_{23}	$2^7.3^2.5.7.11.23$	
M_{24}	$2^{10}.3^3.5.7.11.23$	
J_1	$2^7.3.5.7.11.19$	Janko [132]
J_2 (HJ)	$2^7.3^5.5^2.7$	Janko [133]
J_3 (HJM)	$2^7.3^5.5.17.19$	Janko [133]
J_4	$2^{21}.3^3.5.7.11^3.23.29.31.37.43$	Janko [135]
HS (HiS)	$2^9.3^2.5^3.7.11$	D. Higman and Sims [125]
Sz (Suz)	$2^{13}.3^7.5^2.7.11.13$	Suzuki [208]
Mc	$2^7.3^6.5^3.7.11$	McLaughlin [150]
Ru	$2^{14}.3^3.5^3.7.13.29$	Rudvalis
He (HHM)	$2^{10}.3^3.5^2.7^3.17$	Held [123]
Ly (LyS)	$2^8.3^7.5^6.7.11.31.37.67$	Lyons [140]
ON	$2^9.3^4.5.7^3.11.19.31$	O'Nan [158]
Co_1 $(\cdot 1)$	$2^{21}.3^9.5^4.7^2.11.13.23$	Conway [43]
Co_2 $(\cdot 2)$	$2^{18}.3^6.5^3.7.11.23$	Conway [43]
Co_3 $(\cdot 3)$	$2^{11}.3^7.5^3.7.11.23$	Conway [43]
$M(22)$ (Fi_{22})	$2^{17}.3^9.5^2.7.11.13$	Fischer [67]
$M(23)$ (Fi_{23})	$2^{18}.3^{13}.5^2.7.11.13.17.23$	Fischer [67]
$M(24)'$ (Fi'_{24})	$2^{21}.3^{16}.5^2.7^3.11.13.17.23.29$	Fischer [67]
F_1 (M)	$2^{46}.3^{20}.5^9.7^6.11^2.13^3$ $\times 17.19.23.29.31.41.47.59.71$	Fischer, Griess [109]
F_2 (BM)	$2^{41}.3^{13}.5^6.7^2.11.13.17.19.23$ $\times 31.47$	Fischer
F_3 (E)	$2^{15}.3^{10}.5^3.7^2.13.19.31$	Thompson [216]
F_5 (F)	$2^{14}.3^6.5^6.7.11.19$	Harada [116]

described as explicit subgroups of A_{12} and A_{24} by giving generators, but there are other descriptions. M_{12} and M_{24} are the automorphism groups of the (unique) Steiner triple systems $S(5,6,12)$ and $S(5,8,24)$, and are also the automorphism groups of Golay codes. (See, for example, Conway's chapter in [FSG].)

Janko's groups. These groups all arise from "centralizer of an involution" problems.

Group	Centralizer of an involution
J_1	$Z_2 \times A_5$
J_2, J_3	$2^{1+4}.A_5$
J_4	$2^{1+12}.(3.M_{22}).2$

(Here, as elsewhere in this subsection, the notation indicates the normal structure of the "defining" centralizer, with a nontrivial action whenever

TABLE 5—*continued*

Existence and Uniqueness	Schur Multiplier	Determination
	1	Burgoyne and Fong [37]
	Z_2	Burgoyne and Fong [37]
	Z_{12}	Burgoyne, Fong, Griess[a]
	1	Burgoyne and Fong [37]
	1	Burgoyne and Fong [37]
Janko [132]	1	Janko [132]
Hall and Wales [113]	Z_2	McKay and Wales [148]
G. Higman and McKay [127]	Z_3	McKay and Wales [148]
Norton *et al.* (1980)	1	Lempken [240]
D. Higman and Sims [125]	Z_2	McKay and Wales [147]
Suzuki [208]	Z_6	Griess [108]
McLaughlin [150]	Z_3	Thompson
Conway and Wales [44]	Z_2	Feit, Lyons
G. Higman and McKay [146]	1	Griess [108]
Sims [178]	1	Lyons [140]
Sims, Andrilli [7]	Z_3	O'Nan [158]
Conway [43]	Z_2	Griess [108]
Conway [43]	1	Griess [108]
Conway [43]	1	Griess [108]
Fischer [67]	Z_6	Griess [108], Fischer, Rudvalis[b]
Fischer [67]	1	Griess [108]
Fischer [67]	Z_3	Griess [108], Norton[c]
Griess (1980), see also Thompson [242]	1	Griess
Leon and Sims [139]	1 or Z_2	Griess[d]
Thompson [216]	1	Thompson
Norton [157]	1	Harada [117]

[a] The proof by Burgoyne and Fong [37] that the 2-part of the Schur multiplier of M_{22} was Z_2 was found to be incorrect by Mazée. The character table of $2.M_{22}$ is known. There is a 210-dimensional ordinary representation which can be realized as a real orthogonal representation; using this, Griess has shown that $2.M_{22}$ lifts to $4.M_{22}$ in $\mathrm{Spin}_{210}(R)$ and that this gives an upper bound for the 2-part of the Schur multiplier.

[b] Griess obtained Z_6 as an upper bound; Fischer and Rudvalis showed that it was attained by embedding $M(22)$ in ${}^2E_6(2)$. (Note: this embedding is the starting point for constructing F_2.)

[c] Griess obtained Z_3 as an upper bound; Norton established it by constructing the triple cover of $M(24)$.

[d] Griess has shown that Z_2 is an upper bound, but it is not known whether it is attained. The construction of the Monster F_1 would prove this.

possible; for example, in the case of J_4, 2^{1+12} indicates a normal extraspecial group of order 2^{13}, and the centralizer consists of this extended by $3.M_{22}$ (in fact, nonsplit), extended by an automorphism of order 2 which acts non-trivially on the composition factor M_{22}.)

The group J_1 was the first of the "modern" sporadic groups to be discovered and was shown to exist uniquely by constructing an explicit subgroup of $GL_7(11)$ of order 175 560. The groups J_3 and J_2 were defined by the given centralizer and the assumption that all involutions were, or were not, conjugate. The group J_2 was the first of these whose existence was established. This was done by M. Hall, and we shall discuss this further below. The group J_3 was shown to exist by searching by hand for suitable generators and relations, and then applying the Todd–Coxeter algorithm on a computer. The existence of J_4 has not yet been established, nor is it known whether it would then be unique‡; a *group of J_4-type* is any group with the properties derived by Janko, and by J_4 one should understand this.

Rank 3 permutation groups. Hall's construction of the group J_2 of order 604 800 was as a rank 3 permutation group on 100 letters, having a point stabilizer isomorphic to $U_3(3)$. Four further groups have been constructed directly as rank 3 permutation groups: in the following table we give the basic information that characterizes them. (The Fischer groups have rank 3 permutation representations, but did not specifically arise in this way; they will be discussed later.)

Group	Point Stabilizer	Degree	Nontrivial subdegrees
Hall–Janko J_2	$U_3(3)$	100	36, 63
Higman–Sims HS	M_{22}	100	22, 77
McLaughlin Mc	$U_4(3)$	275	112, 162
Suzuki Sz	$G_2(4)$	1782	416, 1365
Rudvalis Ru	${}^2F_4(2)$	4060	1755, 2304

Centralizer problems: phase 1. Three more groups were obtained directly from abstract centralizer conditions: all were shown to exist only with the aid of a computer. Held's group He was the first and arose out of his attempt to characterize the groups M_{24} and $L_5(2)$ which contain involutions with isomorphic centralizers; he found one possible configuration which could not be eliminated. Lyons' group Ly arose starting with a centralizer isomorphic to $\hat{A}_{11}$, and O'Nan's group ON, although now characterized by its Sylow 2-subgroup alone, was originally predicted from an involution centralizer of the form $(4.L_3(4)).2$. (The uniqueness of O'Nan's group has only recently been established; Sims' original construction and uniqueness theorem was under the assumption that there was an automorphism of order 2. See also Chapter 17 for a discussion of Sims' general methods.)

Conway's groups. For some time it had been known that there was a 24-dimensional real lattice, the *Leech lattice*, corresponding to an unusually dense sphere packing. Conway showed that it had a large automorphism group which he called ·0, whose central quotient group Co_1 was a new simple group, and which contained new simple groups, Co_2 and Co_3, as the

‡ Now both established.

stabilizers of certain short vectors in the lattice. Many of the previously discovered sporadic groups occur as sections of $\cdot 0$. (See Conway [FSG].)

Fischer's groups. The transpositions in the symmetric groups have the property that they form a single conjugacy class of involutions, any two distinct members of which have product of order 2 or 3. A conjugacy class D of involutions in an abstract group having this property is called a class of *3-transpositions*. There exist such classes in the groups $Sp_{2n}(2)$, $O_{2n}^{\pm}(2)$, $U_n(2)$ and $O_n^{\varepsilon}(3)$. (In the final case, they do not generate the whole group.) In attempting to characterize these groups by this property, Fischer considered groups which were generated by a class of 3-transpositions, with a minor restriction (essentially to keep near simplicity), and in the process of classifying such groups discovered three new groups, $M(22)$, $M(23)$ and $M(24)$. The first two are simple; $M(24)$ has a simple subgroup of index 2.

If D was a class of 3-transpositions in a group G under consideration, Fischer showed that if $d \in D$, then G acted by conjugation on D with suborbits $\{d\}$, D_d and E_d, where

$$D_d = \{e \in D \mid o(de) = 2\}$$

and

$$E_d = \{e \in D \mid o(de) = 3\},$$

and that D_d was a class of 3-transpositions in $\langle D_d \rangle$. This enabled him to work inductively; every "known" case for $H = \langle D_d \rangle / Z(\langle D_d \rangle)$ gave rise to a known group, except for $H = U_6(2)$. This led to $M(22)$, leading in turn to $M(23)$ and $M(24)$, but Fischer showed that no further possibilities could arise. The existence was established by the construction of the so-called Fischer triple graphs. The rank 3 permutation group structures of the Fischer groups are as follows.

Group	Stabilizer	Degree	Nontrivial subdegrees	
$M(22)$	$2.U_6(2)$	3510	693,	2816
$M(23)$	$2.M(22)$	31 671	3510,	28 160
$M(24)$	$Z_2 \times M(23)$	306 936	31 671,	275 264

The notation denotes a connection with the Mathieu groups. If E is a maximal commuting subset of D in each of the three Fischer groups, then $|E| = 22, 23$ or 24 and $N(E)/C(E) \cong M_{22}, M_{23}$ or M_{24} respectively.

The Monster and its subgroups. Fischer next considered groups generated by a class of $\{3,4\}$-transpositions (with the obvious definition, allowing also products of order 4). These have not been classified. However, Fischer was led to consider a simple group with this property, in which the centralizer of some involution has the form $(2.{}^2E_6(2)).2$, and was able to obtain almost complete "internal" information. This group, the *Baby Monster* (BM or F_2), has now been shown to exist uniquely.

Before F_2 had been shown to exist, Fischer and Griess (independently) studied a possible simple group which would contain an involution whose centralizer was $2.F_2$. Together with other judiciously guessed information, they were able to present evidence for yet another simple group, called the *Monster* (M or F_1). This has not yet been shown to exist‡. However, a lower bound of 196 883 was found for the degrees of the nonprincipal ordinary irreducible characters. More recently, under the assumption that there is a character of this degree, Livingstone and others have determined the character table and Thompson has shown that the group, if it exists, is unique [242]. (A group satisfying all the known local information, but *not* necessarily possessing a character of degree 196 883, is said to be a *group of F_1-type.*)

More remarkably, it has been shown that F_1 contains elements of orders 3 and 5 having centralizers of the form $Z_3 \times F_3$ and $Z_5 \times F_5$ where F_3 and F_5 are simple groups containing involutions with specified centralizers. These previously unknown simple groups have been obtained by Thompson and Harada, respectively, starting only from the predicted centralizers of $2^{1+8}.A_9$ and $(2.HS).2$.

4. Nonconnected Groups

For an arbitrary group G, let $\mathscr{V}(G)$ denote the set of *four-groups* in G, i.e. subgroups isomorphic to $Z_2 \times Z_2$. Let Γ be the graph whose vertices are the elements of $\mathscr{V}(G)$ and whose edges are the joins of vertices V_1 and V_2 for which $[V_1,V_2] = 1$. Then G is said to be *connected* if the associated graph Γ is connected, and *nonconnected* otherwise. (Note that this differs from the original definition given by Gorenstein in [FSG] which required only a Sylow 2-subgroup to be connected, as also in [104].)

All nonconnected simple groups have now been determined. This work falls into two parts.

(I) The determination of simple groups with a nonconnected Sylow 2-subgroup; and

(II) The characterization of nonconnected groups having a connected Sylow 2-subgroup.

In order to describe the solution of (I), we first introduce some standard notation and terminology that will be required throughout this book. Let p be any prime. If A is an abelian p-group, let $m(A)$ denote the minimal number of generators of A. Then, if P is an arbitrary p-group, define

$$m(P) = \max\{m(A) \mid A \text{ is an abelian subgroup of } P\}$$

‡ Griess has described F_1 as a group of automorphisms of a 196883-dimensional commutative nonassociative algebra over Q, having an associative form. The full uniqueness question remains unsettled.

and

$$r(P) = \max\{m(Q/Q') \mid Q \subseteq P\}.$$

$m(P)$ is called the *rank* of P and $r(P)$ the *sectional rank*; if G is a group, then by the *p-rank* and *sectional p-rank* of G one means, respectively, the rank and sectional rank of a Sylow p-subgroup of G. The p-rank of G will be denoted by $m_p(G)$.

Again, let P be a p-group. It is easily proved that a normal subgroup of P which is maximal subject to being abelian is then precisely its own centralizer. Define

$$SCN(P) = \{A \mid C_P(A) = A \trianglelefteq P\},$$

$$SCN_3(P) = \{A \in SCN(P) \mid m(A) \geqslant 3\},$$

and, for a group G,

$$SCN_3(p) = \{A \in SCN_3(P) \mid P \in Syl_p(G)\}.$$

Here the group G will always be understood by the context.

LEMMA 4.1 (cf. [FSG: p. 109]). *If P is a 2-group and $SCN_3(P) \neq \emptyset$, then P is connected.*

Proof. Let $V \in \mathscr{V}(P)$ and let $Q \in SCN_3(P)$. We first claim that V can be connected to an element of $\mathscr{V}(Q)$. Certainly $C_Q(V) \neq 1$. If $m(C_Q(V)) \geqslant 2$, we may join V to an element of $\mathscr{V}(C_Q(V))$. If $m(C_Q(V)) = 1$, pick $U \in \mathscr{V}(VC_Q(V))$ with $U \cap Q \neq 1$ and choose $u \in U - Q$. Then the Jordan canonical form for the action of u on $\Omega_1(Q)$ must have at least two blocks, and so $|C_{\Omega_1(Q)}(u)| \geqslant 4$. Then U can be joined to an element of $\mathscr{V}(C_Q(u))$. Thus our claim holds.

Now any two elements of $\mathscr{V}(Q)$ are joined. So any two elements of $\mathscr{V}(P)$ are connected.

The relevance of this is that MacWilliams [141] has shown that if P is a 2-group with $SCN_3(P) = \emptyset$, then $r(P) \leqslant 4$. Thus the simple groups with a nonconnected Sylow 2-subgroup will be determined by the classification of simple groups whose sectional 2-rank is at most four. This is a major result of Gorenstein and Harada [99]; we shall list the relevant infinite families after the statement.

THEOREM 4.2. *Let G be a simple group of sectional 2-rank at most four. Then either G is a group of Lie type over a field of odd characteristic and "low" Lie rank (including the possibility that G is a group of Ree type), or else G is isomorphic to one of the following 19 groups:*

$$L_2(8), L_2(16), L_3(4), U_3(4), Sz(8), A_7, A_8, A_9, A_{10}, A_{11},$$

$$M_{11}, M_{12}, M_{22}, M_{23}, J_1, J_2, J_3, Mc \text{ or } Ly.$$

It had previously been shown that the only possible Sylow 2-subgroups for a simple group of 2-rank 2 were dihedral, quasihedral, a wreathed product $Z_{2^n} \wr Z_2$, or a Sylow 2-subgroup of $U_3(4)$, and that the only such simple groups were $L_2(q)$, $L_3(q)$, $U_3(q)$, q odd, A_7, M_{11} and $U_3(4)$ ([2],[3]). Gorenstein and Harada were thus able to concentrate on the case $m_2(G) \geqslant 3$ and showed that the only groups with this property, not listed explicitly in Theorem 4.2, are, for q odd, $G_2(q)$, ${}^3D_4(q)$, $PSp_4(q)$, and $L_4(q)$, $q \not\equiv 1 \pmod 8$, $U_4(q)$, $q \not\equiv -1 \pmod 8$, $L_5(q)$, $q \equiv -1 \pmod 4$, and $U_5(q)$, $q \equiv 1 \pmod 4$. Their proof is extremely long, over 400 pages not counting the many prior characterization theorems they apply, and requires many delicate calculations.

We now turn to the solution of (II). We shall need a definition which we give for arbitrary primes, though our interest here will only be for the prime 2.

Definition. Let G be a group, p a prime, $P \in Syl_p(G)$ and k a positive integer such that $k \leqslant m_p(G)$. Then the subgroup

$$\Gamma_{P,k}(G) = \langle N_G(Q) \mid Q \subseteq P \text{ and } m(Q) \geqslant k \rangle$$

is called the *k-generated p-core* of G.

Note that $\Gamma_{P,k}(G)$ depends on the choice of P, but only up to conjugacy. For arbitrary p, this subgroup has been used by Gorenstein and Lyons in their recent work (see Chapter 4). For $p = 2$, we refer simply to the *k-generated core*, a concept introduced by Gorenstein and Walter [105]; the idea of a 2-generated core appeared in their early work on balanced groups. (See [FSG; Chapter II] and [104].)

Let G be a group and let $S \in Syl_2(G)$. The situation where G has a *proper* 2-generated core, namely $\Gamma_{S,2}(G) \subset G$, is particularly important. This is a generalization of the concept of a strongly embedded subgroup, and it is readily seen from the definitions that group G will have a strongly embedded subgroup if and only if $\Gamma_{S,1}(G) \subset G$. Aschbacher [9] has generalized Bender's theorem and has determined the structure of all groups possessing a proper 2-generated core. For our purposes, we need list only the simple groups that appear in the conclusion of his theorem.

THEOREM 4.3. *Let G be a simple group having a proper 2-generated core. Then G is isomorphic to $L_2(q)$ for suitable $q > 3$, $Sz(2^{2n+1})$, $n \geqslant 1$, $U_3(2^n)$, $n \geqslant 2$, M_{11} or J_1.*

If $G \cong L_2(q)$ and $S \in Syl_2(G)$, the possible values for q and the corresponding structures for the proper 2-generated core H are

$q = 2^n$	$H = N_G(S)$ is strongly embedded,
$q \equiv 3,5 \pmod 8$	$H = N_G(S) \cong A_4$,

or

$q = q_0^{2r+1}$ where $r = 0$ and $\qquad H \cong L_2(q_0)$.

$q_0 = p^{2^n} \equiv 1,7 \pmod 8$ (p prime)

We note the following characterization of proper 2-generated cores. If G is a group and H a subgroup containing a Sylow 2-subgroup S of G, then

$$\Gamma_{S,2}(G) \subseteq H \subset G$$

if and only if

$$m_2(H \cap H^g) \leqslant 1 \quad \text{whenever} \quad g \in G - H.$$

If the inequality were strict, then H would be strongly embedded. In fact, this is the generic case. The groups of 2-rank two, being known, have only to be checked, and J_1 occurs as a special case. The bulk of Aschbacher's proof consists of obtaining a suitable criterion for a strongly embedded subgroup.

Aschbacher's theorem is enough to complete the classification of nonconnected simple groups in view of the following [98: p.65].

PROPOSITION 4.4. *Let G be a group with a connected Sylow 2-subgroup. Then G is nonconnected if and only if G has a proper 2-generated core.*

Proof. Let Γ be the graph associated with $\mathscr{V}(G)$ and let Γ_0 be a connected component of Γ. If $T_0 \in \Gamma_0$ and S is a Sylow 2-subgroup of G containing T_0, then $\mathscr{V}(S) \subseteq \Gamma_0$ since S is connected. Since conjugation by an element of G induces an automorphism of Γ, it follows that the action of G on Γ is transitive on the set of connected components and, if $T \subseteq S$ with $m(T) \geqslant 2$, that

$$N_G(T) \subseteq \langle g \in G \mid \Gamma_0^g = \Gamma_0 \rangle.$$

Thus, if Γ is disconnected, $\Gamma_{S,2}(G) \subset G$.

Conversely, if G has a proper 2-generated core and $S \in Syl_2(G)$, put

$$\Gamma_1 = \mathscr{V}(\Gamma_{S,2}(G)).$$

Then $\Gamma_1 \neq \Gamma$. Now $\Gamma_{S,2}(G) = \Gamma_{S',2}(G)$ for any Sylow 2-subgroup S' of $\Gamma_{S,2}(G)$; hence, if $V \in \Gamma - \Gamma_1$ and V is joined in Γ to a vertex U of Γ_1, we have $[V,U] = 1$ and we may assume that $U \subseteq S$. Now

$$V \subseteq N_G(U) \subseteq \Gamma_{S,2}(G),$$

contrary to assumption. Thus Γ_1 is a union of connected components of Γ (in fact, a single component) and, in particular, Γ is not connected.

As a consequence of these results, further study of simple groups may be restricted to connected groups where certain lines of argument, in particular the signalizer functor methods, work more smoothly. However, the actual division between the general situation and the nonconnected case is not quite as clear cut as this might suggest. The Gorenstein–Harada theorem on groups of sectional 2-rank at most four uses fusion and local arguments to construct Sylow 2-subgroups and explicit centralizers in many situations. These methods are also applied in some low rank problems for connected groups, for example in the work of Foote related to Aschbacher's component theorem which we shall discuss in the next section, and in the work of Harris and Solomon which will be discussed in Section 5 of Chapter 2 and also in Sections 7 and 9 of Chapter 3. These "basic" methods have also been used in the determination of all simple groups with Sylow 2-subgroups of order at most 2^{10} by Beisiegel, Fritz, and Stingl ([26], [72] and [200]).

5. Groups of Component Type

The attack on the general connected simple group begins with a study of the structure of the centralizer of an involution. There is a natural dichotomy which was observed at the outset by Gorenstein and Walter and which was described by Gorenstein in [FSG]; either the centralizer of *every* involution is 2-constrained, or some such centralizer is not. (Recall that a group X is p-*constrained* if $C_X(P) \subseteq O_{p',p}(X)$ when $P \in Syl_p(O_{p',p}(X))$; this is the assertion that the conclusion of the Hall–Higman centralizer lemma holds. Also, write $O(X)$ for $O_{2'}(X)$, the *core* of X.)

This dichotomy is nicely reflected in the known simple groups. In the groups of Lie type defined over fields of characteristic 2, the centralizer of an involution is always 2-constrained; except for some small cases, for groups of Lie type defined over fields of odd characteristic and alternating groups, this is not the case.

In the case of 2-constraint, Gorenstein and Walter were able to prove the following "balanced" theorem [104].

THEOREM 5.1. *Let G be a connected group with $O(G) = 1$. Suppose that the centralizer of every involution of G is 2-constrained. Then $O(C(t)) = 1$ for each involution t of G.*

In fact, under these hypotheses, *all* 2-local subgroups are 2-constrained and corefree. Groups satisfying these conditions are said to be of *characteristic 2 type* and we shall return to this case in the next section. We mention this result here because groups in which the centralizer of some involution is not 2-constrained do *not* satisfy the conclusion of Theorem 5.1 in general. So this is an additional major difference.

In this section, we consider groups in which the centralizer of some involution is not 2-constrained; such a group is of *component type*. (We shall justify this term later.) The attack on these groups falls into three areas.

(I) *The $B(G)$-conjecture.* The object here is to show that if G is a corefree group and t is an involution in G, then $C_G(t)$ is somehow "well-behaved" with respect to $O(C_G(t))$. The attempt to prove this has generalized to the determination of all *unbalanced* groups; a simple group G is unbalanced if it contains an involution t or has an involutory outer automorphism t with $O(C_G(t)) \neq 1$. This determination is known as the *U-conjecture.*

(II) *Aschbacher's component theorem.* This assumes the $B(G)$-conjecture and shows that in general a simple group of component type will have an involution t whose centralizer $C_G(t)$ is "in standard form"; i.e. $C(t)$ has a general structure and certain embedding properties in G which resemble those of a suitably chosen involution in a known simple group. The proof of this does *not* depend on any general properties of the known simple groups.

(III) *Specific characterizations*—the determination of all simple groups having an involution whose centralizer is in a standard form of "known type".

The only stage which is complete is part (II). Most of the problems in (III) have now been solved; in particular, only two remain to be completed to give a proof of (I). (This may seem surprising since the solution of (I) forms the hypothesis for (II) and (III); however, this comes out of the inductive nature of the proofs and the generalization involved.)

Before describing the above problems more carefully, we should observe the relationship between the classification of simple groups of component type and that of groups of characteristic 2 type. Technically they should be considered simultaneously since one takes a minimal counterexample to the list of known simple groups. However, in the case of groups of component type, this really only affects part (III) where a new simple group would just give rise to additional standard form problems; once (I) is proved, (I) and (II) would be unaffected. On the other hand, large parts of the work on groups of characteristic 2 type do assume that every proper subgroup is a $\mathscr{K}$-group, and additional work would be needed if a new simple group did not enjoy those properties of the known simple groups used in the proofs. So, strictly speaking, neither case is absolutely complete without the other.

We now return to groups of component type. To justify this term, we must define components, and prove an important property.

A group X is *quasisimple* if $X = X'$ and $X/Z(X)$ is simple; X is *semisimple* if it is a central product $X_1 \circ \cdots \circ X_n$ of quasisimple groups X_i, $i = 1, \ldots, n$, or if $X = 1$. A *component* of a group X is a subgroup which is subnormal in X and quasisimple.

LEMMA 5.2. *Let A be a component of a group X. If A normalizes a soluble subgroup N of X, then A centralizes N.*

Proof. We shall show that $A = (AN)^{(\infty)}$, for then

$$[A,N] \subseteq A \cap N \subseteq Z(A),$$

and the three subgroup lemma [G; 2.2.3] forces

$$[A,N] = [A,A,N] \subseteq [A,N,A][N,A,A] = 1.$$

Now $A \trianglelefteq\trianglelefteq AN$. If $A \trianglelefteq AN$, there is nothing to prove, so suppose that $M \lhd AN$ with $A \subseteq M \subset AN$. Then $M = A(M \cap N)$ and, by induction, we have $A = M^{(\infty)}$. Hence $A = (AN)^{(\infty)}$.

Definition. A group G is said to be of *component type* if, for some involution t of G, the group $C_G(t)/O(C_G(t))$ has a component; otherwise G is of *noncomponent type.*

By Lemma 5.2, the condition that $C_G(t)/O(C_G(t))$ should have a component is equivalent to the assertion that $C_G(t)$ is *not* 2-constrained, and it is the attention that must now be given to such components that gives rise to this terminology. Characteristic 2 type is a special case of noncomponent type, though it is all we need consider in view of the results discussed in the previous section. However, one must be careful with nonsimple groups; the extension of a group of Lie type over a field of characteristic 2 by an automorphism of order 2 will usually be of component type.

In its simplest form, the object of part (I) of the analysis is to show that if G is a corefree group and t is an involution in G, then any component of $C_G(t)/O(C_G(t))$ is the image of a component of $C_G(t)$ and hence, by Lemma 5.2, acts trivially on $O(C_G(t))$; it is precisely in this form that it is used. In order to state it in its final form, we will need some further standard notation which will also be used elsewhere in this book.

For an arbitrary group X, let $E(X)$ denote the subgroup generated by the components of X. If A is a component of X and $A \neq X$, then $A \trianglelefteq\trianglelefteq M$ for some proper normal subgroup M of X. An easy inductive argument now shows that any component of X will commute with each of its distinct conjugates, and it follows that $E(X)$ is a central product of the components of X (if any), and so is semisimple. (Gorenstein and Walter call this subgroup the *layer* of X and denote it by $L(X)$.) An equivalent definition has been given by Bender [28]; if $F(X)$ is the Fitting subgroup of X, he defines the *generalized Fitting subgroup* $F^*(X)$ to be that subgroup of X containing $F(X)$ for which

$$F^*(X)/F(X) = \text{socle}\,(F(X)C_X(F(X))/F(X)),$$

and puts $E(X) = F^*(X)^{(\infty)}$.

Thus a group G is of component type if

$$E(C_G(t)/O(C_G(t))) \neq 1$$

for some involution t of G. In the *known* simple groups of component type—namely, the alternating groups A_n for $n \geqslant 9$, the groups of Lie type defined over fields of odd characteristic except for $L_2(q)$ and certain small rank groups defined over $GF(3)$, and some of the sporadic groups—it happens that

$$(5.3) \qquad E(C(t)/O(C(t))) = E(C(t))O(C(t))/O(C(t))$$

for every involution t. (This statement is, of course, true whenever $C(t)$ is 2-constrained!) However, *a priori*, there is no reason why this should be the case for an *arbitrary* simple group.

Following Gorenstein and Walter [105], for an arbitrary group X, let $L_{2'}(X)$ denote the perfect normal subgroup of X for which

$$L_{2'}(X)O(X)/O(X) = E(X/O(X));$$

$L_{2'}(X)$ is uniquely determined and is called the 2-*layer* of X. (Bender introduced a subgroup $O_{2',E}(X)$ which is $L_{2'}(X)O(X)$ in this notation.) The conjecture that (5.3) should hold in an arbitrary simple group was part of Gorenstein's original programme and can be restated as

$$(5.3') \qquad L_{2'}(C(t)) = E(C(t));$$

as such, it was referred to as the "semisimplicity of the 2-layer". This was reformulated by Thompson into a conjecture about arbitrary groups and it is in this form, the $B(G)$-conjecture, that we now consider it.

The first step is to generalize the notion of a component. If X is an arbitrary group, then a subnormal subgroup Y is called a 2-*component* of X if $Y = Y'$ and $Y/O(Y)$ is quasisimple. Let $B(X)$ denote the subgroup of X generated by those 2-components of X which are *not* components. An argument similar to that for $E(X)$ shows that $B(X)$ is perfect and that $B(X)/O(B(X))$ is semisimple; indeed,

$$L_{2'}(X) = E(X)B(X).$$

With these definitions, we can now state Thompson's conjecture.

$B(G)$-CONJECTURE. *If G is a group, then $B(C_G(t)) \subseteq B(G)$ for every involution t of G.*

Notice that in a group G for which $O(G) = 1$, necessarily $B(G) = 1$. Thus (5.3) and (5.3′) are immediate consequences of the $B(G)$-conjecture. In fact, the assertions are equivalent (see Corollary 1.17 in Chapter 3). A group for which the $B(G)$-conjecture holds is said to satisfy the *B-property*.

The $B(G)$-conjecture is central to the study of groups of component type, but before going on to the U-conjecture and their relationship, we turn to Aschbacher's *Component Theorem* [10], a result proved before the $B(G)$-conjecture had really been attacked. In effect, Aschbacher's theorem has the

B-property as its sole hypothesis and reduces the classification of simple groups of component type to the proof of the $B(G)$-conjecture and the solution of a "standard form" problem for each quasisimple group. We shall not state his theorem in its original form; we need only the "generic" conclusion in view of subsequent work of Foote [70]. (See Chapter 2, (1.3) and (1.5), and the discussion there.)

THEOREM 5.4. *Let G be a group of component type with $F^*(G)$ simple. Assume that G satisfies the B-property. Then either $F^*(G)$ is a known simple group (of Lie type and low rank) or else G has an involution t such that $C_G(t)$ has a component A which is "standard".*

To say that a subgroup A of a group G is *standard* means that

(i) A is quasisimple,
(ii) $[A,A^g] \neq 1$ for all $g \in G$,

and, if $K = C_G(A)$, then

(iii) $N_G(K) = N_G(A)$,
(iv) K has even order, and
(v) $K \cap K^g$ has odd order for all $g \in G - N_G(K)$.

The term *standard subgroup* was introduced by Aschbacher in [10], where he also defined a *tightly embedded* subgroup as a subgroup K for which (iv) and (v) hold. This concept is closely linked to that of *standard form* for the centralizer of an involution as defined by Gorenstein and Walter [105]. However, their definition required that $m_2(K) = 1$, which is not true in general; when $m_2(K) > 1$, the tight embedding of K has proved a very powerful tool in the subsequent analysis. Also, we note that a standard subgroup A is a component of the centralizer of any involution of $C(A)$.

In the case that there was no standard subgroup, Aschbacher gave some very precise information. Starting with this as hypothesis, Foote showed that the only possibilities for $F^*(G)$ in Theorem 5.4 were, with q odd and $q > 3$,

$$PSp_4(q),\ L_4(q),\ q \not\equiv 1\ (\mathrm{mod}\ 8),\ U_4(q),\ q \not\equiv 7\ (\mathrm{mod}\ 8), \text{ and } G_2(q),\ q \equiv 0\ (\mathrm{mod}\ 3).$$

These groups illustrate rather well why it is not enough just to consider G itself simple. In the first three families, there is no standard subgroup in the simple group itself, but the fixed points under a suitable outer automorphism of order 2 will, in general, yield a standard subgroup in the extension. The situation for $G_2(q)$ is more interesting. For any odd $q > 3$, the simple group $G_2(q)$ does possess a standard subgroup; the centralizer of an involution has two components isomorphic to $SL_2(q)$ which are not conjugate. However, if $q = 3^n$, then the group $G_2(q)$ has a graph automorphism which interchanges these two components, so they cease to be standard in the extension. On the other hand, if n is *odd*, then the Ree group ${}^2G_2(3^n)$ appears as a standard subgroup in a suitable extension!

In general, then, one must solve a series of "standard form" problems, where each known quasisimple group is considered as a standard subgroup. Here, unlike the component theorem, it has been necessary to use explicit detailed knowledge of the known groups. In addition, one may assume for the purposes of the classification programme that the B-property holds in all sections of the group. This assumption is made in Chapter 2 where an account of the standard form problems is given. Most of them have now been solved; in some cases, it was not necessary to make the additional assumption about the B-property. In all the remaining unsolved cases one can assume that $C(A)$ has cyclic Sylow 2-subgroups, and that these can all be completed by solutions to the following standard form problems (in some cases, just the minimal case needed to handle an entire family)‡:

$$A \cong {}^2F_4(2^n), n \geqslant 2, {}^2F_4(2)', 2.{}^2E_6(2), 2.F_2 \text{ and } 2.M(22),$$

and

$$A/Z(A) \cong U_3(3), G_2(3), L_4(3), U_4(3), PSp_4(3), U_6(2) \text{ and } \Omega_8^+(2).$$

(Recall that $n.X$ means a perfect central extension of a simple group X by the cyclic group Z_n.)

Although the B-property is a fundamental hypothesis for the component theorem, the inductive nature of the present approach to its proof actually means that certain standard form problems must be solved first; indeed, all that now remains to be done is to solve the standard form problems for $A/Z(A)$ isomorphic to $L_4(3)$ or $U_4(3)$ *only*, provided that these yield no new simple groups. To prove the $B(G)$-conjecture inductively, one can immediately reduce to the case where, if G is a minimal counterexample, then $F^*(G)$ is simple. Unfortunately one cannot reduce to the case where G is simple because the nonquasisimple 2-component may appear in the centralizer of an *outer* automorphism of $F^*(G)$. Also, one has no inductive hold over the proper subgroups of G. However, it is certainly the case that $O(C_G(t)) \neq 1$ for some involution t of G, and it turns out that one can use this condition inductively. The $B(G)$-conjecture will be an immediate corollary of the determination of all *unbalanced* groups, and this is the route that has been taken.

U-CONJECTURE. *Let G be a group such that $F^*(G)$ is simple, and suppose that G contains an involution t such that $O(C_G(t)) \neq 1$. Then one of the following holds:*

(i) *$F^*(G)$ is an alternating group;*
(ii) *$F^*(G)$ is a group of Lie type defined over a field of odd characteristic;*
(iii) *$F^*(G) \cong L_3(4)$; or*
(iv) *$F^*(G) \cong He$.*

‡ It has been reported that many of these problems have now been eliminated, but details are not yet available.

The attempt to prove the U-conjecture, and hence the $B(G)$-conjecture, is discussed in Chapter 3; its completion depends only on the solution of the two particular standard form problems mentioned earlier. Walter has indicated an alternative approach, and that also is discussed. Together with the completion of the standard form problems, we will have a classification of simple groups of component type, subject to no further groups of *noncomponent* type existing.

6. Groups of Noncomponent Type

In view of the previous two sections, in particular Theorem 5.1, the only remaining phase of the classification programme to be discussed is that of connected groups which satisfy the following definition.

Definition. A group G is of *characteristic 2 type* if $F^*(H) = O_2(H)$ for every 2-local subgroup H of G.

Notice that the groups that have yet to be accounted for are most of the groups of Lie type defined over fields of characteristic 2 (excluding only small cases which appeared as nonconnected groups) and just four of the sporadic groups—M_{24}, Co_2, F_3 and type J_4.

As previously, certain "small" cases have to be dealt with separately. Here we shall mean "small" in two distinct senses: both arose in the N-group paper which, of course, dealt solely with groups of noncomponent type.

First, for an arbitrary group X, define the *2-local p-rank* $m_{2,p}(X)$ by

$$m_{2,p}(X) = \max\{m_p(H) \mid H \text{ a 2-local subgroup of } X\}.$$

Thompson introduced the invariant $e(X)$, the *odd 2-local rank*, defined by

$$e(X) = \max\{m_{2,p}(X) \mid p \text{ any odd prime}\}.$$

Since Frobenius' normal complement theorem implies that a group X with $e(X) = 0$ is nilpotent, we may assume that $e(G) \geqslant 1$. The N-group paper gave three cases, according as

$$e(G) = 1, \quad e(G) = 2, \quad \text{or} \quad e(G) \geqslant 3,$$

and these must be reflected here too, though some special arguments are necessary for $e(G) = 3$. Simple groups G for which $e(G) = 1$ are called *thin*, and those with $e(G) = 2$ *quasithin*. (Here, G need not be of characteristic 2 type.)

Before discussing these, we must consider the other type of "small" problem that arose in the N-group paper. Thompson needed special arguments to handle the case where, for some involution t, the subgroup $O_2(C(t))$ had no noncyclic characteristic abelian subgroup. Such a 2-group is said to be of *symplectic type*; by a theorem of P. Hall, it is a central product of an extra special 2-group and a group which is cyclic, dihedral, generalized quaternion, or quasihedral. This case has to be handled separately in the general characteristic 2 type group also, and cuts across the cases for $e(G)$. Most of the groups of Lie type over $GF(2)$ have this property, motivating the following definition.

Definition. A simple group G is of *$GF(2)$-type* if $F^*(C_G(t))$ is of symplectic type for some involution t of G.

We can now give the precise meaning of part (C) of Gorenstein's classification programme:

Determine all connected simple groups of characteristic 2 type for which either $e(G) \leqslant 2$ or G is of $GF(2)$-type.

Notice that the definition of $GF(2)$-type does not require G to be of characteristic 2 type, but makes an assertion only about a *single* centralizer. Thus some groups of $GF(2)$-type are even of component type. All simple groups of $GF(2)$-type have now been determined; apart from the appropriate groups defined over $GF(2)$, there are some small rank groups of Lie type defined over fields of odd characteristic and many of the known sporadic groups, including the four that had not been characterized previously. A full account of these results is given in Chapter 6, with the actual groups listed by cases, (3.1)–(3.10).

Similarly, one aims to classify all thin and quasithin simple groups by these properties alone. The thin group case was perhaps the hardest in the N-group paper, and subsequent work on thin groups has been modelled on Thompson's approach, even though the problems have been harder. First Janko classified thin simple groups in which all 2-local subgroups were soluble [134], and then Aschbacher gave a complete classification of the thin simple groups [16]. Aschbacher was able to assume that some 2-local subgroup was insoluble in a minimal counterexample, and that any non-abelian composition factor in such a 2-local had cyclic Sylow subgroups for all odd primes; the object of his long and difficult proof was to eliminate all the possibilities. We shall not state his full result here, but just the necessary special case.

THEOREM 6.1. *Let G be a connected thin simple group of sectional 2-rank at least five. Then G is isomorphic to ${}^3D_4(2)$.*

Work is still in progress on the classification of quasithin groups. In Chapter 7, Mason gives an account of his work. He includes thin groups, and although he follows Aschbacher's approach, he has not yet needed to quote his result. By assuming that any necessary standard form problems can be handled in that context, Mason can take his minimal counterexample G to be of characteristic 2 type. Also, by using Janko's result, and the analogous result of F. Smith for $e(G) = 2$, he can assume that some 2-local subgroup of G is insoluble. He is now close to proving the following‡.

(6.2) *Let G be a simple group with $e(G) \leqslant 2$, of minimal order subject to not being a $\mathscr{K}$-group. If G is of characteristic 2 type and every proper subgroup of G is a $\mathscr{K}$-group, then the nonabelian composition factors of the 2-local subgroups of G are amongst the following: $L_2(4)$, $L_3(2)$, $Sp_4(2)'$, $G_2(2)'$, $L_4(2)$, $L_5(2)$, and A_7.*

We now turn to the generic case where $e(G) \geqslant 3$. Here one expects to find groups of Lie type in characteristic 2 plus the second Conway group Co_2 and Thompson's group F_3 (which are both of $GF(2)$-type). There is an extensive survey in the first two sections of Chapter 4, to which the reader is referred; we shall therefore be less precise than previously in our discussion here.

The main work is due to Gorenstein and Lyons [102] and to Aschbacher ([19a], [19b]). Their analysis depends heavily on the assumption that every proper subgroup of the group G under consideration is a $\mathscr{K}$-group, and some of the properties they require still await complete verification. Apart from this and the additional checking that would be required if a new simple group should arise in an earlier phase of the classification programme, part (D) of the programme has now been completed. We shall describe the main results of the complete work here: the reader is reminded that Chapters 4 and 5, as written, represent the position as of mid-1978. In the following discussion, then, assume that G is a simple group of characteristic 2 type with $e(G) \geqslant 3$, and that every proper subgroup of G is a $\mathscr{K}$-group. Also, put

$$\beta_4(G) = \{p \mid p \text{ an odd prime}, m_{2,p}(G) \geqslant 4\}.$$

In Chapter 4, Gorenstein and Lyons describe their work. For the most part they assume that $e(G) \geqslant 4$; we shall discuss the analogous results obtained by Aschbacher in the case $e(G) = 3$ at the end of this section. Their first goal is to obtain an "odd" analogue of Aschbacher's component theorem under suitable hypotheses. In particular they must give conditions which will locate suitable odd primes, and these must reflect the situation that occurs in the groups of Lie type. To state these in full, we would need several technical definitions which are given in Chapter 4; to avoid repeating them, and to concentrate on the conclusions, we shall paraphrase their theorems in the following statements.

THEOREM A. *Suppose that $e(G) \geqslant 4$. Suppose, also, that certain conditions are given which, if satisfied, define a nonempty subset π of $\beta_4(G)$. If $p \in \pi$, then either*

(I) *$p = 3$ and G is of $GF(2)$-type for the prime 3, or*

(II) *G is of standard type for the prime p.*

‡ Mason has now eliminated *all* possible nonabelian composition factors. Thus a minimal counterexample is of component type.

Here, $GF(2)$-type *for the prime* 3 is a particular case of $GF(2)$-type, so the groups that arise in this case are known. *Standard type for the prime* p is the analogue of standard form in the case of groups of component type; namely, G contains an element x of order p such that

(i) $C_G(x)$ has a normal quasisimple subgroup L which is a covering group of a simple group of Lie type of characteristic 2,
(ii) $C_G(L)$ has cyclic Sylow p-subgroups,
(iii) either p divides the order of a Cartan subgroup of L or $p = 3$ and L is defined over $GF(2)$, and
(iv) many other conditions are given.

In their full definition, Gorenstein and Lyons give further information about the prime p depending on the type of L; also, the additional conditions will imply, in particular, that the element x is not weakly closed in a Sylow p-subgroup of $C_G(x)$. This is important in the absence of odd analogues of Glauberman's Z^*-theorem. At first sight it may seem surprising that apparently stronger information is available than as a consequence of Aschbacher's component theorem, but this is a result of the fact that G is of characteristic 2 type and of the assumption that proper subgroups are $\mathscr{K}$-groups.

The possibilities that arise under (II) with the *full* strength of the Gorenstein–Lyons definitions are referred to as *odd standard form problems.* These are discussed in Chapter 8; in particular, Gilman and Griess now have a single theorem which handles all possibilities for $e(G) \geqslant 4$.

This, in a sense, covers one of two cases; the remaining results of Gorenstein and Lyons explain the dichotomy. First they extend the idea of a 2-generated p-core and define, for $P \in Syl_p(G)$, the *weak* 2-generated p-core

$$\Gamma^0_{P,2}(G) = \langle N_G(Q) \mid Q \subseteq P, \quad m_p(Q) \geqslant 2, \quad m_p(QC_P(Q)) \geqslant 3\rangle;$$

they say that $(\Gamma^0_2)_p$ *holds* if $\Gamma^0_{P,2}(G)$ is contained in some 2-local subgroup of G.

THEOREM B. *Suppose that $e(G) \geqslant 4$. Then either the hypotheses of Theorem A can be satisfied, or else $(\Gamma^0_2)_p$ holds for all $p \in \beta_4(G)$.*

Aschbacher has introduced a distinguished subset $\sigma(G)$ of those odd prime divisors p of $|G|$ for which $m_{2,p}(G) \geqslant 3$; in particular, if $e(G) \geqslant 4$, then $\sigma(G) = \beta_4(G)$. The third result of Gorenstein and Lyons is obtained for $e(G) \geqslant 3$ in order to match up with Aschbacher's work.

THEOREM C. *Assume that $e(G) \geqslant 3$ and that $(\Gamma^0_2)_p$ holds for all $p \in \sigma(G)$. Then G possesses an almost strongly p-embedded maximal 2-local subgroup for every $p \in \sigma(G)$.*

Here, to say that a 2-local subgroup M of G is *almost strongly p-embedded* means that, for some $P \in Syl_p(G)$, either $\Gamma_{P,1}(G) \subseteq M$ (in which case M is strongly p-embedded), or else M is soluble, $\Gamma_{P,2}(G) \subseteq M$, and further tight restrictions are given. The conclusion of Theorem C is known as the *Uniqueness Case*. The techniques that must now be applied are very different. Aschbacher has now completed the analysis in this case and has shown that no group can arise. In Chapter 5, he describes the various techniques that he introduced to handle this situation, and he also sketches a proof for the special case $m_{2,3}(G) \geqslant 4$ which he handled first.

This leaves the case of $e(G) = 3$ where one must obtain analogues of Theorem A and B. This has now been completed by Aschbacher.

THEOREM 6.3. *Let G be a simple group of characteristic* 2 *type with $e(G) = 3$, and assume that every proper subgroup of G is a $\mathscr{K}$-group. Then one of the following holds:*

(i) *G is a $\mathscr{K}$-group;*

(ii) *$(\Gamma_2^0)_p$ holds for every $p \in \sigma(G)$; or*

(iii) *G is of standard type for $p = 3$, with odd standard component L isomorphic to $Sp_6(2)$, $L_4(2)$ or $L_5(2)$.*

In case (ii), Theorem C and Aschbacher's work on the Uniqueness Case combine to show that G cannot exist, while the standard form problems in (iii) have been handled by Finkelstein and Frohardt (see Chapter 8). Thus this phase of the classification is now complete.

2

Standard Subgroups in Finite Groups

GARY M. SEITZ

1. Introduction

As a result of Aschbacher's Component Theorem discussed in Chapter 1, one of the key notions in the study of finite groups of component type is that of a standard subgroup. During the last several years a tremendous amount of work has been carried out with the goal of determining all finite groups having a standard subgroup of "known" type. This is a necessary step in the current approach to the classification problem.

In this chapter we shall collect the existing results, and indicate what remains to be done. In several instances we describe general arguments used in proofs of the results. There is not sufficient space for complete proofs, but if one is willing to forget about certain "small" cases (by which we might be referring to order, rank, or the size of the defining field), then many of the arguments are generic and easily understood. The complete results are due to many authors, and the statements given are often combined results. However, references are given for all the original results.

We start by recalling two definitions.

(1.1) *Definition.* A subgroup X of a group G is *tightly embedded* in G if $|X|$ is even, but $|X \cap X^g|$ is odd for each $g \in G - N(X)$.

(1.2) *Definition.* A subgroup A of a group G is a *standard subgroup* of G if A is quasisimple, $N(A) = N(C_G(A))$, $[A, A^g] \neq 1$ for each $g \in G$, and $C_G(A)$ is tightly embedded in G.

It is an easy consequence of the above definitions that if A is a standard subgroup of G, then A is a component of $C_G(t)$ for each involution t in $C_G(A)$. The fact that A commutes with none of its conjugates indicates that, in some sense, A is large.

(1.3) Component Theorem. (Aschbacher [10], Foote [70]). *Let G be a finite group of component type, with $O(G) = 1$. Suppose the $B(G)$-conjecture holds for G. Then, with known exceptions, G contains a standard subgroup.*

More precisely, Aschbacher shows in Theorem 1 of [10] that there exists an involution $t \in G$ and a component A of $C_G(t)$ such that A satisfies certain maximality conditions together with one of the following:

(i) A is standard in G;
(ii) for some component K of G, either $A = K$ or $K \neq K^t$ and

$$A = (C(t) \cap KK^t)';$$

or
(iii) A has 2-rank 1 and there exists a unique conjugate A^g of A such that $[A,A^g] = 1$.

In [70], Foote starts with situation (iii), together with other information available from [10], and determines the possibilities for G. He shows that $E(G) \cong PSp_4(q), L_4(q), U_4(q)$ or $G_2(q)$, for certain odd values of q.

To illustrate the notion of a standard subgroup, let $G = SL(n,q)$ with q and n both odd, and $n \geqslant 7$. Let $t \in G$ be an involution which induces -1 on a hyperplane of the natural module for G. It is clear that

$$E(C_G(t)) = A \cong SL(n-1,q),$$

and one easily checks that A is standard in G. On the other hand, if we let $j \in G$ be an involution with eigenspaces of dimension 3 and $n-3$, then

$$E(C_G(j)) \cong SL(3,q) \times SL(n-3,q).$$

Here, neither component of $C_G(j)$ is standard in G. To see this, just choose an involution z in one of the components with z trivial on an $(n-2)$-space of the natural module for G. Then $C_G(z)$ does not normalize the other component; so the latter is not standard.

For future reference we record a fundamental result on L-balance.

(1.4) (Gorenstein–Walter, (3.1) of [105]). *If X is a 2-subgroup of a group Y, then $L_{2'}(C_Y(X)) \subseteq L_{2'}(Y)$.*

The following basic result indicates how a standard subgroup influences the overall structure of G. We provide a proof of the result in order to indicate how some of the above notions are used.

(1.5) *Let A be a standard subgroup of G. Suppose $O(G) = 1$ and t is an involution in $C_G(A)$. Then one of the following holds:*

(i) $A \trianglelefteq G$;
(ii) $E(G) = X \times X^t$, where $X \cong A$ is simple and $C_G(A) = \langle t \rangle$; or
(iii) $A \subseteq E(G)$ and $E(G)$ is simple.

Proof. By (1.4), $A \subseteq L_{2'}(G) = E(G)$. We first observe that (i) holds if $C_G(E(G)) \neq 1$. For suppose $C_G(E(G)) \neq 1$. As $O(G) = 1$, we may choose an involution

$$z \in C_G(E(G)) \subseteq C_G(A).$$

The tight embedding property coupled with the fact that $N(A) = N(C_G(A))$ shows that $C(z) \subseteq N(A)$. But then $E(G) \subseteq N(A)$ and so A is a component of $E(G)$. Also, $[A,A^g] \neq 1$ for each $g \in G$. Thus, (i) holds. So assume $C_G(E(G)) = 1$ and that (i) is false. In particular,

$$E(G) = G_1 \times \cdots \times G_k,$$

a direct product of simple groups.

For $i = 1,\ldots,k$, let A_i be the projection of A to G_i. Suppose $A_i = 1$ for some i. If z is an involution in G_i, then $z \in C_G(A)$ and, as above, $A \trianglelefteq C_G(z)$. Again this leads to $A \trianglelefteq E(G)$ and then $A \trianglelefteq G$, against our supposition. Therefore, $A_i \neq 1$ for $i = 1,\ldots,k$.

Next, suppose $G_i^t = G_i$, for $i = 1,\ldots,k$. Then $A_i \subseteq C(t) \subseteq N(A)$, for $i = 1,\ldots,k$. Since $A \subseteq A_1 \times \cdots \times A_k$, we conclude that $A = A_i$, for some i. In view of the previous paragraph, $E(G) = G_i$, proving (iii).

Finally, assume $G_i^t = G_j \neq G_i$, for some $i,j \in \{1,\ldots,k\}$. Then, $C = C_{G_iG_j}(t)$ is a maximal subgroup of G_iG_j and $C \subseteq N(A)$. By the above, $A_i \neq 1$, so $[C,A] \neq 1$. By the three subgroup lemma [G;2.2.3], $A = [C,A] \subseteq G_iG_j$; thus $k = 2$ and $E(G) = G_1 \times G_2$. Let $K = C_G(A) \cap N(G_1)$. Then K has index 2 in $C_G(A)$ and $[K,A_1] = [K,A_2] = 1$. However, $A \trianglelefteq C(t)$ implies that $A_i = G_i$ for $i = 1,2$, and so $K \subseteq C_G(E(G)) = 1$. This gives (ii) and completes the proof of (1.5).

In the above result, the first case is regarded as a degenerate case, case (ii) will be referred to as the *wreathed case*, and case (iii) will be called the *quasisimple case*.

The wreathed case causes some difficulties in the study of groups of component type. For, given any simple group A, if we set $G = A \wr Z_2$, the wreathed product, then G is a group having a standard subgroup isomorphic to A. So this configuration is ever present and must be considered in all standard form problems when the standard subgroup is a simple group.

It appears that the notion of a standard subgroup is the correct object of study for groups of component type. The meaning of this remark is twofold. Firstly, we have the Component Theorem that guarantees the existence of such subgroups. Secondly, it has been shown, in nearly all cases, that given a quasisimple group A with $A/Z(A)$ one of the currently known simple groups, one can determine all finite groups G such that $O(G) = 1$, the $B(G)$-conjecture holds for G, and A is a standard subgroup of G. In fact, $O(G) = 1$ is often an unnecessary hypothesis and, in many cases, the $B(G)$-conjecture is not required.

In the context of an inductive proof of the classification of all finite simple groups, one can assume that the simple sections of a minimal counter-example are of known type. In particular, standard subgroups that arise will be of known type. The fact that all but a few known quasisimple groups have been successfully treated as standard subgroups, together with recent results concerning groups that are not of component type, indicates that perhaps the current list of simple groups is near completion (or already complete).

In the remainder of this chapter we will describe the status of the standard form problem. From now on, A denotes a standard subgroup of a finite group G with $O(G) = 1$. In addition, set $K = C_G(A)$ and fix $R \in Syl_2(K)$. We will assume throughout this chapter that the $B(G)$-conjecture holds in all sections of G. This hypothesis is not required for all the results that follow, but it does simplify some statements and is required in a number of results.

In Section 2 we describe results that reduce us to the case where $C_G(A)$ has cyclic Sylow 2-subgroups. The remaining sections consider the possibilities for the group $\tilde{A} = A/Z(A)$. In Section 3 we take $\tilde{A}$ to be a group of Lie type in characteristic 2. Section 4 deals with alternating groups, while in Section 5 $\tilde{A}$ is a group of Lie type in odd characteristic. The sporadic groups are considered in Section 6.

All groups considered here are assumed to be finite. We will use the notational convention that given any group X, $\tilde{X}$ denotes $X/Z(X)$. Let $\mathscr{K}$ denote the class of those groups isomorphic to one of the currently known simple groups. So a group is in $\mathscr{K}$ if it is isomorphic to one of the following:

(i) an alternating group, A_n, $n \geqslant 5$;
(ii) a group of Lie type, $G(q)$;
(iii) a Mathieu group, $M_{11}, M_{12}, M_{22}, M_{23}, M_{24}$;
(iv) a Janko group, J_1, J_2, J_3, J_4;
(v) a Conway group, Co_1, Co_2, Co_3;
(vi) a Fischer group, $M(22), M(23), M(24)'$;
(vii) F_1, F_2, F_3, F_5;
(viii) $HS, Mc, He, Ly, Sz, ON, Ru$.

In the above list, we have included groups of types F_1 and J_4 as known groups; the arguments discussed in this chapter will also carry over to groups of Ree type, though we do not include these explicitly.

The author would like to thank M. Aschbacher, L. Finklestein, R. Solomon and J. Walter for participating in the seminar on standard form problems and for helpful suggestions with regard to this account.

2. Reduction to the Case of R Cyclic

In this section we indicate how the general problem is reduced to the case of R cyclic. The first case to consider is where $m_2(R) > 1$. Shortly after proving the component theorem, Aschbacher considered this situation with $\tilde{A} \cong A_n$ (see [12]). In [22] his result was extended to the following:

(2.1) (Aschbacher–Seitz [22]). *Let $\tilde{A} \in \mathscr{K}$ and suppose $m_2(R) > 1$. Set $X = \langle A^G \rangle$. Then either $X = A$ or $R \cong Z_2 \times Z_2$ and the pair (A,X) is given in the following table:*

	A	X
	A_n	A_{n+4}
	$L_2(4)$	M_{12} $(R \cap X \cong Z_2)$
	$L_2(4)$	J_2
(2.2)	$L_3(4)$	Sz
	$\hat{L}_3(4)$ (4-fold cover)	He
	$Sz(8)$	Ru
	$G_2(4)$	Co_1

Several remarks are in order concerning the above result. The case of $Sz(8)$ was treated in [110]. At the time (2.1) was proved, evidence for the existence of J_4 was not available, so this group was not considered. However, Finkelstein [59] has now shown that only the wreathed configuration can occur when $\tilde{A} \cong J_4$. There is a difficulty regarding the case $\tilde{A} \cong L_3(4)$. The authors appealed to the main result in [154], but it appears that there is an error there. However, Aschbacher and Seitz have subsequently carried out the necessary analysis that they require [22:II]. Also, in their original paper they proved in the case $A \cong G_2(4)$ only that X was "of type Co_1"; this has now been completed.

The following is a very brief outline of the ideas involved in the proof of (2.1). Except for the case $\tilde{A} \cong A_n$, much of the proof is concerned with the elimination of groups as standard subgroups, and we describe only that portion of the proof. The arguments in the A_n case are more along the lines that will be described in the next two sections. First, for a quasisimple group B, we consider the following condition.

Hypothesis (*). Whenever a noncyclic elementary abelian 2-group T acts faithfully on B, with T a Sylow 2-subgroup of a 2-nilpotent tightly embedded subgroup of BT, then $T \subseteq BC(B)$.

Hypothesis (*) is a technical condition which holds for most of the known quasisimple groups. Its importance for the proof of (2.1) is based on the following result, proved in [11].

(2.3) *Suppose that A satisfies Hypothesis* (*), $A \ntrianglelefteq G$, *and $m_2(R) > 1$. Then R is elementary abelian. Moreover, if $g \in G - N(A)$ and*

$$1 \neq T \in Syl_2(K^g \cap N(R)),$$

then either

(a) $|R| = 4$, *or*

(b) $T \in Syl_2(K^g)$ *and* $[T,R] = 1$.

It is possible to choose such an element g such that (b) *holds and* $T \subseteq$ AK.

Suppose now that the pair (A, G) satisfies the hypothesis of (2.1) and $A \subset X$. We assume that A is not an alternating group. The first step is to determine whether or not A satisfies Hypothesis (*). If $\tilde{A} = A/Z(A)$ is a group of Lie type in odd characteristic, this is not particularly difficult (see (4.9) of [22]), and consequently we may choose $T \subseteq AK$ as in (2.3). On the other hand, A cannot admit such a group T unless $A \cong L_2(q)$, $5 \leqslant q \leqslant 9$ ((4.9) of [22]). From here we reduce to the case $q = 5$ or 9, and appeal to [12].

Similar arguments work if $\tilde{A}$ is a sporadic group, other than M_{12}, J_2, He, Sz, Ru, or Co_1. For the latter groups, *ad hoc* arguments are used to obtain a contradiction.

If $\tilde{A}$ is a group of Lie type in characteristic 2, the situation is more complicated. It is true that A satisfies Hypothesis (*) (see (20.1) in [21]), although the proof requires a careful analysis of centralizers of involutory outer automorphisms of A. But, unlike the situation in odd characteristic, the group A usually admits groups such as T. For example, in most cases, if $t \in A$ is an involution, then $O_2(Z(C_A(t))) = Y$ is a TI-set in A, and $|Y| = q$, where $\tilde{A}$ is defined over a field of order q. So $T \subseteq AK$ may project onto Y, if $q \geqslant 4$. In this case the argument is roughly this. Using properties of the groups $\langle Y, Y^a \rangle$, for certain $a \in A$, it is shown that T can be replaced by $T^g \subseteq AR$ such that T^g projects into a root subgroup U_r of A (usually, r is a long root), and from here induction can be applied as follows. Let $J_r = \langle U_r, U_{-r} \rangle$ and $D = E(C_A(J_r))$. Then one shows that D is a standard subgroup of $C_G(J_r)$,

$$R \in Syl_2(C_G(J_r) \cap C_G(D)) \quad \text{and} \quad DO(C_G(J_r)) \ntrianglelefteq C_G(J_r).$$

Therefore, we may apply induction, and in all but a few cases this yields a contradiction. The exceptional cases either lead to an acceptable configuration or are eliminated by special arguments.

In view of Finkelstein's work on J_4, (2.1) remains valid when $\tilde{A}$ is an arbitrary group in $\mathscr{K}$. It is, therefore, permissible to assume $m_2(K) = 1$ from now on.

The case of R a generalized quaternion group is handled by the following beautiful result of Aschbacher.

(2.4) (Aschbacher, Corollary II in [15]). *Let G be a finite group with $F^*(G)$ simple, and suppose K is a tightly embedded subgroup of G having generalized quaternion Sylow 2-subgroups. Then $F^*(G)$ is isomorphic to a group of Lie type in odd characteristic, M_{11}, or M_{12}.*

Note that (2.4) only assumes the existence of K. The standard subgroup is not required. This indicates the force of the tight embedding property when the Sylow 2-subgroups of K are not cyclic.

We will not discuss the proof of (2.4) in this chapter. However, we remark that, except for the cases leading to M_{11} and M_{12}, it is shown that $F^*(G)$ is generated by particular subgroups of G, with relations among those subgroups similar to those in a group of Lie type. We shall discuss this method of identifying groups of Lie type in the discussion of (3.5) in the next section.

In view of the results of this section we will assume the following hypotheses for the remainder of the chapter:

(1) $O(G) = 1$;
(2) the $B(G)$-conjecture holds in all sections of G; and
(3) R is cyclic.

3. $\tilde{A} \cong G(2^a)$

In this section we take $\tilde{A} \cong G(2^a)$, a group of Lie type defined over a field of characteristic 2. Recall that R is now assumed to be cyclic; so let $\langle t \rangle = \Omega_1(R)$. Moreover, we have the standing hypotheses that $O(G) = 1$ and the $B(G)$-conjecture holds in all sections of G.

One would expect that, except for a few small exceptions, either we are in the wreathed case ($E(G) \cong A \times A$) or $E(G)$ is also a group of Lie type in characteristic 2. Notice that in the latter case t must necessarily induce an outer automorphism of $E(G)$—a graph, field, or graph–field automorphism. (The possible pairs $(A, \langle A^G \rangle)$ are described in Section 19 of [21].)

In describing the results there is a natural subdivision, according to whether $\tilde{A}$ has Lie rank 1 (i.e., a Bender group), Lie rank 2, or Lie rank at least 3.

(3.1) (Griess, Mason and Seitz [110]). *Suppose $\tilde{A}$ is of Lie rank 1 and $A \ntrianglelefteq G$. Then one of the following holds:*

(i) $E(G) \cong A \times A$;
(ii) *$E(G)$ is a group of Lie type in characteristic 2 and t induces an outer automorphism on $E(G)$;*
(iii) *$\tilde{A} \cong L_2(4)$ and $E(G) \cong M_{12}, A_9, J_1, J_2, A_7, L_2(25), L_3(5), U_3(5), G_2(5)$, or ${}^3D_4(5)$;*
(iv) *$A \cong Sz(8)$ and $E(G) \cong Ru$; or*
(v) *$A \cong L_2(8)$ and $E(G) \cong G_2(3)$.*

The idea of the proof of (3.1) is as follows. The basic goal is to find a Sylow 2-subgroup of G, transfer out portions of this 2-group, and then quote an appropriate characterization theorem. In cases (i) and (ii) we appeal to Goldschmidt [91] or Gilman–Gorenstein [74], while a number of other cases can be handled using Gorenstein–Harada [99]. The groups of characteristic 3 and 5 result as a consequence of the isomorphisms $L_2(8) \cong {}^2G_2(3)'$ and $L_2(4) \cong L_2(5)$.

To determine the Sylow 2-subgroups of G, a "pushing-up" argument is used. Start with $S \in Syl_2(A)$ and consider $Y = N_G(\Omega_1(Z(SR)))$. The group

$\Omega_1(Z(SR))$ is elementary abelian of order 2^{a+1} and, typically, one shows that Y acts as a 2-transitive permutation group on $\Omega = t^G \cap \Omega_1(Z(SR))$, that $\Omega \neq \{t\}$, and that in this permutation representation the one point stabilizer contains a normal subgroup regular on the remaining points of Ω. The main result of [124] applies and, in most cases, shows that Y^Ω contains a regular normal 2-subgroup. Sylow 2-subgroups of Y are manageable, and either they are Sylow groups of G or the process of pushing up can be repeated. Eventually, a Sylow 2-subgroup of G is obtained and fairly delicate arguments are used to determine the structure of this 2-group as well as the fusion pattern. At this point, transfer arguments are possible, and the characterization theorems apply.

The above arguments are, for most part, carried out within 2-local subgroups. To a large extent this is also true when $\tilde{A}$ has Lie rank 2, but as the Lie rank increases these arguments are only relevant in the early stages of the proofs.

Suppose $\tilde{A}$ has Lie rank 2. Then $\tilde{A}$ is one of the groups $L_3(2^a)$, $Sp_4(2^a)'$, $U_4(2^a)$, $U_5(2^a)$, $G_2(2^a)'$, ${}^3D_4(2^a)$, or ${}^2F_4(2^a)'$. Most cases have now been considered, the major exception being the last case. The results usually deal with one family at a time, and often separate papers exist for the cases $q = 2$ and $q \geqslant 4$. This last subdivision is just a technicality for the groups $U_5(2^a)$ and ${}^3D_4(2^a)$, but for several cases there is a genuine difference, owing to the fact that the groups $L_3(2)$, $Sp_4(2)'$, $G_2(2)'$ and $U_4(2)$ ($\cong PSp_4(3)$) all have dual roles—as groups of both even and odd characteristic. For our purposes it is convenient to regard $L_3(2)$ as $L_2(7)$, $Sp_4(2)'$ as A_6, and $G_2(2)'$ as $U_3(3)$. These groups will be discussed in the corresponding section. The following theorem summarizes the existing results (with the exception of $A \cong U_4(2)$, which is considered later) when $Z(A)$ has odd order. Later in this section, we shall discuss the situation where $Z(A)$ has even order.

(3.2) *Suppose $A \ntrianglelefteq G$, $|Z(A)|$ is odd, and one of the following holds:*

(i) $\tilde{A} \cong L_3(2^a)$, $a \geqslant 2$ (*Seitz* [171]);
(ii) $\tilde{A} \cong Sp_4(2^a)$, $a \geqslant 2$ (*Gomi* [94]);
(iii) $\tilde{A} \cong U_4(2^a)$ *or* $U_5(2^a)$, *with* $a \geqslant 2$ (*Miyamoto* [152], [153]); *or*
(iv) $\tilde{A} \cong U_5(2)$, $G_2(2^a)$ $a \geqslant 2$, ${}^3D_4(2^a)$ (*Yamada* [234], [235], [236]).

Then either $E(G) \cong A \times A$ or $E(G)$ is a group of Lie type in characteristic 2 and t induces an outer automorphism on $E(G)$.

The methods used in dealing with the various cases of (3.2) are often quite similar and it would be nice if, eventually, these groups could be dealt with simultaneously. The basic idea is this. Let A_1, A_2 be the distinct maximal parabolic subgroups of A that contain a fixed Sylow 2-subgroup of A. For each of these 2-local subgroups one applies a pushing up procedure that eventually results in 2-local subgroups, P_1 and P_2, of G such that, for $i = 1,2$,

$$A_i \subset P_i, \qquad O_2(A_i) \subseteq O_2(P_i),$$

and P_i contains a Sylow 2-subgroup of G. The connections existing between A_1 and A_2 force restrictions on the groups P_1 and P_2, and (3.1) can be used to identify $P_i/O_2(P_i)$ for $i = 1,2$. For the cases where $E(G)$ should turn out to be a group of Lie rank 2, the idea is to construct a (B,N)-pair in G, in such a way that P_1 and P_2 are maximal parabolic subgroups of G. The identification is then made by appealing to [69]. In the case where $E(G)$ should be a direct product of two copies of A, this is reflected in the groups P_1 and P_2. Namely, P_1 and P_2 are direct products and the construction of a subgroup $G_0 \subseteq G$ with $G_0 \cong A \times A$ is not too difficult. However, showing that $G_0 \trianglelefteq G$ (hence $G_0 = E(G)$) can be quite complicated. It is toward this end that it sometimes is convenient to use the $B(G)$-conjecture.

(3.3) (Gomi [95]). *Suppose $A \cong U_4(2)$ and $A \ntrianglelefteq G$. Then one of the following holds*‡:

(i) *$E(G) \cong PSp_4(9)$, $U_4(3)$, $L_4(3)$, or $L_5(3)$;*

(ii) *$E(G) \cong L_4(4)$ or $U_4(2) \times U_4(2)$; or*

(iii) *there is a 2-central involution $z \in A$ such that $C_G(z)$ contains a quasisimple subgroup L satisfying*

  (a) *$z \in L$,*

  (b) *$W = O_2(L) \cong Z_4$,*

  (c) *$[L,O(C_G(z))] = 1$,*

  (d) *$L/\langle z\rangle$ is a standard subgroup of $C_G(z)/\langle z\rangle$ and W is a Sylow 2-subgroup of $C_G(L/\langle z\rangle)$, and*

  (e) *either $L/O(L) \cong SU_4(3)$ or $L/Z(L)$ has a Sylow 2-subgroup isomorphic to a Sylow 2-subgroup of $L_6(q)$, $q \equiv 3$ (mod 4).*

We remark that Foote has shown that the latter case in (e) does not occur. However, the case of $L/O(L) \cong SU_4(3)$ is a possibility, and should lead to $E(G) \cong U_5(3)$. For the moment, this remains open.

The next result deals with the case where $\tilde{A}$ has Lie rank at least 3. The result is not quite inclusive since the cases $\tilde{A} \cong Sp_6(2)$, $U_6(2)$, and $\Omega_8^{\pm}(2)$ require different arguments.

CONJECTURE (3.4). *Suppose $\tilde{A} \cong Sp_6(2)$, $U_6(2)$, or $\Omega_8^{\pm}(2)$, $|Z(A)|$ is odd, and $A \ntrianglelefteq G$. Then one of the following holds:*

(i) *$E(G) \cong A \times A$;*

(ii) *$E(G)$ is a group of Lie type in characteristic 2 and t induces an outer automorphism on $E(G)$; or*

(iii) *$A \cong \Omega_8^{+}(2)$ and $G \cong Aut(M(22))$.*

‡ The argument of [95] used an incorrect knowledge of the Schur multiplier of M_{22}. However, Gomi has carried out the necessary additional arguments under the assumption that the $B(G)$-conjecture holds.

Work on (3.4) is nearly complete. Alward [6] has completed the case $\Omega_8^-(2)$ and Egawa [49] has handled $\Omega_8^+(2)$. Gomi has just announced a solution of the case $Sp_6(2)$ and Yamada has completed the case $U_6(2)$ [237].

(3.5) (Seitz, [172]). *Assume $\tilde{A}$ has Lie rank at least 3, $|Z(A)|$ is odd, $A \ntrianglelefteq G$, and Conjecture (3.4) holds. Then one of the following holds:*

(i) *$E(G) \cong A \times A$;*

(ii) *$E(G)$ is a group of Lie type in characteristic 2 and t induces an outer automorphism on $E(G)$; or*

(iii) *$A = \Omega_8^+(2)$ and $G = Aut(M(22))$.*

Before outlining the proof of (3.5), we remark that Conjecture (3.4) is required only when $\tilde{A}$ is defined over the field of 2 elements. The proof proceeds along the following lines. The case $\tilde{A} \cong L_n(2^a)$ is treated in [171], where the arguments are basically those of the rank 2 configurations. So assume $\tilde{A} \ncong L_n(2^a)$. The first step is to find a nice conjugate t^g of t, with $t^g \in C(t) - \{t\}$.

Let Σ be the root system associated with A, and let U_α, $U_{-\alpha}$ be opposite root subgroups of A for α a long root in Σ. Set $J_\alpha = \langle U_\alpha, U_{-\alpha} \rangle$. The fusion analysis is aimed at showing that there exists $t^g \neq t$ with $t^g \in A^\# t \cap C(J_\alpha)$ (for the orthogonal groups, things are a bit different). Suppose false. By the Z^*-theorem, we immediately have

$$t \neq t^g \in C(t) \subseteq N(A),$$

and it is not difficult to show that we may choose $t^g \in A^\# \langle t \rangle$. Say x is the projection of t^g to A. Using the results of [21], is usually possible to conclude $t^g \in A^\# t$. For example, this will happen if $x \in C_A(x)^{(\infty)}$. Also, the results of [21] show that x centralizes an A-conjugate of J_α, unless A is a classical group and x has large Jordan rank. The procedure is now similar to that in (3.1). Let

$$Q = \Omega_1(Z(O_2(C(\langle t, t^g \rangle))))$$

and consider the action of $Y = N_G(Q)$ on $t^G \cap Q$. It is shown that Y^Ω is a 2-transitive or rank 3 permutation group but, unlike the situation in (3.1), various results on permutation groups lead to a contradiction. The point is that the 1-point stabilizer is known (from the results in [21]), as is the 2-point stabilizer, and the configuration is impossible.

Once we have $t \neq t^g \in A^\# t \cap C(J_\alpha)$, we set $D = E(C_A(J_\alpha))$ and $E = E(C_G(J_\alpha))$. It is straightforward to see that D is a standard subgroup of $C_G(J_\alpha)$ and

$$R \in Syl_2(C_G(J_\alpha) \cap C_G(D)).$$

The existence of t^g leads to the statement that $DO(C_G(J_\alpha)) \ntrianglelefteq C_G(J_\alpha)$. Therefore, we are in a position to apply induction, Conjecture (3.4), or (3.2).

We find that either $E \cong D \times D$ or E is a group of Lie type in characteristic 2 and t induces an outer automorphism. In the first case we expect $E(G) \cong A \times A$, and in the second case $E(G)$ should be simple.

Consider the following example. Say $A \cong Sp_{2n}(2^a)$, $n \geqslant 4$. Then $D \cong Sp_{2n-2}(2^a)$ and we suppose $E \cong Sp_{2n-2}(2^{2a})$. So t induces a field automorphism on E, and we aim at proving $E(G) \cong Sp_{2n}(2^{2a})$. What follows is a general method used to construct groups of Lie type. We begin by introducing the Dynkin diagram of A:

$$\underset{n}{\circ} \Rightarrow \underset{n-1}{\circ} - \circ \cdots \underset{2}{\circ} - \underset{1}{\circ}.$$

Letting $\{\alpha_1, \ldots, \alpha_n\}$ be a system of fundamental roots for Σ, we have $A = \langle J_{\alpha_1}, \ldots, J_{\alpha_n} \rangle$, and we may choose α so that $D = \langle J_{\alpha_2}, \ldots, J_{\alpha_n} \rangle$. Now, E also has Dynkin diagram of type C_{n-1} and there exist unique root subgroups $\hat{U}_{\alpha_i}$, $\hat{U}_{-\alpha_i}$ of E such that $U_{\alpha_i} \subset \hat{U}_{\alpha_i}$ and $U_{-\alpha_i} \subset \hat{U}_{-\alpha_i}$, for each $i \in \{2, \ldots n\}$.

Next we choose $s \in J_{\alpha_i}$, for $i = 1, \ldots, n$, such that s_i is an involution in $N - H$, where H is a fixed Cartan subgroup normalizing each U_α, $\alpha \in \Sigma$, and the Tits system of A is given by

$$\left(\left(\prod_{\alpha > 0} U_\alpha\right) H, N\right).$$

Then,

$$U_{\pm\alpha_2}^{s_1 s_2 s_3} = U_{\pm\alpha_1} \quad \text{and} \quad U_{\pm\alpha_3}^{s_1 s_2 s_3} = U_{\pm\alpha_2}.$$

Setting $w = s_1 s_2 s_3$, we have $D^w = \langle J_{\alpha_1}, J_{\alpha_2}, J_{\alpha_4}^w, \ldots, J_{\alpha_n}^w \rangle$. The group to look at is $G_0 = \langle E, E^w \rangle$. Define $\hat{U}_{\alpha_1} = \hat{U}_{\alpha_2}^w$ and $\hat{U}_{-\alpha_1} = \hat{U}_{-\alpha_2}^w$. The idea is to show, first, that $G_0 = \langle \hat{U}_{\pm\alpha_1}, \ldots, \hat{U}_{\pm\alpha_n} \rangle$, and to then show $G_0 \cong I = Sp_{2n}(2^{2a})$. The former is fairly easy, the latter requires some work. We apply a result of Curtis (Theorem 1.4 of [46]). That result shows that if for each i, $j \in \{1, \ldots, n\}$ there is a (natural) isomorphism between the groups $\langle \hat{U}_{\pm\alpha_i}, \hat{U}_{\pm\alpha_j} \rangle$ and the corresponding subgroups in Y, and if these isomorphisms agree on common domains, then G_0 is isomorphic to a central extension of Y.

In our situation there is not too much to check. If $i,j \in \{2, \ldots, n\}$ we have nothing to check. Suppose $i = 1$. For $j = 2$,

$$\langle \hat{U}_{\pm\alpha_1}, \hat{U}_{\pm\alpha_2} \rangle = \langle \hat{U}_{\pm\alpha_2}, \hat{U}_{\pm\alpha_3} \rangle^w,$$

although it is not obvious that $\hat{U}_{\pm\alpha_3}^w = \hat{U}_{\pm\alpha_2}$. For the last item, we must show that the Weyl group of A permutes the root subgroups of G just as it permutes those of A. There is some work involved here, but once this is achieved it is not difficult to obtain the missing relations:

$$[\hat{U}_{\pm\alpha_1}, \hat{U}_{\pm\alpha_j}] = 1, \quad \text{for} \quad j > 2.$$

The above configuration is, by far, the nicest, since the root system for D was the same as that for E, and the root system of E was well placed in that of G_0. This is not always the case, and constructions can be much more difficult. We remark that starting from $A \cong Sp_{2n}(2^a)$, any of the following are candidates for $E(G)$: $Sp_{2n}(2^{2a})$, $L_{2n}(2^a)$, $L_{2n+1}(2^a)$, $U_{2n}(2^a)$, $U_{2n+1}(2^a)$, $\Omega^+_{2n+2}(2^a)$, $\Omega^-_{2n+2}(2^a)$, and $A \times A$.

Having constructed a group G_0, the rest of the proof is concerned with showing $G_0 \trianglelefteq G$. This implies $G_0 = E(G)$. This part of the proof is fairly complicated and differs considerably in the quasisimple and wreathed cases.

Let $\Omega = \{G_0^g \mid g \in G\}$ and let $\alpha = G_0$. We wish to show that $\Omega = \{\alpha\}$. In the quasisimple case one uses properties of involutions in $Aut(G_0)$ to show that, with a few exceptions, $G_\alpha = N_G(G_0)$ controls G-fusion of its involutions and t fixes just the point α. This information (and the arguments leading to it) are then used to determine $C_G(z)$ when $z \in G_0$ is a 2-central involution. It is shown that z fixes just α, at which point we get $G_0 \trianglelefteq G$ by appealing to a theorem of Holt [129]. In the wreathed case one uses the hypothesis concerning the $B(G)$-conjecture together with induction to prove that $C_G(X) \subseteq G_\alpha$ for each 2-group $X \neq 1$ lying in one of the components of G_0. Once this is achieved, a result of Aschbacher (Theorem 5 of [10]) yields $G_0 \trianglelefteq G$.

A final remark concerning the last part of the proof of (3.5)—the proof that $G_0 \trianglelefteq G$. The arguments outlined above required detailed information about the group G_0. In the quasisimple case, certain properties of involutions in $Aut(\tilde{G}_0)$ were required, and this information was available from the work in [21]. In the wreathed case we needed information on $Aut(U)$, where

$$U = \prod_{\alpha > 0} U_\alpha$$

is a Sylow 2-subgroup of A. This information was developed during the course of the proof of (3.5). At the same time, one has the impression that knowing a great deal about $C_G(t)$ and having the existence of the group G_0, it should be possible to prove general results giving the normality of G_0 in G. Unfortunately, there are obstructions to proving such a result. For consider the following example. Consider the embeddings

$$Sp(2n,2^a) \subset SU(2n,2^a) \subset SU(2n+1,2^a),$$

and denote the groups by A, B, C, respectively. The involutory field automorphism of C (which may also be viewed as a graph automorphism) has A as a standard subgroup and stabilizes B. So if we let $B \subseteq C$ take the place of $G_0 \subseteq G$, we do not have $B \trianglelefteq C$, even though $A = E(C_G(t)) \subseteq B$ and A is standard in B.

To complete the discussion of groups of Lie type over fields of order a power of 2 we must consider the cases where $\tilde{A} \cong G(2^a)$ and $Z(A)$ is of even order. This only occurs for twelve choices of $\tilde{A}$ and in each case $a = 1$ or 2

TABLE 1

$\tilde{A}$	$\widetilde{E(G)}$	Reference
$L_2(4)$	$L_3(5)$, $U_3(5)$	Gorenstein–Harada [99]
$L_3(2)$	$L_3(7)$, $U_3(7)$	Gorenstein–Harada [99]
$L_3(4)$	ON	Nah [154]
$Sp_4(2)'$	$L_3(9)$, $U_3(9)$	Aschbacher [15]
$U_4(2)$	None	Seitz [173]
$G_2(4)$	None	Solomon [190] or Seitz [173]
$Sp_6(2)$	Co_3	Seitz [173]
$U_6(2)$	?	
$L_4(2)$	Mc	Solomon [186]
$\Omega_8^+(2)$	?	
$F_4(2)$	None	Seitz [173]
${}^2E_6(2)$	?	

The methods involved vary and are somewhat *ad hoc*. Recall that we are still under the assumption that R is cyclic. It is usually the case that t is a 2-central involution, so a Sylow 2-subgroup of G is visible. In small cases this Sylow group has sectional 2-rank at most 4 and G is determined by Gorenstein–Harada [99]. In some cases one is led to a contradiction fairly easily. In configurations that do, in fact, lead to a simple group, one can often reduce to the exact centralizer of an involution, or at least the correct Sylow 2-subgroup of G. At this point it is usually possible to quote a previous characterization theorem. For example, in [173] it is easily shown that if $A \cong \hat{Sp}_6(2)$, then A contains a Sylow 2-subgroup of G, so G is determined by Solomon [185].

In Table 1, we list the possibilities for $\tilde{A}$ (we have included $L_2(4)$, $L_3(2)$, and $Sp_4(2)'$ here), followed by the existing information (if any) available, and a reference. In each case the cyclic group R is the centre of A. For the case $\widetilde{E(G)} \cong ON$, $R \cong Z_4$. In all other cases $R \cong Z_2$.

4. $\tilde{A} \cong A_n$, $n \geqslant 5$

In this section we assume $\tilde{A} \cong A_n$, $n \geqslant 5$, and continue the assumptions that R is cyclic and that the $B(G)$-conjecture holds in all sections of G. Recall, $\langle t \rangle = \Omega_1(R)$. The results here are complete, except that one of the results assumes that the U-conjecture holds in all proper sections of G.

There are similarities between this case and the situation in Section 3. For small values of n, there are complexities due to the exceptional isomorphisms

$$A_5 \cong L_2(4) \cong L_2(5), \quad A_6 \cong L_2(9) \cong Sp_4(2)', \quad \text{and} \quad A_8 \cong L_4(2).$$

In addition, A_n has a covering group $\hat{A}_n$, where $Z(\hat{A}_n) \cong Z_2$, except for $n = 6,7$, where $Z(\hat{A}_n) \cong Z_6$. For some small values of n, these coverings give rise to exceptional situations. On the other hand, for $n \geqslant 12$ things smooth out, and using arguments having features similar to those described for the proof of (3.5), it is shown that the expected occurs. That is, $E(G) \cong A$, $A \times A$, or A_{n+2}, where in the last case t acts as a transposition on $E(G)$.

(4.1) (Solomon [188]). *Suppose that* $A \cong A_n$, $n \geqslant 8$, *and* $A \ntrianglelefteq G$. *Suppose, also, that the U-conjecture holds in all proper sections of G. Then one of the following holds:*

(i) $E(G) \cong A \times A$ *and t interchanges the components of G*;
(ii) $E(G) \cong A_{n+2}$ *and t induces a transposition on* $E(G)$;
(iii) $n = 8$ *and* $E(G) \cong L_4(4)$ *or HS*; *or*
(iv) $n = 10$ *and* $E(G) \cong F_5$.

We indicate, below, some of the ideas used in the proof of (4.1). Suppose $n \geqslant 12$. Choose $X \subseteq A$, with X the pointwise stabilizer of an $(n-4)$-set. Then $X \cong A_4$ and $D = C_A(X) \cong A_{n-4}$. The idea is to show that D is a standard subgroup of $C_G(X)$, with

$$R \in Syl_2(C_G(X) \cap C_G(D)) \quad \text{and} \quad DO(C(X)) \ntrianglelefteq C(X).$$

This involves showing that $t^g \in C(X)$ for some $t^g \in C(t)$-$\{t\}$. But once this has been achieved, induction can be applied to the group $C_G(X)$. (However, we must allow for the cases $8 \leqslant n-4 \leqslant 11$).

Typically, $E = \cdot E(C_G(X) \cong D \times D$ or A_{n-2}. The former leads to the wreathed case and the latter to the case $E(G) \cong A_{n+2}$. This involves two steps. First, construct a group G_0 of the desired type, and then show that $G_0 = E(G)$.

The first step involves a careful pasting together of certain subgroups of G. For example, suppose $E\langle t\rangle \cong S_{n-2}$ with t a transposition. Then

$$C(t) \cap E\langle t\rangle \supseteq S_{n-4},$$

and there is a conjugate $t^g \neq t$, such that $A\langle t^g\rangle \cong S_n$. Consider the set $\{1,\ldots,n\}$ and regard t as $(n+1, n+2)$. Then label

$$A\langle t^g\rangle = \langle (1,2),\ldots,(n-1,n)\rangle.$$

The missing transposition is $(n,n+1)$. By looking at $C_G((1,2))$ one can find this element and determine the necessary relations to conclude that

$$G_0 = \langle (1,2),\ldots,(n+1,n+2)\rangle \cong S_{n+2}.$$

Suppose $E = D_1 \times D_2 \cong D \times D$. Choose $a \in A$ such that a is the product of four 2-cycles, $X^a \subseteq D$, and $C(a) \supseteq D \cap D^a \cong A_{n-8}$. Clearly $E^a = D_1^a \times D_2^a$ and one shows that

$$E \cap E^a \cong (D \cap D^a) \times (D \cap D^a).$$

Replacing a by at, if necessary, we may assume that

$$D_1 \cap D_1^a \cong D \cap D^a \cong D_2 \cap D_2^a.$$

Set $X_i = \langle D_i, D_i^a \rangle$, for $i = 1,2$. Then

$$C(D_1 \cap D_1^a) \supseteq \langle D_2, D_2^a \rangle = X_2,$$

and additional arguments yield that $[X_1, X_2] = 1$. However, $X_1^t = X_2$ and $A \subseteq X_1 \circ X_2$. It follows that $G_0 = X_1 X_2 \cong A \times A$.

The proof that $G_0 = E(G)$ differs in the quasisimple and wreathed cases (as in (3.5)). The quasisimple case is the more straightforward of the two.

When $8 \leqslant n \leqslant 11$, the arguments are more complicated. This is to be expected in view of the exceptions that occur when $n = 8$ or 10. In addition to the above arguments, Solomon considers the group $N_G(E)$, where $E \cong Z_2 \times Z_2 \times Z_2$ and is regular on $\{1, \ldots, 8\}$.

For $n = 5,6,7$ different techniques are used. More emphasis is placed on finding properties of the Sylow 2-subgroups of G. Combining several results we have

(4.2) *Suppose $A/O(A) \cong A_5, A_6$, or A_7 and $A \ntrianglelefteq G$. Then one of the following holds:*

(i) *$E(G) \cong A \times A$;*

(ii) *$E(G)$ has sectional 2-rank at most 4 (hence is determined in [99]); or*

(iii) *$E(G) \cong U_4(3), PSp_4(3), Sp_4(4), L_5(2), L_4(3), HS$, or He.*

The proof of (4.2) is due to Harada [115], Harris–Solomon [121], Harris [118] and Fritz [72]. Suppose (i) does not hold. Then (1.5) implies that $E(G)$ is simple. Since R is assumed to be cyclic, the structure of $C(t)/O(C(t))$ is severely restricted, which in turn restricts the possibilities for a Sylow 2-subgroup of G. Let $S \in Syl_2(C(t))$. If S contains a self-centralizing elementary abelian subgroup of order 8, then Theorem 2 of [115] implies that (ii) holds. For example, this occurs if $R = \langle t \rangle$ and $S \subseteq AR$. Assuming (ii) false, the results of [118] and [121] show that $E(G)$ has Sylow 2-subgroups of order at most 2^{10}, and one can apply the results of Beisiegel [25], Fritz [72] and Stingl [200].

The next result handles the case where $Z(A)$ has even order.

(4.3) (Solomon [186], Aschbacher [15]). *Suppose $|Z(A)|$ is even and $A \ntrianglelefteq G$. Then one of the following holds:*

(i) *$n = 8$ and $E(G) \cong Mc$;*

(ii) *$n = 11$ and $E(G) \cong Ly$;*

(iii) *$n = 5$ and $E(G) \cong L_3(5)$ or $U_3(5)$; or*

(iv) *$n = 6$ and $E(G) \cong L_3(9)$ or $U_3(9)$.*

The cases $n = 5,6,7$ are consequences of Corollary 2 of [15], while Solomon deals with larger values of n. The cases $8 \leqslant n \leqslant 11$ are exceptional and were, for the most part, settled as centralizer of involution results at the time when the existence of Mc and Ly became apparent. See Theorem 4.4 of [186] for the appropriate credits.

The main result of [186] is proved in a context much more general than our present situation, so as to be applicable to the proof of the U-conjecture. But it is not too difficult to illustrate the ideas involved in dealing with our configuration, for n suitably large. Suppose that $A = \hat{A}_n$ and $A \ntrianglelefteq G$. In order to make the ideas more transparent we will take $n \geqslant 17$, although roughly the same arguments work when $n \geqslant 12$.

By the Z^*-theorem we may choose $t \neq t^g = s \in C(t)$. Then s induces an involution $\tilde{s}$ on $\tilde{A} \cong A_n$ and, since $Aut(A_n) = S_n$, we regard $\tilde{s}$ as an involution in S_n. We may assume

$$\tilde{s} = (1,2)\ldots(2m-1,2m).$$

If $l = n-2m \geqslant 4$, then $C_A(s)$ contains $I \cong \tilde{A}_l(\hat{A}_4 \cong SL(2,3))$, and $\langle t\rangle = Z(I)$. But viewing I as a subgroup of $C_G(s)$, it is not difficult to see that t must centralize A^g—a contradiction. Therefore, $n-2m \leqslant 3$, and $m \geqslant 7$.

Let $z \in A$ satisfy $\tilde{z} = (1,2)\,(3,4)$. Then $o(z) = 4$ and $C_A(s) \cap C_A(z)$ contains a subgroup $J \cong A_{m-2}$. Now consider $z \in C(s)$, where we may assume z induces

$$\tilde{z} = (1,2,3,4)\ldots(4j-3,\ldots,4j)\,(4j+1,4j+2)\ldots(4k+1,4k+2)$$

on $\tilde{A}^g$. Then t induces $(1,2)\ldots(4j-1,4j)$ on $\tilde{A}^g$ and, by the above, $n-4j \leqslant 3$. So the only simple composition factor of $C(s) \cap C(z)$ is A_j. This gives $j \geqslant m-2$, whereas $j \leqslant n/4 \leqslant (2m+3)/4$. This is a contradiction.

5. $\tilde{A} \cong G(q)$, q odd

In this section $\tilde{A}$ is taken to be a finite group of Lie type defined over a field of odd characteristic. Write $\tilde{A} \cong G(q)$, q odd. The $B(G)$-conjecture is assumed to hold in all groups considered.

There is a strong connection between the results mentioned in this section and the results required for the proof of the U-conjecture and the $B(G)$-conjecture. For this reason, and in order to facilitate proofs, the results are stated more generally than is required for the standard form problem. Of course, in proving the $B(G)$-conjecture one must drop the above assumption and prove theorems for 2-components instead of components.

We need the following definition. An involution j of a group X is called a *classical involution* if $C_X(j)$ contains a 2-component (or solvable 2-component) having 2-rank 1 and containing j. Classical involutions exist in simple groups of Lie type defined over fields of odd characteristic, with the exception of the groups $L_2(q)$ and ${}^2G_2(q)$. Indeed, if L is such a group, let J be generated by two opposite root subgroups for a long root in the root system of L. Then $J \cong SL_2(q)$ and j is a classical involution of L, where $\langle j \rangle = Z(J)$ (see (4.2) of [22]).

The following major result of Aschbacher determines those simple groups that contain a classical involution.

(5.1) (Aschbacher [15]). *Let X be a finite group with $F^*(X)$ simple, and let j be a classical involution in X. Then $F^*(X)$ is isomorphic to a finite group of Lie type in odd characteristic or $F^*(X) \cong M_{11}$.*

The above theorem is fundamental for the results of this section. For, if $\tilde{A} \not\cong L_2(q)$ or ${}^2G_2(q)$ (the latter existing only for q an odd power of 3), then $\tilde{A}$ contains a classical involution and one tries to show that G must also contain a classical involution. Once this is achieved, (5.1) yields the identification of $F^*(G)$.

In the following we will describe some recent work of Walter [230] and Harris [119], [120]. There is considerable overlap in their results, although the approaches are different. At the time of this writing some of the results are not in final form. Nevertheless, the work is significant as it relates to both the standard form problem and the $B(G)$-conjecture. We take the liberty of combining some of the results, and refer the reader to the references for more precise and more general statements.

Let $\mathscr{I}$ be a class of simple groups. By $\bar{\mathscr{I}}$ we will denote the smallest class of groups that contains $\mathscr{I}$ and contains all simple groups L for which there is an involution $j \in Aut(L)$ such that $E(C_L(j))$ has a composition factor in $\bar{\mathscr{I}}$. Let

$$\mathscr{I} = \{L_2(q) \mid q > 3, q \text{ odd}\} \text{ and}$$

$$\mathscr{I}_0 = \{L_2(q) \mid q \text{ a power of a prime } p \geqslant 5\}.$$

Also, $Chev(p)$ denotes the simple groups of Lie type defined over extension fields of $\boldsymbol{F}_p$, and

$$Chev^*(p) = Chev(p) - \{L_2(p^n) \mid n \geqslant 1\}.$$

The fields of characteristic 3 are especially troublesome so we start with results for larger characteristics.

(5.2) (Walter [230], Harris [120]). *If $L \in \bar{\mathscr{I}}_0$, then L is isomorphic to one of the following groups:*

(i) *$G(q)$, a group of Lie type in odd characteristic;*

(ii) *A_n, $n \geqslant 7$; or*

(iii) $M_{12}, J_1, J_2, J_3, HS, Mc, Ly, Sz, He, ON, Co_3, F_5$.

The above should also hold when $\mathscr{I}_0$ is replaced by $\mathscr{I}$, but at present this is dependent on certain hypotheses which we describe later.

(5.3) (Walter [229]). *Let L be a finite group with $E(L)$ quasisimple. Suppose j is an involution in L and $E(C_L(j))$ has a composition factor in Chev*(p), for $p \geqslant 5$ a prime. Then one of the following holds:*

(i) $E(\tilde{L}) \in$ *Chev**(p);

(ii) $p = 5$ *and* $E(\tilde{L}) \cong A_n$; *or*

(iii) $p = 7$ *and* $E(\tilde{L}) \cong L_3(4)$.

Although (5.2) and (5.3) are sufficient for our purposes, we note that both Harris and Walter have extensions of these results that hold for 2-components rather than components (see [230] and [120]). At the moment Harris' results are dependent on the verification of certain properties of groups of Lie type, although work on these properties is in progress. Also, both authors have results for fields of characteristic 3, although in each case the results have an extra hypothesis. We discuss the results later.

In the following we sketch a proof of (5.3), reducing to (5.2). Later we will make a few remarks about the proof of (5.2). The arguments below are due to Walter. We simply combined ideas appearing in [229] and [230].

Let L be a minimal counterexample to (5.3). Choose an involution $j \in L$ such that $C_L(j)$ has a component B and $\tilde{B} \in$ *Chev**(p), $p \geqslant 5$. Then B contains a classical involution t_1, with

$$\langle t_1 \rangle = Z(L_1) \quad \text{and} \quad SL_2(q) \cong L_1 \trianglelefteq E(C_B(t_1)).$$

By (1.4), $L_1 \subseteq E(C_G(t_1))$, and an easy argument shows that there is a component L_2 of $C_G(t_1)$ such that either $L_1 \subseteq L_2$ or $L_1 \subseteq L_2 L_2^j \neq L_2$. We claim that there exist involutions s, x and subgroups X, Y of L such that the following conditions hold:

(5.4)

(i) $\langle x \rangle = Z(X)$ and $X \cong SL_2(q)$;

(ii) $X \subseteq Y$ and Y is a component of $C_G(x)$;

(iii) $s \in C(x) \cap N(Y)$; and

(iv) X is a component of $C_L(\langle x,s \rangle)$.

In fact, if $L_1 \subseteq L_2$, we can set $x = t_1$, $s = j$, $X = L_1$, and $Y = L_2$. So suppose $L_1 \subseteq L_2 L_2^j \neq L_2$. Since $L_1 \trianglelefteq\trianglelefteq C(\langle t_1, j \rangle)$, we have $L_2 \cong SL_2(q)$. Set $\langle t_2 \rangle = Z(L_2)$. As above, there exists a component L_3 of $C_G(t_2)$ with $L_2 \subseteq L_3$ or $L_2 \subseteq L_3 L_3^{t_1} \neq L_3$. Assume the latter holds. As $L_1 \subseteq C_G(t_2)$ and $[L_2, L_1] = L_2$, we necessarily have $L_1 \subseteq N(L_3 L_3^{t_1})$. But $L_1 = L_1'$ forces $L_1 \subseteq N(L_3)$, and this contradicts $t_1 \in L_1$. Therefore $L_2 \subseteq L_3$ and we set

$$(s, x, X, Y) = (t_1, t_2, L_2, L_3).$$

This proves the claim.

Choose x, s, X, and Y satisfying (5.4). Then s acts on $\tilde{Y}$ and $C_{\tilde{Y}}(s)$ contains an $L_2(q)$ component, with preimage in Y being $X \cong SL_2(q)$. We may now apply (5.2) to obtain the structure of Y. For convenience, we assume $\tilde{Y}$ is a group of Lie type in odd characteristic. We remark that this is necessarily the case if $p \neq 5, 7, 17$.

By L-balance, $Y \subseteq E(L)$. If $x \in Z(E(L))$, then $E(L) = Y$ and we are done. So assume $x \notin Z(E(L))$. If $\tilde{Y} \cong L_2(q_1)$ for q_1 a power of q, then $\tilde{x}$ is a classical involution in $E(\tilde{L})$ and we apply (5.1) to get the structure of $E(L)$. So suppose $\tilde{Y} \ncong L_2(q_1)$. Then properties of centralizers of involutions in groups of Lie type imply that $\tilde{Y} \cong P\Omega_n^{\pm}(q)$, for $n > 5$.

Choose $X_1 \cong X_2 \cong SL_2(q)$ in Y such that $[X_1,X_2] = 1$ and $\tilde{X}_1\tilde{X}_2 \cong \Omega_4^+(q)$ in $\tilde{Y}$. Let $\langle z_i \rangle = Z(X_i)$, for $i = 1,2$. From the structure of Y (which is isomorphic to an image of $\mathrm{Spin}(n,q)$), we conclude that $X_1X_2 = X_1 \times X_2$ and $x = z_1z_2$.

Next, we look at $C_G(z_1)$. By (1.4), $X_1X_2 \subseteq E(C_G(z_1))$, and an easy argument shows that there exist components D_1 and E_1 of $E(C_G(z_1))$ such that $X_1 \subseteq D_1$ and $X_2 \subseteq E_1$. Suppose $D_1 \neq E_1$. As $z_1 \in Z(D_1)$ and $z_2 \in E_1 \subseteq C(D_1)$, we have $x = z_1z_2 \in C(D_1)$. But X_1 is a component of $C_G(\langle x,z_1 \rangle)$ and this implies $X_1 = D_1$. Applying (5.1) to $\widetilde{E(L)}$ we have the structure of $E(L)$. So assume $X_1 \neq D_1$. Therefore, $X_1X_2 \subseteq D_1$. The same argument shows that $z_2 \notin Z(D_1)$ and so $\tilde{D}_1$ contains $\tilde{z}_2$ as a classical involution. By (5.1), $\tilde{D}_1$ is a Chevalley group (M_{11} does not occur as $q \geqslant 5$). Since $C_{\tilde{D}_1}(\tilde{z}_2)$ contains both $\tilde{X}_1 \cong L_2(q)$ and $\tilde{X}_2 \cong SL_2(q)$ as components, we have $\tilde{D}_1 \cong P\Omega_7(q)$. In particular, $C_{\tilde{D}_1}(\tilde{z}_2)$ contains a third component $\tilde{X}_3$ with

$$\tilde{X}_2\tilde{X}_3 \cong SL_2(q) \circ SL_2(q), \quad X_3 \cong SL_2(q) \quad \text{and} \quad Z(X_3) = \langle z_1,z_2 \rangle.$$

Exactly the same arguments apply to $C_G(z_2)$. There is a component D_2 of $C_L(z_2)$ with $\tilde{D}_2 \cong P\Omega_7(q)$ and $X_1X_2 \subseteq D_2$. Moreover, the same argument shows $X_3 \subseteq D_2$. In $\tilde{D}_2$ we have

$$\tilde{X}_2 \cong L_2(q) \cong \Omega_3(q), \quad \text{and} \quad \tilde{X}_1\tilde{X}_3 \cong \Omega_4^+(q).$$

So there is an element $u \in D_2$ such that $\tilde{u}$ is an involution inducing an outer diagonal automorphism on each of $\tilde{X}_1$ and $\tilde{X}_3$, while centralizing $\tilde{X}_2$. Clearly, $u \in N(X_i)$, $i = 1,2,3$, so $u \in C(\langle z_1,z_2 \rangle)$ and u acts on D_1. However, in the group $D_1\langle u \rangle$, the relation $[u, X_2] = 1$ forces $[u, X_3] = 1$. This is a contradiction.

We have seen how the proof of (5.3) utilized a special case of (5.2). It was necessary to identify a group Y for which there exist involutions s and x satisfying (5.4). From (5.1) we could assume $Y\langle s \rangle \subset L$, so, by minimality of L, (5.3) holds in all sections of $Y\langle s \rangle$. So to obtain a proof of (5.3), one only needs to prove (5.2) under the additional assumption that (5.3) holds in all sections of the group being considered. Once this has been achieved, the inductive hypothesis on (5.2) is no longer required and both results are verified.

Suppose then that $L \in \mathscr{I}_0$ is a minimal counterexample to (5.2) and that (5.3) holds in all sections of $L\langle j \rangle$, for $j \in Aut(L)$. There is some involution

$j \in Aut(L)$ such that $E(C_L(j))$ contains a component $B \subset L$ with $\tilde{B}$ in $\mathscr{I}_0$. The Component Theorem is proved in such a way that we may assume B to be a standard subgroup of $L\langle j \rangle$. By minimality of L, B is of type (i), (ii) or (iii) in the statement of (5.2).

The results of Section 4 handle the case $\tilde{B} \cong A_n$, and our inductive hypothesis allows us to assume $\tilde{B} \notin Chev^*(p)$, $p > 3$. We are reduced to $\tilde{B} \cong L_2(q)$ or $\tilde{B}$ a sporadic group. Sporadic standard subgroups will be discussed in Section 6, so assume $\tilde{B} \cong L_2(q)$. In view of (2.1) and (2.4) we may assume that $C(B) \cap L\langle j \rangle$ has a cyclic Sylow 2-subgroup, say J.

Let $J \subseteq S \in Syl_2(C(j))$. In many cases we can show that $L\langle j \rangle$ has sectional 2-rank at most 4, whereupon the main theorem of [99] determines the structure of L. For example, this happens if $S \in Syl_2(L\langle j \rangle)$. Also, if there exists a subgroup S_1 of S such that

$$Z_2 \times Z_2 \times Z_2 \cong S_1 = C_S(S_1),$$

then a result of Harada (Theorem 2 of [115]) implies that $L\langle j \rangle$ has sectional 2-rank at most 4.

The latter configuration does occur in a number of cases where $S \notin Syl_2(L\langle j \rangle)$. For example, suppose $g \in N(S) - S$ and $g^2 \in S$. Then $J \cap J^g = 1$ and $J^g \subseteq S$. If $S \subseteq B \times J$, then

$$Z(S) = J \times \Omega_1(Z(S \cap B)),$$

and this forces $J \cong Z_2$. Setting $S_1 = S_2 \times J$ for S_2 a Klein subgroup of $S \cap B$, we have such a subgroup.

The results of Harris [118] and Harris–Solomon [121] reduce the general case to the case $J = \langle z \rangle \cong Z_2$, q is a square, $C(z)$ contains an involution u acting as a field automorphism on $B \cong L_2(q)$, and $|B|_2 > 8$. Rather technical arguments are used to complete the proof of (5.2).

To describe the situation in characteristic 3, we begin by defining $\mathscr{C}(3)$ to be the union of the following families of simple groups:

$Chev(3)$
$\{A_{2n} \mid n \geqslant 4\}$
$\{G_2(2^n), Sp_4(2^n), L_4(2^n), U_4(2^n) \mid n = 2^m \geqslant 2\}$
$\{U_5(2^n) \mid n = 2^m \geqslant 1\}$
$\{Mc, HS, Co_1, F_5\}$.

CONJECTURE (5.5). *For X a group, $F^*(X)$ is isomorphic to a group in $\mathscr{C}(3)$ if the following hold:*

(i) *$F^*(X)$ is simple;*
(ii) *the $B(G)$-conjecture holds in all sections of X; and*
(iii) *X contains a standard subgroup I with $\tilde{I} \cong L_3(3)$, $U_3(3)$, $G_2(3)$, $L_4(3)$, $U_4(3)$, $PSp_4(3)$, $P\Omega_7(3)$, or $P\Omega_8^{\pm}(3)$.*

(5.6) (Harris [120], Walter [230]). *Assume that Conjecture* (5.5) *holds. If G is a finite group with $F^*(G)$ simple and $G \in \overline{\mathscr{C}(3)}$, then $F^*(G)$ is isomorphic to a group in $C(3)$.*

Conjecture (5.5) is stronger than necessary for the proof of (5.6). Both Harris and Walter require only certain of the possible coverings of the groups described in (iii). Walter's approach in [230] deals with the cases $\tilde{I} \cong P\Omega_7(3)$ and $P\Omega_8^{\pm}(3)$, so these groups can be deleted from the list. We have been informed by Harris that $L_3(3)$ is no longer necessary, and Gomi's work in [95] handles most of the cases for $\tilde{I} \cong PSp_4(3)$. Once Conjecture (5.5) is verified, the results of this section will be complete and, in particular, the standard form problem for $\tilde{A} \cong G(q)$, q odd, will be completely settled.

6. Sporadic Groups

In this section $\tilde{A}$ denotes a sporadic simple group. As usual we assume that the $B(G)$-conjecture holds in all sections of G and that R is cyclic. In view of (1.5) we may assume that $F^*(G)$ is simple.

For X a simple group and n an integer, the notation nX will be used to denote a quasisimple group I with $Z(I) \cong Z_n$ and $I/Z(I) \cong X$.

While there are some general methods available (see [58] and [172]), there is a great deal of case analysis. Consequently, we will be content with simply listing the results in tabular form.

In Table 2 we indicate the cases that have already been treated. The first column gives the choice of A. The second column indicates the possibilities for G (if any). The third column gives references for the particular case, while in the fourth column we indicate any additional hypotheses needed for the proof. For certain of the Fischer groups the following conjecture has been assumed to hold.

CONJECTURE (6.1). *Let X be a finite group with $F^*(X)$ quasisimple and containing a subgroup Y such that $|Z(Y)|$ is even and $YZ(X)/Z(X)$ is standard in $\bar{X} = X/Z(X)$. Suppose $Y/Z(Y) \cong U_6(2)$, $M(22)$, ${}^2E_6(2)$, or F_2. Then one of the following holds:*

(i) *$Y/Z(Y) \cong U_6(2)$ and $F^*(\bar{X}) \cong M(22)$.*
(ii) *$Y/Z(Y) \cong M(22)$ and $\bar{X} \cong M(23)$ or $M(24)'$.*
(iii) *$Y/Z(Y) \cong {}^2E_6(2)$ and $\bar{X} \cong F_2$.*
(iv) *$Y/Z(Y) \cong F_2$ and $\bar{X}$ is of F_1 type.*

In view of the results of Table 2 (p. 62) the only cases left to consider are A isomorphic to $2.F_2$ or $2.M(22)$.

TABLE 2

A	G	Reference	Extra Hypothesis
M_{11}	$Aut(Mc)$	[99]	
M_{12}	Co_3	[238]	precise centralizer
$2.M_{12}$	$Aut(Sz)$	[64]	U-Conjecture holds
M_{22} $3.M_{22}$		[57]	
$2.M_{22}$ $6.M_{22}$		Harada	
$4.M_{22}$ $12.M_{22}$		methods of [58]	
M_{23}		[55]	
M_{24}		[49]	
J_1	$Aut(ON)$	[54]	
J_2	$Aut(Sz)$	[56]	
$2.J_2$		[64]	U-Conjecture holds
J_3 $3.J_3$		[56]	
J_4		[59]	
HS $2.HS$	F_5 $Aut(F_5)$	[116], [188]	U-Conjecture holds
Sz $3.Sz$		[190]	$L_3(4)$ standard form problem solved
$2.Sz$ $6.Sz$		[64]	U-Conjecture holds
Mc $3.Mc$		[186]	
Ru		[190]	U-Conjecture holds
$2.Ru$		[58]	
He		[111]	U-Conjecture holds in proper sections and $L_3(4)$ standard form problem solved
Ly		[186]	
ON		[190]	$L_3(4)$ standard form problem solved
Co_1		[190]	
$2.Co_1$		[58]	
Co_2		[142]	
Co_3		[64]	U-Conjecture holds
$2.M(22)$	$M(23)$	[130]	$A = C_G(Z(A))$
$M(23)$; $M(22)$ $M(24)'$ $3.M(24)'$; F_1 F_2	$M(24)$	[190]	Conjecture 6.1 holds in all proper sections of G

3

The $B(G)$-Conjecture and Unbalanced Groups

RONALD SOLOMON

1. A Brief Prehistory

Following Brauer's suggestion that simple groups of even order should be characterized by the structure of the centralizer of an involution, and Feit and Thompson's proof that a simple group does necessarily have even order, there was a parallel development of two lines of research.

(1) The characterization of finite simple groups G, given the precise structure of $C_G(t)$ for some involution t of G.
(2) The attempt to force $C_G(t)$ to have one of the precise structures treated in (1).

Much of the work in (1) was finally subsumed under the headings of "standard form problems" and the "$O_2(C_G(t))$ symplectic type problem" which are discussed in Chapters 2 and 6. It was recognized early that a major problem arising under heading (2) was the determination of $O(C_G(t))$.

The problem may be illustrated by the following easy example. The group $L = A_5$ has a faithful representation on the 3-dimensional vector space V over $\boldsymbol{F}_{19}$. Let $G = VL$ and let

$$T = \{1,t_1,t_2,t_3\} \in Syl_2(L).$$

Clearly $C_G(t_i) \cong Z_2 \times D_{38}$ for all i. Now assume that H is a simple group with Sylow 2-subgroup T and $C_H(t_i) \cong Z_2 \times D_{38}$ for all i. We would like to derive a contradiction by exhibiting a normal subgroup $W$of H isomorphic to V. We also know what W should be, namely

$$W = \langle O(C_H(t_i)) \,|\, 1 \leqslant i \leqslant 3\rangle.$$

We have a start towards proving that $W \trianglelefteq G$. Indeed, as $N_H(T)$ permutes $\{t_i \,|\, 1 \leqslant i \leqslant 3\}$, we have

(1) $N_H(T) \subseteq N_H(W)$.

Moreover, as $C_H(t_i) = O(C_H(t_i))T$, we have

(2) $C_H(t_i) \subseteq N_H(W)$ for $1 \leqslant i \leqslant 3$.

Thus we are in position to quote a celebrated theorem of Bender, of which we state a corollary here.

THEOREM 1.1 (Bender [29]). *Let G be a finite simple group having a proper subgroup M satisfying*

(a) *$N_G(T) \subseteq M$ for $T \in Syl_2(M)$, and*

(b) *$C_G(t) \subseteq M$ for all involutions t of M.*

Then G is isomorphic to $L_2(2^n)$, $U_3(2^n)$ or $Sz(2^{2n-1})$ for some $n \geqslant 2$.

In our context, taking $G = H$ and $M = N_H(W)$, we see, by comparison of $Z_2 \times D_{38}$ with the involution centralizers in $L_2(2^n)$, $U_3(2^n)$ and $Sz(2^{2n-1})$, that M cannot be a proper subgroup of H. Thus $M = H$ and, as $\langle 1 \rangle \neq W \trianglelefteq M$, we have in fact $W = H$. Unfortunately, this contradicts nothing. However, if we could prove that W had odd order, then we would be done. This became the focal problem for the first line of attack on $O(C(t))$. The first major contributions were made by Thompson. We set up his notation.

Definition 1.2. Let A be a 2-subgroup of the finite group G. The set of *A-signalizers* $И(A)$ is the set of all A-invariant subgroups of G of odd order. $И^*(A)$ is the set of maximal elements of $И(A)$ under inclusion.

From Thompson's perspective, the key problem was to show, for suitably large 2-subgroups A of G, that $И^*(A) = \{O(G)\}$. Thompson's setting was the class of N-groups; however his line of reasoning transfers verbatim to a larger class defined by Gorenstein.

Definition 1.3. A finite group G is of *characteristic* 2 *type* if every 2-local subgroup of G is 2-constrained and $SCN_3(2) \neq \emptyset$‡.

The first observation is that the result holds 2-locally for $A \in SCN(2)$.

LEMMA 1.4. *Let H be a 2-constrained group with $S \in Syl_2(H)$ and $A \subseteq S$. Then the following hold.*

(1) *$O(C(A)) \subseteq O(H)$.*

(2) *If $A \in SCN(S)$, then $И^*(A) = \{O(H)\}$.*

Proof. We easily reduce to $O(H) = \langle 1 \rangle$. Let $R = O_2(H)$ and $X \in И(A)$. Assume that either $X = O(C(A))$ or $A \in SCN_3(S)$. In either case

$$[X, C_R(A)] \subseteq X \cap R = \langle 1 \rangle.$$

Thus by Thompson's $(P \times Q)$-Lemma [G:(5.3.4)], $[C_X(A), R] = \langle 1 \rangle$. Then, by definition of 2-constraint, $C_X(A) = \langle 1 \rangle$. In particular, (1) holds and we

‡ This definition, required only up to Theorem 1.9 of this section, is more general than that taken elsewhere in this book. Cf. Chapter 1, Sections 5 and 6.

may assume $A \in SCN(S)$ and $X = [X,A]$. As $A \trianglelefteq S$, we have

$$[A,R] \subseteq C_R(A) \subseteq C_R(X)$$

and hence, by the three subgroup lemma,

$$[X,R] = [X,A,R] \subseteq [A,R,X][R,X,A] \subseteq [R,A].$$

Thus X stabilizes the chain $R \supseteq [R,A] \supseteq \langle 1 \rangle$. Hence X centralizes R and, as above, $X = \langle 1 \rangle$.

Using this fact and his celebrated Transitivity Theorem, Thompson was able to prove (for N-groups) the following Signalizer Theorem.

THEOREM 1.5. *Let G be a finite group of characteristic 2 type and let $A \in SCN_3(2)$. Then $И^*(A)$ contains a unique element*

$$W_A = \langle O(C(a)) \mid a \in A^\# \rangle.$$

COROLLARY 1.6. *Let G be a finite group of characteristic 2 type and let $S \in Syl_2(G)$. Then there is an odd order subgroup W of G with*

(1) *$W = \langle O(C(b)) \mid b \in B^\# \rangle$ for all $B \subseteq S$ with $m(B) \geqslant 2$, and*

(2) *$N_G(W) \supseteq \langle N_G(B) \mid B \subseteq S, m(B) \geqslant 2 \rangle$.*

Proof. (1) is an easy consequence of (1.5) and the "balance property",

(1.7) *$O(C(a)) \cap C(b) \subseteq O(C(b))$ for a, b commuting elements of $S^\#$.*

This, in turn, is immediate from (1.4(1)). Then (2) is immediate from (1).

Walter first explicitly noted the importance of studying the subgroup

$$\Gamma_{S,k}(G) = \langle N_G(B) \mid B \subseteq S, m(B) \geqslant k \rangle,$$

calling it the k-generated core of G. Particularly important was the case $k = 2$. Various special results of Gorenstein and Walter were finally subsumed in the following theorem of Aschbacher [9: Theorem 1].

THEOREM 1.8. *Let G be a finite simple group with $m_2(G) \geqslant 3$ having a proper 2-generated core. Then G is isomorphic to $L_2(2^n)$, $U_3(2^n)$ or $Sz(2^{2n-1})$ with $n \geqslant 2$ or to Janko's group J_1. In particular, $O(\Gamma_{S,2}(G)) = \langle 1 \rangle$.*

Combining (1.6) and (1.8), we obtain the following theorem of Gorenstein and Walter [104] (due to Thompson for N-groups).

THEOREM 1.9. *Let G be a finite simple group of characteristic 2 type. Then $O(C(t)) = \langle 1 \rangle$ for all involutions t of G.*

Now, many of the known simple groups which are not of characteristic 2 type do in fact have an involution t with $O(C(t)) \neq \langle 1 \rangle$. Thus in trying to extend this line of attack, two questions arise:

(1) What properties prevent one from proving that $O(C(t)) = \langle 1 \rangle$?

(2) What analogous theorems might be provable in general?

Both questions were addressed by Gorenstein and Walter during the period following Walter's work on groups with abelian Sylow 2-subgroups. They were led to the consideration of an important subgroup called the layer. We recall the following definitions.

Definition 1.10. Let G be a finite group. A 2-*component* of G is a perfect, subnormal subgroup H of G such that $H/O(H)$ is quasisimple. If H itself is quasisimple, H is called a *component* of G. The product of all 2-components of G is $L_{2'}(G)$, the *layer* of G. The product of all components of G is denoted by $E(G)$. The product $E(G)F(G) = F^*(G)$ is called the *generalized Fitting subgroup* of G.

These subgroups have the following basic properties.

PROPOSITION 1.11. *Let G be a finite group. Then the following hold.*

(1) *$E(G)$ and $L_{2'}(G)$ are characteristic subgroups of G.*
(2) *$E(G)$ is semisimple (i.e., a central product of the components).*
(3) $C_G(F^*(G)) = Z(F(G))$.
(4) $L_{2'}(G)O(G)/O(G) = E(G/O(G))$.

We sketch a proof of the following important extension of (1.4).

LEMMA 1.12. *Let G be a finite group, t an involution of G and $D \subseteq O(C(t))$. Let $L_1, L_2, \ldots, L_r$ be the 2-components of G. Then the following hold.*

(1) *D normalizes L_i for $1 \leqslant i \leqslant r$ and $[O_{2',2}(G),D] \subseteq O(G)$.*
(2) *If $(L_i)^t \neq L_i$, then $[L_i,D] \subseteq O(L_i)$.*
(3) *If $D \nsubseteq O(G)$, there exists a $\langle t\rangle$-invariant 2-component L_i with*

$$L_i = O(L_i)[L_i,t] = O(L_i)[L_i,D].$$

Proof. We may assume that $O(G) = \langle 1\rangle$ and $D \neq \langle 1\rangle$. By the $(P\times Q)$-lemma, $[O_2(G),D] = \langle 1\rangle$. If $(L_i)^t \neq L_i$, let $K = C_{L_i(L_i)^t}(t)$. Then K is maximal in $L_i(L_i)^t$ and $[K,D]$ has odd order. It follows that D centralizes $L_i(L_i)^t$. Suppose $L_i = (L_i)^t$ and let $S \in Syl_2(\langle L_i,t\rangle)$ and $R = N_S(\langle Z(L_i),t\rangle)$. As $D \subseteq O(C(\langle Z(L_i),t\rangle))$, $[R,D]$ has odd order. As $R \nsubseteq Z(L_i)$, $(L_i)^D = L_i$.

Finally, as $D \neq \langle 1\rangle$, D does not centralize $E(G)$. Hence for some i,

$$[L_i,D] = [L_i,t] = L_i.$$

Definition 1.13. Let G be a finite group. An *unbalancing triple* for G is a triple (a,x,L) where a and x are commuting involutions and L is an $\langle x\rangle$-invariant 2-component of $C_G(a)$ such that

$$L = O(L)[L, O(C(x)) \cap C(a)].$$

The proof of (1.9) in fact proves the following result.

THEOREM 1.14. *Let G be a finite simple group with $SCN_3(2) \neq \emptyset$. Then one of the following holds:*

(1) $O(C(t)) = \langle 1 \rangle$ *for all involutions* t *of* G; *or*
(2) G *has an unbalancing triple.*

This theorem is the point of departure for the attack on the Unbalanced Group Conjecture, which we shall state later. Before we do that, let us remind ourselves that the original objective was to get $C_G(t)$ into the shape of the centralizer of an involution in a known simple group. Gorenstein and Walter made the following observations.

(1) In all known simple groups, $L_{2'}(C(t)) = E(C(t))$ for every involution t.
(2) If one can show that $L_{2'}(C(t)) = E(C(t))$ for every involution t of a simple group G and if $L_{2'}(C(t)) \neq \langle 1 \rangle$ for some involution t of G, then one should be able to prove that for some involution s of G, $C_G(s)$ looks "very much" like the centralizer of an involution in a known simple group.

Assertion (2) was made precise by Aschbacher in his Component Theorem [10: Theorem 1], discussed in Chapter 1. Gorenstein and Walter launched, in the late 1960s, a major effort to prove

(GW) Let G be a finite group with $F^*(G)$ simple. Then $L_{2'}(C(t)) = E(C(t))$ for every involution t of G.

This effort was joined by John Thompson in 1973 and he reformulated (GW) as the $B(G)$-Conjecture.

Definition 1.15. Let G be a finite group. $B(G)$ is the product of all 2-components of G which are *not* quasisimple.

Thus $L_{2'}(G) = E(G)B(G)$ with $E(G) \cap B(G) \subseteq Z(E(G))$.

$B(G)$-CONJECTURE. *Let* G *be a finite group and* t *an involution of* G. *Then* $B(C(t)) \subseteq B(G)$.

We shall refer to this conjecture henceforth as (B). The equivalence of (GW) and (B) is an easy consequence of the important L-balance Theorem of Gorenstein and Walter [105: (4.2)].

THEOREM 1.16. *Let* G *be a finite group and* t *an involution of* G. *Let* K *be a* 2-*component of* $C(t)$. *Then there is a* 2-*component* L *of* G *such that one of the following holds:*
(1) $K \subseteq L = L^t$, *or*
(2) $K \subseteq LL^t$ *and* $L/Z^*(L) \cong K/Z^*(K)$.
In particular, $L_{2'}(C(t)) \subseteq L_{2'}(G)$.

Corollary 1.17. *(B) is equivalent to (GW).*

Proof. Assume (B) and let G be a finite group with $F^*(G)$ simple. Then $B(G) = \langle 1 \rangle$, so $B(C(t)) = \langle 1 \rangle$ for all involutions t. Thus $L_{2'}(C(t)) = E(C(t))$ for all involutions t.

Now assume (GW) and let G be a finite group and t an involution of G. Let K be a 2-component of $B(C(t))$. By L-balance, $K \subseteq LL^t$ for some 2-component L of $L_{2'}(G)$. If $L \subseteq B(G)$, we are done. Thus we may assume $LL^t \subseteq E(G)$. If $L \neq L^t$, then

$$K \subseteq M = (C_{LL^t}(t))' \quad \text{and} \quad M/Z(M) \cong K/Z^*(K).$$

Thus $O(K) \subseteq Z(K)$, contrary to assumption. Thus $L = L^t$ and K is a 2-component of $B(C_L(t))$. Let $\bar{H} = \langle L,t \rangle / Z(L)$. Then $F^*(\bar{H}) = \bar{L}$ is simple and $\bar{K}$ is a 2-component of $B(C_{\bar{L}}(\bar{t}))$, contrary to (GW).

In attempting to prove (GW), Gorenstein and Walter developed a variant of Thompson's signalizer method, called the signalizer functor method. We shall describe this in the next section.

2. Signalizer Functors

Thompson's signalizer arguments in the N-group paper were designed to prove that $\mathcal{U}^*(A) = \{O(G)\}$ for $A \in SCN_3(2)$. To prove (B), we would like to prove a similar assertion for $O(B(G))$. However $\mathcal{U}(A)$ is too large a set. Gorenstein's idea was to restrict attention to a subset $\mathcal{U}_\theta(A)$ defined by

(2.1) $\mathcal{U}_\theta(A) = \{X \in \mathcal{U}(A) \mid C_X(a) \subseteq \theta(C(a)) \text{ for } a \in A^\#\}$

where θ is a function assigning to $C(a)$ an odd order subgroup of $C(a)$ for each $a \in A^\#$. It will be more convenient for us to think of the domain of θ as $A^\#$ and write $\theta(a)$ for $\theta(C(a))$. It is desirable to have $\theta(a) \in \mathcal{U}_\theta(A)$ for all $a \in A^\#$. This motivates the following definition.

Definition 2.2. Let A be an abelian 2-subgroup of the finite group G. An A-*signalizer functor* is a function θ on $A^\#$ such that $\theta(a)$ is an A-invariant odd order subgroup of $C(a)$ and

$$\theta(a) \cap C(b) \subseteq \theta(b) \quad \text{for} \quad a,b \in A^\#.$$

Note that (1.7) says that if G is a group of characteristic 2 type and A an abelian 2-subgroup of G, then O is an A-signalizer functor and $\mathcal{U}(A) = \mathcal{U}_O(A)$. The "justification" for Definition 2.2 is the Signalizer Functor Theorem of Gorenstein [FSG] and Goldschmidt [89].

THEOREM 2.3. *Let A be an abelian 2-subgroup of the finite group G with $m(A) \geqslant 3$. Let θ be an A-signalizer functor. Then*

$$W_A = \langle \theta(a) \mid a \in A^\# \rangle$$

is the unique maximal element of $\mathcal{U}_\theta(A)$.

The slickest proof of this is due to Bender [30]. Note that this generalizes Thompson's Signalizer Theorem (1.5). Moreover the same proof as sketched above gives the following generalization of (1.9).

THEOREM 2.4. *Let G be a finite group with $O(G) = \langle 1 \rangle$ and $SCN_3(2) \neq \emptyset$. Suppose that θ is an A-signalizer functor for all abelian 2-subgroups A of G with $m(A) \geqslant 3$. Suppose further that*

$$\theta(a)^g = \theta(a^g) \quad \textit{for} \quad a \in A^{\#},\ g \in G.$$

Then $\theta = 1$; i.e., $\theta(a) = 1$ for all $a \in A^{\#}$.

Now in trying to prove (B) an apparently desirable choice for θ would be

$$\theta_B(a) = O(B(C(a))).$$

Unfortunately an easy example shows the infelicity of this choice. Let V be the standard permutation module for S_7 over $GF(3)$ with basis $\{v_1, v_2, \ldots, v_7\}$. Let G be the semidirect product of V with S_7 and

$$A = \langle (1,2), (3,4), (5,6) \rangle.$$

It is easily calculated that

$$\theta_B((i,i+1)) = \langle v_j - v_k \,|\, j,k \in \{1,2,\ldots,7\} - \{i,i+1\} \rangle.$$

Then $v_3 + v_4 + v_5 \in \theta_B((1,2)) \cap C((3,4))$ but $v_3 + v_4 + v_5 \notin \theta_B((3,4))$. Thus θ_B is not an A-signalizer functor. Nevertheless, the possibility that some variant of θ_B could be concocted which would give a quick proof of (B) remains a tantalizing thought! However, for the present, we must content ourselves with describing those functors which have proved most useful in work on this problem. There are essentially two.

Historically, the first functor (besides O itself) to be put to use was the 2-*balanced functor* of Gorenstein and Walter.

Definition 2.5. For E a 4-subgroup of the finite group G, define

$$\Delta(E) = \bigcap_{e \in E^{\#}} O(C(e)).$$

For A an elementary 2-subgroup of G with $m(A) \geqslant 4$, define for $a \in A^{\#}$

$$\theta_\Delta(a) = \langle C(a) \cap \Delta(E) \,|\, E \subseteq A, \quad E \cong E_4 \rangle.$$

The motivation for this definition is the following proposition about known simple groups.

PROPOSITION 2.6. *Let H be a known simple group, G a subgroup of $\mathrm{Aut}\, H$ containing H, and E a 4-subgroup of G. Then one of the following holds.*

(1) $\Delta(E) = \langle 1 \rangle$.

(2) *$H \cong A_{4n+3}$ for some $n \geqslant 1$, E is semi-regular on $4n$ letters and $\Delta(E) \cong Z_3$.*

(3) *$H \cong L_2(q)$ for some odd q and $\Delta(E)$ is a cyclic group of field automorphisms of H.*

Using this proposition one can prove, for example, the following result.

PROPOSITION 2.7. *Let G be a finite group. Suppose that if (a,x,J) is an unbalancing triple in G, then $J/Z^*(J)$ is a known simple group not isomorphic to A_{4n+3} or $L_2(q)$ for q odd, $n > 2$. Then θ_Δ is an A-signalizer functor for all abelian 2-subgroups A with $m(A) \geqslant 4$.*

COROLLARY 2.8. *Let G be as in (2.7). Suppose that $SCN_5(2) \neq \emptyset$ and $\Gamma_{S,3}(G) = G$ for $S \in Syl_2(G)$. Then $\Delta(E) = \langle 1 \rangle$ for all 4-subgroups E of G.*

Ignoring the connectivity questions and the problems of A_{4n+3} and $L_2(q)$, we may reasonably ask how having $\Delta(E) = \langle 1 \rangle$ helps us towards proving (B). We give a partial answer in the next lemma.

LEMMA 2.9. *Let G be a finite group with $\Delta(E) = \langle 1 \rangle$ for all 4-subgroups E of G. Suppose that (B) holds in all proper sections of G. Let t be an involution of G and J a 2-component of $C(t)$. Suppose there exists an 8-subgroup E of $C(t) \cap N(J)$ with $L_{2'}(C_J(E)) \neq \langle 1 \rangle$. Then J is quasisimple.*

Proof. Let $X = O(J)$ and, for $F \subseteq E$, let $X_F = C_X(F)$. Without loss, $t \in E$. Let K be a 2-component of $C(E)$ contained in J.

Let F be a 4-subgroup of E and L any 2-component of $\langle K^{L_{2'}(C(F))} \rangle$. For $f \in F^\#$, let $M_f = \langle L^{L_{2'}(C(f))} \rangle$. If M_f is a product of two 2-components of $C(f)$, then clearly $[X_F,L] \subseteq O(C(f))$. If M_f is a single 2-component then, as (B) holds in $M_f F/\langle f \rangle$, we again have $[X_F,L] \subseteq O(C(f))$. Thus, in all cases, $[X_F,L] \subseteq O(C(f))$ for all $f \in F^\#$. Hence $[X_F,L] \subseteq \Delta(F) = \langle 1 \rangle$ for all 2-components L of $\langle K^{L_{2'}(C(F))} \rangle$. In consequence, $[X_F,K] = \langle 1 \rangle$ for all 4-subgroups F of E. As $X = \langle X_F \mid |E:F| = 2 \rangle$, we have $[X,K] = \langle 1 \rangle$. Thus $XC_J(X) \trianglelefteq J$ and $XC_J(X) \nsubseteq Z^*(J)$, whence $XC_J(X) = J$. As $J = J'$, $X \subseteq Z(J)$, as desired.

Results such as (2.9) show that if $\Delta = \langle 1 \rangle$, then attention may be focused on 2-components J for which $m_2(C(J/O(J))) \leqslant 2$ and J is "small". Here the methods seem to be *ad hoc*. The $L_3(4)$ case, for example, is treated in [111: Section 5].

The second important type of functor was introduced by Goldschmidt in [90] and [92].

Definition 2.10. Let G be a finite group, $V = E_1E_2$ an elementary 2-subgroup of G with $m(E_1) \geqslant 3$ and $m(E_2) = 2$. For $e \in E_1^\#$, define

$$\theta_C(e) = [O(C(e)), E_2](O(C(e)) \cap O(C(E_2))).$$

The key result for applying θ_C is Goldschmidt's "hidden commutator lemma," [92: (2.5)].

LEMMA 2.11. *Suppose that U, V, W are subgroups of the p-group E such that $U \subseteq Z(E) \cap V$ and $W \subseteq N_E(V)$. If E acts on the solvable p'-group X and $X = [X,V]$, then*

$$C_X(W) = \langle C_X(W) \cap [C_X(U_0),V] \mid m(U/U_0) = 1 \rangle.$$

This lemma may be applied to prove the following result, which may be extracted from Foote's proof of Lemmas 4.1 and 4.2 in [71].

PROPOSITION 2.12. *Let G, V and θ_C be as in* (2.10). *Then one of the following holds.*

(1) *θ_C is an E_1-signalizer functor.*

(2) *There exist elements $e_i \in E_i^{\#}$, a subgroup X of $O(C(e_1))$ and a 2-component J of $C(e_2)$ such that*

$$X = [X,E_2] \subseteq O(C(e_1)) \cap C(e_2),$$

and

$$J = O(J)[J,X'] = O(J)[J,e_1] = O(J)[J,E_2].$$

(3) *There exist $f,f_0 \in E_1^{\#}$ and a 2-component K of $C(f)$ with*

$$K = O(K)[K,f_0] = O(K)[K,\theta(f_0) \cap C(f)] = O(K)[K,E_2].$$

In particular, we have the following corollary.

COROLLARY 2.13. *Let G, V and θ_C be as in* (2.10). *Then θ_C is an E_1-signalizer functor provided the following condition holds:*

(CS) *For all $t \in E_1^{\#} \cup E_2^{\#}$ and all 2-components J of $C(t)$, either*

$$[J,E_1] \subseteq O(J) \quad \text{or} \quad [J,E_2] \subseteq O(J).$$

Condition (CS) is called *core-separation* by Goldschmidt and holds, for example, if E_1 and E_2 lie in disjoint strongly closed 2-subgroups of G. Another context in which (CS) holds relates to maximal 2-components, which we now define.

Definition 2.14. Let G be a finite group. $\mathscr{L}(G)$ is the set of all 2-components of centralizers of involutions of G. For $K,L \in \mathscr{L}(G)$, we say $L < K$ if there exist commuting involutions s,t of G with $K \lhd\lhd C(s)$, $L \lhd\lhd C(t)$, $s \in C(L/O(L))$ and $\langle L^{L_{2'}(C(s))} \rangle = KK^t$. We let $\ll$ be the transitive extension of $<$ on $\mathscr{L}(G)$ and say L is *maximal* in $\mathscr{L}(G)$ if $L \ll M$ implies that $L/Z^*(L) \cong M/Z^*(M)$. We let $\mathscr{L}^*(G)$ be the set of maximal elements of $\mathscr{L}(G)$.

Remarks. (1) If $L \in \mathscr{L}(G)$ and $|L/Z^*(L)| \geqslant |M/Z^*(M)|$ for all $M \in \mathscr{L}(G)$, then $L \in \mathscr{L}^*(G)$. In particular, if $\mathscr{L}(G) \neq \emptyset$, then $\mathscr{L}^*(G) \neq \emptyset$.

(2) If $L \in \mathscr{L}^*(G)$, $M \in \mathscr{L}(G)$ and $L \ll M$, then $M \in \mathscr{L}^*(G)$.

The following lemma illustrates the relevance of maximal 2-components to θ_C.

LEMMA 2.15. *Let G be a finite group with $L_1,L_2 \in \mathscr{L}^*(G)$ and $L_1L_2 = L_2L_1$. Suppose that E_i is an elementary 2-subgroup of L_i with $m(E_1) \geqslant 3$ and $m(E_2) = 2$. Then θ_C is an E_1-signalizer functor.*

If $W_C(E_1,E_2) = \langle \theta_C(e_1) | e_1 \in E_1^\# \rangle$, then a short argument shows that $W_C(E_1,E_2)$ is uniquely determined by (F_1,f_2) for any 4-subgroup F_1 of E_1 and any $f_2 \in E_2^\#$. This gives precisely the mobility needed to prove that $N_G(W_C(E_1,E_2))$ is a strongly embedded subgroup and hence to prove the following result.

THEOREM 2.16 ([187; Theorem 1.1]). *Let G be a finite group with $F^*(G)$ simple. Let $\{L_1,L_2,\ldots,L_r\}$ be a non-empty set with $L_i \in \mathscr{L}^*(G)$ and $L_iL_j = L_jL_i$ for $1 \leqslant i,j \leqslant r$. Suppose that $m_2(L_i) \geqslant 3$ for all i. Then $r = 1$.*

Sharper results than (2.15) have been obtained by Foote under the following hypothesis.

(F) Let G be a finite group. Whenever $H \subseteq G$ and s,t are commuting involutions of H, L is a 2-component of $C_H(t)$ and $X = O(C_H(s)) \cap C_H(t)$, then either $[L,X'] \subseteq O(C_H(t))$ or $L/Z^*(L) \cong L_2(q)$.

Foote's hypothesis is motivated by the observation that (F) holds in any finite group all of whose simple sections are known. Its usefulness resides in the fact that it excludes possibility (2) of (2.12) unless $J/Z^*(J) \cong L_2(q)$, in which case *ad hoc* arguments may be applied.

Using (F), Foote is able to prove the following result [71: (4.1), (4.2)].

THEOREM 2.17. *Let G, V, θ_C be as in* (2.10). *Assume that* (*F*) *holds and that $L \in \mathscr{L}^*(G)$ with $E_1 \subseteq C(L/O(L))$ and $E_2 \subseteq L$. Then θ_C is an E_1-signalizer functor.*

Indeed Foote's results, which we shall state more precisely in Section 7, suggest that one might be able to prove under fairly weak hypotheses:

(C) Let G be a finite group with $F^*(G)$ simple. If $L \in \mathscr{L}^*(G)$ with $m_2(C(L/O(L))) \geqslant 3$, then $\theta_C(e) = 1$ for every involution $e \in C(L/O(L))$.

THEOREM 2.18. *Let G and L satisfy the hypotheses and conclusion of* (*C*). *Then L is quasisimple.*

Proof. Let $X = O(L)$, $E_2 \subseteq L$, $E_1 \subseteq C(L/O(L))$. By (2.11),

$$[X,E_2] = \langle [C_X(e_1),E_2] | e_1 \in E_1^\# \rangle.$$

As $E_2 \subseteq L$ and $L \in \mathscr{L}^*(G)$, it follows easily that

$$[C_X(e_1),E_2] \subseteq O(L^{L_{2'}(C(e_1))}) \subseteq O(C(e_1)).$$

Hence

$$[C_X(e_1),E_2] = [C_X(e_1),E_2,E_2] \subseteq [O(C(e_1)),E_2] \subseteq \theta_C(e_1) = \langle 1 \rangle.$$

Thus $[X,E_2] = \langle 1 \rangle$, whence $XC_L(X) = L$ and, as in (2.9), L is quasisimple.

It is interesting to note the similarity between (2.9) and (2.18). Use of either θ_Δ or θ_C leads one, under suitable hypotheses on 2-components, to the conclusion that $B(C(t)) = 1$ if $m_2(C(L/O(L))) \geqslant 3$ for L a 2-component of $C(t)$, and thus 2-components of $C(t)$ are *all* components. To get further using either functor seems to require detailed knowledge of L and certain "neighbouring" 2-components in $\mathscr{L}(G)$. However in attempting to prove (B), induction gives one no detailed information about any element of $\mathscr{L}(G)$. This fact precipitated the next phase in the history of (B).

3. The Unbalanced Group Conjecture

In order to make better use of induction in proving (B), the idea crystallized in 1974 to attempt instead to prove the Unbalanced Group Conjecture, which we now state and henceforth refer to as (U).

(U) *Let G be a finite group with $F^*(G)$ simple. Suppose that $O(C(t)) \neq \langle 1 \rangle$ for some $t \in G^{\#}$ with $t^2 = 1$. Then $F^*(G)$ is isomorphic to one of the following:*
  (1) *a group of Lie type over $GF(q)$ for some odd q,*
  (2) *A_{2n+1} for some $n \geqslant 2$, or*
  (3) *$L_3(4)$ or Held's group, He.*

We shall refer to a group G as *balanced* if $O(C(t)) \subseteq O(G)$ for all involutions $t \in G$; otherwise, G is *unbalanced*.

We shall often have occasion to assume that "(U) holds in proper sections of G," by which we shall mean that the conclusions of (U) hold in all proper sections of G which satisfy the hypotheses of (U).

First let us note that (GW) follows from (U) by inspection of the listed groups. Thus (B) follows as well, by (1.17).

Next observe the following corollary to (1.14) and the classification of simple groups of sectional 2-rank at most 4 by Gorenstein and Harada [99].

PROPOSITION 3.1. *Let G be a minimal counterexample to (U). Then G has an unbalancing triple (a,x,L). Setting $D = O(C(x)) \cap C(a)$, the quotient group*

$$H = \langle L,D,x \rangle / C_{LD}(L/O(L))$$

is a proper section of G with $F^*(H) \cong L/Z^*(L)$ simple and $O(C_H(\bar{x})) \neq \langle 1 \rangle$ where $\bar{x}$ is the image of x in H. In particular $L/Z^*(L)$ is isomorphic to one of the groups listed in the conclusion of (U).

Thus (U) gives us an inductive handle on some $L \in \mathscr{L}(G)$. The proof of (U) then proceeds by a case-by-case treatment of the possibilities for L. This is the subject of the rest of this chapter.

Before we close this section it is wise to put the "change from (B) to (U)" in better perspective. Without doubt, (U) as well as (GW) were prominent in the minds of Gorenstein, Walter, Thompson and others long before 1974. Moreover the inductive advantage of (U) over (B) was also clear. However, equally clear was the great disadvantage of (U). Recall that the original reason for wanting to prove (B) was to provide a tool for characterizing simple groups G with $\mathscr{L}(G) \neq \emptyset$. However a proof of (U) entails a characterization of almost every simple group G with $\mathscr{L}(G) \neq \emptyset$. Thus most of the problems for which (B) would be useful would have to be solved before (B) was proved. A partial compensation is the fact that (B) holds in every *proper* section of a minimal counterexample to (U).

The factor which decisively tipped the argument in favor of (U) was the announcement by Aschbacher in 1974 of a major theorem characterizing almost all groups of Lie type over $GF(q)$ for q odd. This is the subject of Section 4.

4. Aschbacher's Theorem and its Consequences

In working on (B), Thompson was led to the problem of finite simple groups G with a tightly embedded subgroup K having a quaternion Sylow 2-subgroup. Typically K "should be" a root $SL(2,q)$ in a group G of Lie type over $GF(q)$ for some odd q. Thompson developed some ideas for getting hold of a "Weyl-type" subgroup of G, which together with K should determine G. These ideas were implemented by Aschbacher in early 1974. Conversations with Walter led to a weakening of the hypotheses to incorporate a characterization of the orthogonal groups over fields of odd order. In its final form, Aschbacher's theorem is the following.

THEOREM 4.1 ([15: Theorem I]). *Let G be a finite group with $F^*(G)$ simple. Let z be an involution in G with $z \in K \lhd\lhd C(z)$. Assume the following.*

(a) *K has quaternion Sylow 2-subgroups.*

(b) *For k a 2-element of $K - \langle z \rangle$, $k^G \cap C(z) \subseteq N(K)$.*

(c) *For $g \in C(z) - N(K)$, $[K, K^g] \subseteq O(C(z))$.*

Then one of the following holds.

(1) *$F^*(G)$ is a group of Lie type over a field of odd order other than $L_2(q)$ or ${}^2G_2(q)$.*

(2) *$F^*(G) \cong M_{11}, M_{12}, Sp_6(2)$ or $\Omega_8^+(2)$.*

Roughly speaking, Aschbacher sets $\Omega = K^G$ and, for $J \in \Omega$, defines $\Omega(J)$ to be those elements of Ω which commute with J modulo core (i.e., $[J^*, J] \subseteq O(J)$ whenever $J^* \in \Omega(J)$). His goal is to reach the following "generic situation".

(1) $X = \langle \Omega(K) \rangle$ is transitive on $\Omega(K)$.

(2) $F^*(X)$ is quasisimple.

(3) $G = \langle X, X^g \rangle$ for $K^g \in \Omega(K)$.

(4) $m_2(C_G(KX)) = 1$.

Once this situation is reached, G can be quickly identified using Curtis–Phan relations, as described in the proof of (3.5) in Chapter 2.

There are three main obstacles to reaching the generic situation:

(A) proving X is transitive on $\Omega(K)$,
(B) proving X is not 2-constrained, and
(C) proving $O(X) \subseteq Z(X)$.

Part (A) is the "low rank" case and, like all low rank cases, it is long, painful and impossible to summarize. Signalizer arguments are not needed by Aschbacher because they were previously carried out by Gorenstein, Harada and others in prior characterization theorems.

Part (B) requires Aschbacher to identify all 2-constrained groups H having a class K^G as in (4.1). It turns out, roughly, that $H = Z_2 \wr A_n$. He then fairly easily determines G if $X \cong Z_2 \wr A_n$.

Part (C) is quite easy for Aschbacher. He uses a slight variant of Goldschmidt's functor θ_C defined on elementary subgroups of $Z^*(K_1K_2K_3)$ where the K_i's are permuting members of K^G. Induction and elementary properties of groups of Lie type show θ_C is balanced and then it follows easily that $\theta_C = 1$ if $\langle\Omega(K_1)\rangle$ is large enough. Otherwise $\langle\Omega(K_1)\rangle$ is known by Part (A) and *ad hoc* arguments again show that $\theta_C = 1$.

Aschbacher's theorem allowed Thompson to deduce the following major reduction of (U), assuming certain facts about the groups of Lie type which were subsequently proved by Burgoyne [36].

THEOREM 4.2. *Let G be a finite unbalanced group with $F^*(G)$ simple having an unbalancing triple (a,x,J) with $J/O(J)$ a group of Lie type over a field of odd order. Suppose that all proper sections of G satisfy (U). Assume that $J/O(J) \cong L_2(q)$. Then either $F^*(G)$ is a group of Lie type over a field of odd order or G has an unbalancing triple (b,y,K) with $b \in K$ and $K/O(K) \cong \hat{A}_n$ for some $n \geqslant 9$.*

The proof of (4.2) is inspired by ideas of Walter which form the heart of his work in [230] on groups of Lie type, as described in Chapter 2. The "unbalanced condition" makes the argument particularly easy. We sketch the argument briefly.

Let $D = O(C(x)) \cap C(a)$. One proves that $\langle x,D\rangle$ normalizes, but does not centralize, some $L \in \mathscr{L}(J)$ with $L/O(L) \cong SL(2,q)$. Let b be an involution of $Z^*(L)$ and set $L_1 = C_L(b)$ and $K = \langle L_1^{L_{2'}(C(b))}\rangle$. If $K/O(K) \cong SL(2,q)$, then Aschbacher's Theorem identifies G. As (b,x,K_1) is an unbalancing triple for any 2-component K_1 of K, induction gives one of the following:

(1) $K/Z^*(K)$ is an orthogonal group of dimension at least 6 over $GF(q)$,
(2) $K/O(K) \cong \hat{A}_n$ for some $n \geqslant 9$,
(3) $K/O(K)$ is the 16-fold covering group of $L_3(4)$, or
(4) $K/O(K) \cong SL(2,q) \times SL(2,q)$.

Case (3) is eliminated by an easy fusion argument. In case (1), one finds a subconfiguration of type (4) in K. If $K = K_1K_2$ with $t_i \in Z(K_i)$, a second application of (4.1) forces $K_1K_2 \subseteq M$, a 2-component of $C(t_1)$ with (t_1,x,M) an unbalancing triple and $t_2 \notin Z^*(M)$. This forces $M/O(M) \cong Spin(7,q)$, but $O(C(x)) \cap C(t_1)$ acts nontrivially on both K_1 and K_2, which is impossible by inspection in $Spin(7,q)$. This leaves only (2), as required.

Prior to all of this, in 1973, Solomon had proved a result which in particular implied the following result.

THEOREM 4.3 ([186: (1.3)]). *Let G be a finite unbalanced group with $F^*(G)$ quasisimple having an unbalancing triple (a,x,J) with $J/O(J) \cong \hat{A}_m$ for some $m \geqslant 9$. Then $F^*(G) \cong \hat{A}_n$ for some $n \geqslant m$.*

The heart of the proof is a fusion argument similar to that outlined for the standard component $\hat{A}_n$ case in Chapter 2. The only added complication arises in the case $m = 11$, when the case $J < K$ with $K/O(K)$ isomorphic to the Lyons group Ly must be handled. As in Aschbacher's work, the functor θ_C defined on elementary subgroups of $Z^*(J_1J_2J_3)$ with $J_i \in J^G$ permits one to reduce to an easy standard form problem.

Thus (4.2) and (4.3) yielded the following reduction.

THEOREM 4.4. *Let G be a minimal counterexample to (U) and let (a,x,J) be an unbalancing triple in G. Then $J/O(J)$ is isomorphic to one of the following groups:*

(1) *A_{2n+1} for some $n \geqslant 3$,*
(2) *$L_2(q)$ for some odd $q \geqslant 5$,*
(3) *a covering group of $L_3(4)$, or*
(4) *He.*

5. The Alternating Groups

In 1975, Solomon completed the following additional reduction of (U).

THEOREM 5.1 ([189: Theorem 1.1]). *Let G be a finite group with $F^*(G)$ simple. Assume that all proper sections of G satisfy (U) and that (a,x,J) is an unbalancing triple of G with $J/O(J) \cong A_{2n+1}$ for some $n \geqslant 4$. Then $F^*(G) \cong A_{2m+1}$ for some $m > n$.*

COROLLARY 5.2. *Let G be a minimal counterexample to (U). If (a,x,J) is an unbalancing triple of G, then one of the following holds:*

(1) *$J/O(J) \cong L_2(q)$ for some odd $q \geqslant 5$,*
(2) *$J/O(J) \cong A_7$ or He, or*
(3) *$J/Z^*(J) \cong L_3(4)$.*

Much of the work in proving (5.1) involves the solution of the standard component problem for A_n, which was accomplished by Aschbacher [12] and Solomon [188]. This is described in Chapter 2.

To reduce to this case, the Goldschmidt functor θ_C again proves effective.

Definition 5.3. Let J be a finite group with $\bar{J} = J/O(J) \cong A_n$. A *root* 4-*subgroup* E of J is a 4-subgroup fixing $n-4$ points in the permutation representation of $\bar{J}$ of degree n. Let $\mathscr{R}(J)$ denote the set of root 4-subgroups of J.

Then, in the notation of (2.10), θ_C is defined on $V = E_1 \subseteq N(J)$ with $E_2 = E_1 \cap J \in \mathscr{R}(J)$. The problem is to verify the following.

(5.4) *There do not exist* $e, e_1 \in E_1^{\#}$ *with* $D = O(C(e_1)) \cap C(e)$ *and* $[D, E_2]$ *acting nontrivially on some 2-component* $\bar{L}$ *of* $C(e)/O(C(e))$.

If $L_{2'}(C_J(E_1)) \subseteq L$, an easy contradiction follows. If not, one may choose $F \in \mathscr{R}(J)$ with $E_2 \subseteq K = L_{2'}(C_J(F))$ and $\langle K, L_{2'}(C_L(F))\rangle$ contained in a 2-component M of $C(F)$ with $[M, C_D(F)] \nsubseteq O(M)$. Then induction gives the structure of M and, again, an easy contradiction follows.

With (5.4), $\theta_C = 1$ follows easily if $n > 11$, and with some effort if $n \in \{9, 11\}$. The results of [12] and [188] complete the proof of (5.1).

At this point, it would appear that the proof of (U) was almost complete. Indeed what remains to be proved is the following special case.

CONJECTURE 5.5. *Let G be a finite unbalanced group all of whose proper sections satisfy (U). Assume that $G = \langle H, a\rangle$ where $H = F^*(G)$ is simple and (a, x, J) is an unbalancing triple of G. Assume that for* every *unbalancing triple (b, y, K) of G, one of the following holds:*

(a) *$K/O(K) \cong A_7$, He or $L_2(q)$ for some odd $q \geqslant 5$, or*

(b) *$K/Z^*(K) \cong L_3(4)$.*

Then one of the following is the case:

(1) *$H \cong A_9$ and $J \cong A_5$ or A_7, or*

(2) *$H = G \cong A_{11}$ and $J \cong A_7$, or*

(3) *$G \cong Aut(He)$ and $J/Z(J) \cong A_7$ or $L_3(4)$.*

In particular we notice that in all the conclusions to (5.5) there is an unbalancing triple (b, y, K) with $K \in \mathscr{L}^*(G)$. However it is not *a priori* clear that any unbalancing component need be maximal in G. A typical situation one might imagine is the following: (a, x, J) is an unbalancing triple with $J \cong L_2(q)$, $P \in Syl_2(C(J))$ with $P^x = P$. J is subnormal in $C(b)$ for all $b \in C_P(x)^{\#}$ but there exists $b_1 \in P - C_P(x)$ with J properly embedded in $K \lhd\lhd C(b_1)$. Now K may no longer be involved in any unbalancing triple, so all inductive grip on K is lost. This obstacle was surmounted by Gilman and Solomon, but they paid a price for it. We describe this in our next section.

6. Pumping up Unbalancing Components

In Thompson's work on unbalancing components of Lie type and Solomon's work on unbalancing $\hat{A}_n$'s, one always had an unbalancing triple (a,x,J) with $a \in J$. This fact made it possible to manouvre so as never to lose track of the unbalancing involution. In Solomon's work on unbalancing A_n's, this problem never mattered because the answers to the standard form problem for A_{2n+1}, $n \geqslant 4$, involved only A_{2m+1} for $m \geqslant n$. However in confronting unbalancing triples (a,x,J) with $J/O(J) \cong L_2(q)$, one faces the nightmare of "pumping up" into a wide variety of 2-components, which should be irrelevant to the proof of (5.5). Gilman and Solomon undertook in [76] to put a lid on this Pandora's box by a more careful examination of the nature of the relation $<$ and its transitive extension $\ll$. They found the following criterion for maximality.

PROPOSITION 6.1. *Let G be a finite group all of whose proper sections satisfy (B). Let J be a 2-component of $C(a)$ for some involution a of G. Let $S \in Syl_2(C(a) \cap N(J))$ and let $P = C_S(J/O(J))$. Then $J \in \mathscr{L}^*(G)$ if the following conditions hold:*

(1) *if b is an involution of P and L is a 2-component of $\langle J^{L_{2'}(C(b))}\rangle$, then $L/Z^*(L) \cong J/Z^*(J)$, and*
(2) *if b is an involution of $C_P(S)$, then $S \in Syl_2(C(b) \cap N(L))$.*

This was an extension of the idea of *nonembedded components* introduced by Gilman in [73], when (B) holds for G. In this context the idea is easy to explain. Hypothesis (2) guarantees that $\langle J^{L_{2'}(C(b))}\rangle \neq LL^a$ with $L \neq L^a$, for any $b \in P$. Thus, by Hypothesis (1), if $J < L$, then $J = L$. Hence if $J \ll L$, then $J = L$ and, in particular, $J \in \mathscr{L}^*(G)$.

Motivated by (6.1), they defined a maximal unbalancing triple.

Definition 6.2. A *maximal unbalancing triple* (a,x,J) of G is an unbalancing triple such that if b is an involution of $C(a) \cap C(J/O(J))$ and L is a 2-component of $\langle (L_{2'}(C_J(b))^{L_{2'}(C(b))}\rangle$, then the following two conditions hold.

(1) If (b,y,L) is an unbalancing triple for some $y \in C(a) \cap N(J)$, then $J/Z^*(J) \cong L/Z^*(L)$.
(2) If $S \in Syl_2(C(a) \cap N(J))$ and $b \in Z(S)$, then $S \in Syl_2(C(b) \cap N(L))$.

The first important result is the following.

PROPOSITION 6.3. *Let G be a minimal counterexample to (U). Then G has a maximal unbalancing triple.*

The key results treat the cases of unbalancing components with

$$J/O(J) \cong A_7, \quad J/Z^*(J) \cong L_3(4), \quad \text{or} \quad J/O(J) \cong L_2(q).$$

THEOREM 6.4. *Let G be a minimal counterexample to (U). Suppose that (a,x,J) is an unbalancing triple of G with $J/O(J) \cong A_7$. Then either $J \in \mathscr{L}^*(G)$ or $J < K$ with (b,y,K) a maximal unbalancing triple and $K/O(K) \cong He$.*

Moreover, if there exists a 4-subgroup E of $C(a) \cap N(J)$ with $\Delta = C(a) \cap \Delta(E)$ and $[J,\Delta] = J$, then $J \in \mathscr{L}^*(G)$.

The next two results require hypotheses concerning the solution of certain standard component problems.

Hypothesis $(GS1)$. Whenever H is a finite group with $F^*(H)$ simple and J is a standard subgroup in H with $J/Z(J) \cong L_3(4)$, then $F^*(H)$ is isomorphic to $L_3(16)$, He, Suzuki's sporadic simple group Sz, or O'Nan's simple group ON.

THEOREM 6.5. *Let G be a minimal counterexample to (U). Assume that $(GS1)$ holds in all proper sections H of G. Assume that (a,x,J) is a maximal unbalancing triple of G with $J/Z^*(J) \cong L_3(4)$. Then $J \in \mathscr{L}^*(G)$.*

Hypothesis $(GS2)$. Whenever H is a finite balanced group with $F^*(H)$ simple and $J \cong L_2(q)$ is a standard component in H for some odd q, then $F^*(H)$ is known.

The list of possibilities for $F^*(H)$ is quite long. It may be found in [76: Section 1].

THEOREM 6.6. *Let G be a minimal counterexample to (U). Assume that $(GS2)$ holds in all proper sections H of G. Assume that (a,x,J) is a maximal unbalancing triple of G with $J/O(J) \cong L_2(q)$ for some odd q. Then there exists $L \in \mathscr{L}^*(G)$ with $L/O(L) \cong L_2(q_1)$ for some odd q_1.*

The upshot of all of this is that, subject to the proof of (GS1) and (GS2), a *maximal* 2-component L is produced with $L/O(L) \cong L_2(q)$, A_7 or He, or with $L/Z^*(L) \cong L_3(4)$.

In order to give a brief notion of part of the proof, we shall assume that the (B)-conjecture holds in G. Although it may seem to beg the central question, this in no way affects the basic thrust of the argument. It merely serves to avoid some notational complications which arise when having to move back and forth frequently between H and $H/O(H)$ for 2-local subgroups H of G. We assume that (a,x,J) is a maximal unbalancing triple of G but that $J \notin \mathscr{L}^*(G)$. The point of departure is the following observation.

LEMMA 6.7. *Let $S \in Syl_2(C(a) \cap N(J))$ and $P = C_S(J)$. There exists a non-identity subgroup B of P with $W = N(B)$ and $\bar{W} = W/B$, satisfying:*

(1) *$\langle J^{L_{2'}(W)} \rangle = KK^a = L_{2'}(W)$ and $B \in Syl_2(C_W(L_{2'}(W)))$,*

(2) *$\bar{J}$ is standard in $\langle \overline{K,a} \rangle$, and*

(3) *if $E \subseteq P$ with either $|E| > |B|$ or $|E| = |B|$ and $|N_S(E)| > |N_S(B)|$, then $J \lhd\lhd N_G(E)$.*

The lemma essentially follows from (6.1) and the Aschbacher–Foote component theorem discussed in the previous chapters, as $J \ncong SL(2,q)$. The definition of a maximal unbalancing triple then forces the following result.

LEMMA 6.8. $B \cap B^x = \langle 1 \rangle$.

Now (6.7) and (6.8) severely restrict the possibilities for the fusion of a in $N(B)$. Indeed, setting $D = S \cap J$, one shows that

(6.9) $a^{N(B)} \cap C_{\langle D,a \rangle}(x) = \{a\}$.

Using (6.9) and Hypotheses (GS1) and (GS2), one quickly proves Theorem 6.5 and reduces in the other cases to the following possibilities, where K is as in Lemma 6.7:

(6.10)

	J	KK^a
(1)	A_7 or $L_2(q)$	$\hat{A}_7 \circ \hat{A}_7$ or $SL(2,q) \circ SL(2,q)$
(2)	$L_2(q)$	$L_2(q^2)$
(3)	$L_2(q^2)$	$PSp_4(q)$
(4)	$L_2(9)$	A_8

Application of Theorem 4.1 then proves Theorem 6.4 and reduces Theorem 6.6 to the following situation.

PROPOSITION 6.11. *Assume the hypotheses of* (6.6). *Then G has a 2-subgroup B_0 with $L \lhd\lhd C(B_0)$ and $L/O(L) \cong A_8, PSp_4(q)$ or $Sp_4(q)$.*

It is worth noting that this situation is achieved fairly rapidly by Gilman and Solomon and almost all of the complication of their paper arises in eliminating the A_8 and $PSp_4(q)$ cases. This is scarcely odd, inasmuch as, for $q > 5$, the single infinite family of groups having unbalancing $L_2(q)$ components is the family $P\Omega_n(q)$, where

$$P\Omega_4^+(q) \cong L_2(q) \times L_2(q), \quad P\Omega_4^-(q) \cong L_2(q^2) \quad \text{and} \quad P\Omega_5(q) \cong PSp_4(q).$$

For $q = 5$, there is, of course, the additional family, A_{2n+1}, but Theorem 5.1 can be invoked to dispose of those groups immediately.

In any case, the results are finally proved leaving the following problems to be solved to complete the proof of (U).

(6.12) The classification of groups G with $F^*(G)$ simple, which are either balanced or minimal counterexamples to (U), and have either

(1) an unbalancing triple (a,x,J) with $J \in \mathscr{L}^*(G)$ and $J/O(J) \cong A_7$ or He, or $J/Z^*(J) \cong L_3(4)$; or

(2) $J \in \mathscr{L}^*(G)$ with $J/O(J) \cong L_2(q)$.

(6.13) The classification of balanced groups G with $F^*(G)$ simple, having a standard subgroup J with $J/Z(J) \cong L_3(4)$.

At this point our problem, which as of (5.5) seemed to be reduced to a characterization of A_9, A_{11} and He, has burst its confines and now entails certain "irrelevant" characterizations. We have the advantage, however, of *knowing* that the groups still to be identified are "small", in the sense of having a maximal 2-component of type A_7, $L_2(q)$, $L_3(4)$ or He. This "smallness" will become even more precise in the next section.

7. The A_7 and $L_2(q)$ Problems, Phase I

In order to complete (6.12) for A_7 and $L_2(q)$, it clearly suffices, in view of Aschbacher's theorem (4.1), to determine all finite groups G satisfying the following hypothesis.

Hypothesis 7.1. G is a finite group with $F^*(G)$ simple such that

(1) (U) holds in every proper section of G,

(2) there exists $J \in \mathcal{L}^*(G)$ with $J/O(J) \cong A_7$ or $L_2(q)$ for some odd $q \geqslant 5$, and

(3) for all $K \in \mathcal{L}^*(G)$, $m_2(K) > 1$.

In 1975, Foote obtained the following reduction of the problem in [71].

THEOREM 7.2. *Let G satisfy Hypothesis* 7.1. *If $m_2(C(J/O(J))) \geqslant 2$, then $F^*(G)$ is isomorphic to A_9, A_{10}, A_{11}, M_{12} or the Hall–Janko group J_2.*

We have discussed Foote's use of the Goldschmidt functor θ_C in Section 2. In the case $m_2(C(J/O(J))) \geqslant 3$, this immediately produces a signalizer functor on a large set of elementary subgroups. However, if $C(J/O(J))$ has $SCN_3(2) = \emptyset$, some delicate fusion analysis is still needed to prove that $\theta_C = 1$. Once this is proved, an easy contradiction follows. When $C(J/O(J))$ has 2-rank 2, the argument is almost entirely fusion-theoretic arriving either at the hypotheses of a theorem of Harada on groups with small self-centralizing 2-subgroups [115: Theorem 4] or at a set of 2-local data very similar to that of $L_2(q) \wr Z_2$. In the latter case, Goldschmidt's methods for handling strongly closed 2-subgroups yield a contradiction.

At this time, Harris and Solomon and, independently, Fritz undertook the case where $m_2(C(J/O(J))) = 1$. In this case, the following result is easily proved.

PROPOSITION 7.3. *Let G satisfy Hypothesis* 7.1 *with $m_2(C(J/O(J))) = 1$. Then either $F^*(G)$ has sectional* 2-*rank at most* 4 *or the following hold:*

(1) *$|C(J/O(J))|_2 = 2$, and*

(2) *there is an involution σ in $N(J)$ inducing a field automorphism on $J/O(J)$.*

Completing this problem is a remarkably tedious and still uncompleted task. Fritz's work was undertaken in the context of a project of Held to determine all simple groups G with $|G|_2 \leqslant 2^{10}$. Fritz's result was the following.

THEOREM 7.4 ([72]). *Let G satisfy Hypothesis* 7.1. *If G is simple and $|G|_2 \leqslant 2^{10}$, then G is known.*

Harris and Solomon, using the full determination of simple groups G with $|G|_2 \leqslant 2^{10}$, obtained the following theorem.

THEOREM 7.5 ([121: Theorem 2]). *Let G satisfy Hypothesis* 7.1. *If $|J|_2 = 8$ and $|N(J)/JC(J)|_2 = 2$, then either $F^*(G)$ has sectional* 2*-rank at most* 4 *or $F^*(G)$ is isomorphic to $Sp_4(4)$, $L_5(2)$, $U_5(2)$ or He.*

COROLLARY 7.6. *Let G satisfy Hypothesis* 7.1 *with $J/O(J) \cong A_7$. Then $F^*(G) \cong A_9, A_{11}$ or He.*

Harris extended (7.5) in [118: Theorem 2].

THEOREM 7.7. *Let G satisfy Hypothesis* 7.1. *If $|J|_2 = 8$, then either one of the conclusions of* (7.5) *holds or $F^*(G)$ is isomorphic to the Higman–Sims group.*

COROLLARY 7.8. *Let G satisfy Hypothesis* 7.1. *Then either $F^*(G)$ is known or the following hold:*

(1) *$J/O(J) \cong L_2(q^2)$ for some odd $q \geqslant 7$,*

(2) *$|C(J/O(J))|_2 = 2$, and*

(3) *there exists an involution σ in $N(J)/C(J/O(J))$ with $L(C_J(\sigma)) \neq \langle 1 \rangle$.*

The primary significance of (7.7) for later work is the elimination of the troublesome case $J/O(J) \cong L_2(9)$. The proofs of (7.4), (7.5) and (7.7) are very painful fusion analyses in the spirit of [99].

8. The $L_3(4)$ and He Cases

In 1976, Griess and Solomon completed the reduction of (U) to the maximal $L_2(q)$ and standard $L_3(4)$ problems by proving the following theorem ([111: Theorem 1.2]).

THEOREM 8.1. *Let G be a minimal counterexample to* (U). *Assume that* ($GS1$) *holds in all sections of G and* ($GS2$) *holds in all proper sections of G. Then there exists $J \in \mathscr{L}^*(G)$ with $J/O(J) \cong L_2(q)$ for some odd q.*

The proof proceeds by contradiction. Thus we assume that there is no $J \in \mathscr{L}^*(G)$ with $J/O(J) \cong L_2(q)$ for any odd q. Then by (5.5), (6.4), (6.5) and (6.6), the following holds.

PROPOSITION 8.2. *Let G satisfy the hypotheses of* (8.1). *Then*

(1) *if (a,x,J) is an unbalancing triple of G, then $J/O(J) \cong L_2(5)$, $L_2(7)$, A_7 or He, or $J/Z^*(J) \cong L_3(4)$,*

(2) *if (b,y,K) is a maximal unbalancing triple of G, then $K/Z^*(K) \cong L_3(4)$ or He, and*

(3) *θ_Δ is an A-signalizer functor for all abelian 2-subgroups A of G with $m(A) \geqslant 4$.*

We remark that $L_2(5)$, $L_2(7)$, $L_3(4)$ and He are 2-balanced groups. Theorem 6.4 guarantees that 2-components of A_7 type in G are locally 2-balanced in G‡. Thus θ_Δ is an A-signalizer functor.

Using (8.2), Griess and Solomon are able to quickly show in the case where $K \in \mathscr{L}^*(G)$ with $K/O(K) \cong He$ that $\theta_\Delta = 1$ and then that K is a standard subgroup of G. They then solve the standard He problem, assuming (GS1) for proper sections of G.

In the case when $K/Z^*(K) \cong L_3(4)$ for every maximal unbalancing triple (b,y,K), the problem divides into two cases.

Case A: The graph of commuting 8-subgroups of $KC(K/O(K))$ is disconnected.

Case B: The graph of commuting 8-subgroups of $KC(K/O(K))$ is connected.

In Case A, careful fusion analysis shows that $O^2(G)$ has a 2-local subgroup H such that $H/O(H)$ is a non-splitting extension of $Z_4 \times Z_4 \times Z_4$ by $GL(3,2)$. Then a theorem of O'Nan [158] and work of Sims and Andrilli [7] yields a contradiction by showing that $O^2(G) \cong ON$, whence G is a balanced group.

In Case B, it is proved that $\Delta(E) = 1$ for all 4-subgroups of $KC(K/O(K))$. Then Lemma 2.9 immediately yields that K is a standard subgroup of G provided that $m_2(C(K/O(K))) > 2$, and (GS1) gives a contradiction.

One is left with a fairly tight configuration with $m_2(C(K/O(K))) \leqslant 2$. The failure of K to be standard yields another centralizer whose layer contains $K_1K_2 \ldots K_r$, $1 \leqslant r \leqslant 4$, with $K_i/O(K_i) \cong L_2(5)$ or $L_2(7)$ for all i, $1 \leqslant i \leqslant r$. Comparison of these two centralizers finally yields a contradiction.

Thus (8.1) reduces (U) to the problems of proving (GS1) and (GS2) and handling certain groups with $J \in \mathscr{L}^*(G)$, $J/O(J) \cong L_2(q)$. Combining this with (7.8), we obtain the following reduction.

THEOREM 8.2. *(U) will follow from the classification of finite groups G with $F^*(G)$ simple such that (U) holds in every proper section of G and either*

(U1) *there is a standard subgroup L of G with $L/Z^*(L) \cong L_3(4)$, or*

(U2) *there is $L \in \mathscr{L}^*(G)$ satisfying*

  (a) *$L/O(L) \cong L_2(q^2)$ for some odd $q \geqslant 7$,*

  (b) *$|C(L/O(L))|_2 = 2$, and*

  (c) *there is an involution $\sigma \in N(L)/C(L/O(L))$ with $L_{2'}(C_L(\sigma)) \neq \langle 1 \rangle$.*

‡ In the notation of Theorem 6.4, this means that, for all 4-subgroups E of $C(a) \cap N(J)$, $[L,\Delta(E)] \subseteq O(C(a))$.

The status of the standard $L_3(4)$ problem is discussed in Chapter 2. In the next section we discuss Phase II of the maximal $L_2(q)$ problem.

9. The $L_2(q)$ Problem, Phase II

By (7.8), the maximal $L_2(q)$ problem had been reduced to a fairly precise "centralizer of an involution" problem. Unfortunately this problem has a large number of solutions. After studying for a while the possibility of extending the 2-group analysis used in [72], [118] and [121] to handle the general case, Harris concluded that a more fruitful approach would be to try to "pump up" $L_{2'}(C_L(\sigma))$ to a maximal 2-component of G. (Here we use the notation of (8.2).) Unfortunately, this approach makes the solution of the maximal $L_2(q)$ problem rely on the solution of the standard component problem for all groups of Lie type over fields of odd order. This may however be the only reasonable method of solving this problem.

As the bulk of Harris' work is discussed in the context of standard Chevalley components in Chapter 2, we shall limit ourselves to the statement of his main result and some discussion of the reduction of its proof to the standard form setting.

Hypothesis H. Whenever H is a finite group with $F^*(H)$ simple, satisfying (U), and L is a standard subgroup of H with $L/Z(L)$ isomorphic to $P\Omega_n^{\pm}(3)$ for some n, $5 \leqslant n \leqslant 8$, then $F^*(H)$ is a known simple group.

THEOREM 9.1 ([120]). *Let G be a finite group with $F^*(G)$ simple and $|G:F^*(G)| \leqslant 2$. Suppose that (U) and (H) hold in all proper sections of G. Suppose that $J \in \mathscr{L}^*(G)$ with $J/O(J) \cong L_2(q^2)$ for some odd $q > 3$. Then one of the following holds:*

(1) *G has sectional 2-rank at most 4, or*
(2) *$J \cong L_2(q^2)$ and $F^*(G) \cong P\Omega_n^{\pm}(q)$ for $n = 5$ or 6, or*
(3) *$J \cong L_2(q^2)$ and $F^*(G) \cong P\Omega_8^-(q^{1/2})$.*

The essential observation is that if G does not have sectional 2-rank at most 4, there exists an involution $\sigma \in N(J)$ with

$$L_{2'}(C_J(\sigma))/O(L_{2'}(C_J(\sigma))) \cong L_2(q).$$

Harris then studies the group $H_1 = C(\sigma)$ in which (U) holds by induction. The aim is to prove that either

$$L_{2'}(H_1)/O(L_{2'}(H_1)) \cong SL(2,q) \circ SL(2,q),$$

with $\sigma \in L_{2'}(H_1)$ or $L_{2'}(H_1)/O(L_{2'}(H_1)) \cong U_4(q^{1/2})$. The former case occurs if $F^*(G) \cong P\Omega_n^{\pm}(q)$ for $n = 5$ or 6, while the latter occurs if $F^*(G) \cong P\Omega_8^-(q^{1/2})$.

As (U) and (H) hold in all sections of H_1, Harris is able to invoke his general results on Lie type components to determine the possibilities for $L_{2'}(H_1)/O(L_{2'}(H_1))$. Fusion arguments then reduce one to the two possibilities mentioned above. If

$$L_2{}'(H_1)/O(L_{2'}(H_1)) \cong SL(2,q) \circ SL(2,q),$$

then $F^*(G)$ is determined immediately by (4.1). If

$$L_{2'}(H_1)/O(L_{2'}(H_1)) \cong U_4(q^{1/2}),$$

more work is required to prove that, for involutions $t \in L_{2'}(H_1)$, there exists $L_t \lhd\lhd L_{2'}(C_t)$ with $t \in L_t$ and $L_t/O(L_t) \cong SL(2,q^{1/2})$. Once this is proved, $F^*(G)$ is again determined by (4.1).

Given the announced completion by Gomi, Nah and Walter of some of the open standard component problems, Harris' result has the following consequence.

COROLLARY 9.2. *The Unbalanced Group Conjecture is true if the following two assertions are valid.*

(1) *Let G be a finite simple group with a standard subgroup L such that $L/O(L) \cong L_3(4)$ and $m_2(C(L)) \geqslant 2$. Then G is isomorphic to Suzuki's sporadic simple group.*

(2) *Let G be a finite group with $F^*(G)$ simple. Assume that (U) holds in all sections of G and G has a standard subgroup L such that $L/Z(L)$ is isomorphic to $L_4(3)$ or $U_4(3)$ and $C(L)$ has cyclic Sylow 2-subgroups. Then $F^*(G)$ is isomorphic to $L_5(3)$, $U_5(3)$, $L_4(9)$, $PSp_8(3)$, $P\Omega_7(3)$ or $P\Omega_8^{\pm}(3)$.*

10. Other Approaches

The principal alternate approach to (U) thus far undertaken is contained in the work of Walter [231]. Walter's point of view is essentially the same as that previously described. Indeed, his ideas and suggestions had an important influence on the investigations we discussed earlier. We shall briefly indicate the principal differences in his approach.

First, Walter undertakes the classification of a somewhat larger set of groups than the unbalanced groups.

Definition 10.1. A finite group G is *$\mathscr{E}$-unbalanced* if $O(C_G(E)) \nsubseteq O(G)$ for some elementary 2-subgroup E of G.

Walter's observation is that the attempt to classify $\mathscr{E}$-unbalanced groups G with $F^*(G)$ simple gives additional inductive leverage, while not significantly extending the list of groups to be identified.

In the previously described work, the Gorenstein–Walter functor θ_Δ only figures in the Griess–Solomon paper described in Section 9. Walter undertakes to utilize θ_Δ at a much earlier stage in the argument. He assumes only Theorems 4.1 and 4.3 and attempts to prove that θ_Δ is an A-signalizer functor for suitably chosen A. The presence of unbalancing components of A_n and $L_2(q)$-types makes this task delicate and requires a "lining-up" lemma to guarantee the existence of any "good" A. If $m(A) \geqslant 5$, the $L_2(q)$ problem can be avoided by replacing θ_Δ by the analogous functor associated with $\Delta(E)$ for $E \cong E_8$. This of course increases the connectivity problems. If $m(A) = 4$, θ_Δ is replaced by θ_{Δ_p} where

$$\Delta_p(E) = \bigcap_{e \in E^\#} O_p(C(e)).$$

This functor, originally suggested by Goldschmidt, has the following advantage.

Proposition 10.2. *Let G be a finite group with $F^*(G)$ a* known *simple group. Then one of the following holds:*

(1) *$\Delta_p(E) = \langle 1 \rangle$ for all odd p and all 4-subgroups E of G, or*
(2) *$p = 3$, $F^*(G) \cong A_{4n+3}$, $\Delta_3(E) \cong Z_3$ and E is conjugate to an elementary 2-group acting semi-regularly on $\{1,2,\ldots,4n\}$.*

The drawback of θ_{Δ_p} is the difficulty of deducing, once $\Delta_p(E) = \langle 1 \rangle$, that 2-layers are semisimple.

This emphasis on the Δ-functors rather than Goldschmidt's commutator functors represents the principal difference between Walter's approach and the approach described in the earlier sections. Of course, his concurrent work on groups with components of Lie type over $GF(q)$ permits him to avoid the complications encountered by Gilman and Solomon in trying the control nonmaximal $L_2(q)$ components.

We conclude this chapter by mentioning two significantly different ideas for treating this problem. Both aim at a direct attack on the B-Conjecture. The first was Thompson's original attack on (B). The idea was to use factorization techniques and methods of Bender to study maximal subgroups of G containing $O(F(B(C_G(t))))$ for a suitably chosen involution t of G. One of the difficulties encountered by Thompson involves a "pushing-up" problem for p-constrained groups H with $H/O_p(H) \cong SL(2,p^n)$, where p is an odd prime. Important results on this type of problem have recently been obtained by Glauberman and Niles. (See Chapter 9.) Thus Thompson's approach might warrant reinspection.

The second idea, which has more of the status of a dream, is Alperin's goal of deducing the B-Conjecture from a major elaboration of Brauer's theory of p-modular characters. Recall that

$$O(G) = \bigcap_{\psi \in B_0(G)} Ker\,\psi$$

where $B_0(G)$ is the principal 2-block of G. Now if H is a 2-local subgroup of

G, then in general, for $\psi \in B_0(G)$, *not* all of the irreducible constituents of $\psi|_H$ lie in $B_0(H)$. However, suppose one could establish the existence of a functor B assigning to each finite group G a subset of $Irr(G)$ such that

(1) $O(B(G)) = \bigcap_{\psi \in B(G)} Ker\,\psi$, and

(2) if $\psi \in B(G)$, H is a 2-local subgroup of G and φ is an irreducible constituent of $\psi|_H$, then $\varphi \in B(H)$.

Then the B-Conjecture would be a trivial corollary, for one would have

$$O(B(H)) \subseteq \bigcap_{\psi \in B(G)} Ker\,\psi = O(B(G)).$$

Indeed, if such a theory could be developed, it would probably work for all primes p and give a simultaneous proof of the "$B_p(G)$-Conjecture" for all primes p. Such a theory might eliminate all need for signalizer functors in the study of finite simple groups. An interesting beginning towards such a goal is to be found in the recent papers of Alperin, Broué and Puig (e.g. [4]).

It would be wonderful to have three completely different proofs of the $B(G)$-Conjecture. More realistically, it would be nice to have one proof *soon*!

4

Finite Groups of Characteristic 2 Type

DANIEL GORENSTEIN and RICHARD LYONS

1. Introduction

In this chapter, we shall outline the results which we have obtained on simple groups G of characteristic 2 type in which all proper subgroups are $\mathscr{K}$-groups and $e(G) \geqslant 4$.

Here, by definition, a *$\mathscr{K}$-group* is a group, all of whose composition factors are among the known simple groups, while $e(G)$ denotes the maximum 2-*local p-rank* $m_{2,p}(G)$ of G as p ranges over all odd primes, and for a given prime p, $m_{2,p}(G)$ is the maximum of $m_p(H)$ as H ranges over all 2-local subgroups of G. Recall also that G is of *characteristic* 2 *type* if $F^*(H) = O_2(H)$ for every 2-local subgroup H of G; equivalently, every such H is 2-constrained with trivial core.

Our reason for focusing on the case $e(G) \geqslant 4$ is that this case can be treated in an essentially uniform manner. (However, considerable special analysis is required to establish our results when $p = 3$ and $m_{2,3}(G) = 4$.) The overall analysis divides into three major parts which give conditions for G to possess the following, respective, internal structures.

(A) Either the existence of standard components in centralizers of elements of odd prime order or the existence of maximal 2-locals with O_2 of symplectic type.
(B) The existence of 2-locals containing weak 2-generated p-cores.
(C) The existence of almost strongly p-embedded 2-locals.

The exact meanings of (A), (B) and (C) will be given in Section 2.

Aschbacher's work in the corresponding case $e(G) = 3$ has led to an entirely analogous three part division of that problem and he has carried through the corresponding analysis of both parts (A) and (B). (Moreover, in addition, under (A) he has treated all but two of the resulting standard form problems: namely, $p = 3$ and $C_G(x) \cong Z_3 \times A_8$ or $Z_3 \times Sp_6(2)$, x an element of G of order 3.) His conclusions in part (B) are identical to those which we reach in the case $e(G) \geqslant 4$. Hence, to avoid duplication of effort, our analysis in part (C) (which has the conclusion of (B) as its hypothesis), covers both the cases $e(G) = 3$ and $e(G) \geqslant 4$ (thus in part (C) we assume only that $e(G) \geqslant 3$).

The proof of our results requires a great many properties of $\mathscr{K}$-groups, the verification of which is itself a long and arduous task. Standard techniques exist for reducing most of the needed assertions to corresponding statements about simple (and quasisimple) $\mathscr{K}$-groups. Hence our analysis depends upon a prior systematic study of certain questions concerning the known simple groups. The principal such questions relate to properties of *local balance* and *generation* with respect to elementary abelian p-groups, p an odd prime, which are entirely similar to those considered in the past for the prime 2. The main results in this direction have been recently established by Seitz [174]; the specific statements we need can be derived from his theorems, in some cases directly, and in others, with some additional analysis.

It is the verification of these required properties of simple $\mathscr{K}$-groups which has slowed completion of our work. That task is now fortunately nearly finished. Manuscripts for parts (A) and (B) have been typed, but with references to the preliminary properties of $\mathscr{K}$-groups omitted. The same has now been done for part (C). Thus our results under (A), (B) and (C) cannot, at the time of writing, be considered to be theorems. This must await completion of the two manuscripts which will incorporate the various properties of $\mathscr{K}$-groups, which we need (hopefully to be accomplished before the appearance of this book).

Furthermore, it should also be mentioned that our results in part (C) depend upon yet another major effort, just completed:

(D) The embedding of 2-local subgroups.

It turned out that, to establish our results on almost strongly p-embedded 2-locals, we required certain facts about the embedding of 2-locals in G which were very similar to properties which Aschbacher needed for his investigations of simple groups of characteristic 2 type which possess a strongly 3-embedded maximal 2-local subgroup (and, more generally, an almost strongly p-embedded maximal 2-local, p any odd prime). Again to avoid duplication of effort, Aschbacher and the present authors have combined forces and are preparing a single paper [20] which covers the various results we need in part (D). These results will be stated explicitly in the next section, but discussion of their proof will be omitted. Hopefully that manuscript, too, will be finished before the appearance of this book.

2. Definitions and Statements of the Main Results

To state our principal results precisely, we require considerable preliminary terminology. Here, G will denote a simple group of characteristic 2 type with all proper subgroups $\mathscr{K}$-groups.

We set

$$\beta_k(G) = \{p | p \text{ an odd prime}, m_{2,p}(G) \geqslant k\}.$$

Thus, in parts (A) and (B), we assume that $\beta_4(G) \neq \emptyset$, while in part (C) we assume only that $\beta_3(G) \neq \emptyset$.

In his work on the case $e(G) = 3$, Aschbacher introduces an ordering in the set $\beta_3(G)$ [19]. This ordering is incorporated into the following definitions:

$$\sigma(G) = \beta_4(G) \quad \text{if} \quad e(G) \geqslant 4;$$

while if $e(G) = 3$, set

$$\sigma(G) = \{p \,|\, p \in \beta_3(G), p \geqslant 7\} \quad \text{if} \quad \beta_3(G) \nsubseteq \{3,5\};$$

$$\sigma(G) = \{5\} \quad \text{if} \quad \beta_3(G) \subseteq \{3,5\}, \quad \text{but} \quad \beta_3(G) \neq \{3\};$$

and

$$\sigma(G) = \{3\} \quad \text{if} \quad \beta_3(G) = \{3\}.$$

For each $p \in \sigma(G)$, the set of elementary abelian p-subgroups B of G which are of maximal rank subject to lying in a 2-local subgroup of G (i.e., $m_p(B) = m_{2,p}(G)$) plays a fundamental role in our analysis. We denote this set by $\mathscr{B}_{\max}(G;p)$.

It is best to define the next set of terms for an arbitrary finite group X and p-subgroup P of X. We set

$$\mathscr{E}^p(X) = \text{set of elementary abelian } p\text{-subgroups of } X,$$

$$\mathscr{E}^p_k(X) = \text{set of elementary abelian } p\text{-subgroups of rank } k,$$

$$\Gamma_{P,k}(X) = \langle N_X(Q) | Q \subseteq P, m_p(Q) \geqslant k \rangle,$$

and

$$\Gamma^0_{P,k}(X) = \langle N_X(Q) \,|\, Q \subseteq P, m_p(Q) \geqslant k, m_p(QC_P(Q)) \geqslant k+1 \rangle.$$

If $P \in Syl_p(X)$, we call $\Gamma_{P,k}(X)$ the *k-generated p-core* of X and $\Gamma^0_{P,k}(X)$ the *weak k-generated p-core* of X. They are clearly determined up to conjugacy by the choice of P. Moreover, if $\Gamma_{P,1}(X) \subset X$ for $P \in Syl_p(X)$, we say that X has a *strongly p-embedded* subgroup.

Weak k-generated p-cores arise in a natural way in the course of arguments which involve questions of "k-connectedness for the prime p" (see Section 5 for definition of 2-connectedness = connectedness).

Finally for any $E \in \mathscr{E}^p(X)$, we set

$$\Delta_X(E) = \bigcap_{e \in E^\#} O_{p'}(C_X(e)).$$

Properties of "balance for the prime p" are expressible in terms of this Δ notation.

The conclusion of part (B) involves weak p-generated 2-cores. To state those of parts (A) and (C) we need three definitions.

Definition 2.1. We say that G is of *$GF(2)$-type for the prime $p \in \sigma(G)$* if, for some maximal 2-local subgroup M of G, $O_2(M)$ possesses no noncyclic characteristic abelian subgroup and M contains an element of $\mathscr{B}_{\max}(G;p)$.

By P. Hall's well-known theorem, $O_2(M)$ is then the central product of an extra-special group with a cyclic or maximal class 2-group. The term "$GF(2)$-type" is suggested by the fact that in many families of groups of Lie type defined over the prime field $GF(2)$, $O_2(C(t))$ is extraspecial for some involution t.

Simple groups of $GF(2)$-type have been completely classified by Aschbacher [13,14], S. Smith, [182,183,184] and Timmesfeld [221]. We could therefore have dropped this alternative from the conclusion of part (A) and replaced it by the assertion that G is known, but for clarity of exposition it is preferable to include the $GF(2)$-type possibility explicitly.

To avoid technicalities, we state our next definition in a less precise form than appears in the actual conclusion of part (A).

Definition 2.2 (slightly simplified). G is of *standard type for the prime $p \in \beta_4(G)$* provided there is $B \in \mathscr{B}_{\max}(G;p)$, $x \in B^{\#}$, and a normal quasisimple subgroup L of C_x $(= C_G(x))$ with the following properties:

(1) L is a covering group of a group of Lie type of characteristic 2;
(2) either p divides the order of a Cartan subgroup of L or $p = 3$ and L is defined over $GF(2)$;
(3) every element of B induces an inner-diagonal automorphism on L;
(4) B does not centralize some B-invariant 2-subgroup of C_x;
(5) $C_G(L)$ has cyclic Sylow p-subgroups; and
(6) if $D \in \mathscr{E}_2^p(B)$ with $x \in D$, and $L_0 = C_L(D)^{(\infty)}$ satisfies the condition $C_B(L_0) = D$ (such D and L_0 exist by the isomorphism type of L and the embedding of B), then there is $x^* \in D - \langle x \rangle$ such that, if L^* denotes the normal closure of L_0 in C_{x^*}, then $L^*/O_{p'}(L^*)$ is a covering group of a group of Lie type of characteristic 2 and x does not centralize $L^*/O_{p'}(L^*)$ (whence $C_B(L^*/O_{p'}(L^*)) = \langle x^* \rangle$).

Remarks. Condition (5) justifies the use of the term "standard". The full definition of standard type involves refinements of both conditions (2) and (6). There is a minor restriction on the isomorphism type of L_0 for (6) to hold. Also, L^* is usually quasisimple in (6). A consequence of the definition and properties of $\mathscr{K}$-groups is that $\langle x \rangle$ is not weakly closed in a Sylow p-subgroup of C_x, so that one has an analogue of the Z^*-theorem. Of course, (6) is a much sharper statement.

We call the group L^* in (6) a "neighbour" of L. As one would anticipate, the structure of G should, in general, be completely determined by L and its neighbour L^*. In fact, one would hope to show first that the group $G_0 = \langle L, L^* \rangle$ has a (B,N)-pair and hence is a group of Lie type of characteristic 2, and then that $G = G_0$. This is the usual procedure for treating standard form problems. Finkelstein, Frohardt, Gilman, Griess and Solomon are presently treating the various possibilities that can arise for L and L^* (including the two residual cases when $e(G) = 3$, mentioned above); and at this writing, their work is nearly completed [61,65,66,75] (see Chapter 8).

Once these standard form problems have indeed been settled, the conclusion of part (A) will be that G is determined and, in fact, is a group of Lie type of characteristic 2. Then the study of simple groups G of characteristic 2 type with $e(G) \geqslant 3$ and all proper subgroups $\mathscr{K}$-groups will be reduced to those which satisfy the conclusion of part (C), for all $p \in \sigma(G)$. This conclusion is embodied in the following definition.

Definition 2.3. Let M be a maximal 2-local subgroup of G and p a prime in $\sigma(G)$. We say that M is *almost strongly p-embedded* in G provided the following conditions are satisfied:

(1) $\Gamma_{P,2}(G) \subseteq M$ for some Sylow p-subgroup P of G;
(2) either $\Gamma_{P,1}(G) \subseteq M$ (in which case M is a strongly p-embedded 2-local subgroup of G) or M is solvable and, according as $p \geqslant 5$ or $p = 3$, condition (3) or (4) holds;
(3) (a) P is abelian;
  (b) there is $P_0 \subseteq P$ of order p such that P_0 is weakly closed in P with respect to G;
  (c) if $C = C_G(P_0)$, then $L = C^{(\infty)} \cong L_2(p^n)$, $n \geqslant 2$, $L \subseteq M$, and $C_C(L)$ has cyclic Sylow p-subgroups.
  (d) $O_{p'}(M)P_0$ is a Frobenius group with kernel $O_2(M)$ and complement $O_{p'}(C)P_0$; and
  (e) if $P_1 \subseteq P$ is of order p and $P_1 \neq P_0$, then $N_G(P_1) \subseteq M$;
(4) (a) $\sigma(G) = \{3\}$, $e(G) = m_{2,3}(G) = 3$, and $P \cong Z_3 \wr Z_3$;
  (b) there is $P_0 \subseteq P$ of order 3 with $P_0 \nsubseteq J(P)$ such that $C = C_G(P_0) \nsubseteq M$;
  (c) M covers $C/O(O_{3'}(C))$ and $F(O(O_{3'}(C))) \cap M = 1$;
  (d) if T is a P-invariant Sylow 2-subgroup of $O_{3'}(M)$, then $C_T(P_0) \cong Q_8 \circ Z_{2^n}$ for some n; and
  (e) if $P_1 \subseteq J(P)$ with P_1 of order p, then $N_G(P_1) \subseteq M$.

Remarks. By the definition, either M is strongly p-embedded in G or there is a single conjugacy class of subgroups of P of order p such that, if P_0 is in the class, then $N_G(P_0) \subseteq M$; in the latter case, the structures of M and of $C_G(P_0)$ are very precisely pinned down. Note that if $p = 3$ and $m_{2,3}(G) \geqslant 4$, then the definition implies that M is, in fact, strongly 3-embedded in G.

Aschbacher is presently studying the situation in which G possesses an almost strongly p-embedded maximal 2-local subgroup for each $p \in \sigma(G)$; his work, too, is nearing completion‡. Moreover, he has already handled the

‡ This work has now been completed.

important special case in which G possesses a strongly 3-embedded maximal 2-local subgroup and $m_{2,3}(G) \geqslant 4$ (proving that no such simple group G exists). This means that for his further analysis, he can assume that $m_{2,3}(G) \leqslant 3$, which puts a severe restriction on the possible proper simple sections involved in G. The goal of that analysis is, as in the special case already treated, to show that no such simple group exists.

As the previous discussion clearly indicates, the conclusions of part (A) lead eventually to the construction of the groups of Lie type of characteristic 2, while those of parts (B) and (C)—the so-called "uniqueness case"—will hopefully always lead to a contradiction. Thus a fundamental dichotomy exists between the conclusions of part (A) versus (B) and (C). This dichotomy is reflected in the corresponding hypotheses of each part, which involve suitable Δ-assumptions. These are formalized in the following terminology:

For any $p \in \sigma(G)$, we say that $(\Delta O)_p$ *holds*, provided for every $B \in \mathscr{B}_{\max}(G;p)$ and every $D \in \mathscr{E}_2^p(B)$, we have

(1) $\Delta_G(D)$ is of odd order if $p \geqslant 5$;
(2) $[\Delta_G(D), B]$ is of odd order if $p = 3$.

The distinction in the definition for $p \geqslant 5$ and $p = 3$ is made for minor technical reasons.

The subdivision in our analysis of G is made according as condition $(\Delta O)_p$ holds for some $p \in \sigma(G)$ or fails for every $p \in \sigma(G)$. In the first case, we argue in part (A) that G is either of $GF(2)$-type or of standard type for some prime in $\sigma(G)$; while in the second case, we show ultimately (part (C)) that G possesses an almost strongly p-embedded maximal 2-local subgroup for each $p \in \sigma(G)$.

The following additional term will help to make this dichotomy even sharper. For any $p \in \sigma(G)$, we say that $(\Gamma_2^0)_p$ *holds*, provided $\Gamma_{P,2}^0(G)$ is contained in a 2-local subgroup of G for any $P \in Syl_p(G)$.

The conclusion of part (B) is then that, for $p \in \sigma(G)$, either condition $(\Delta O)_p$ holds or condition $(\Gamma_2^0)_p$ holds. However, for this conclusion to be correct, we must place a restriction on p, based on the fact that our overall analysis is carried out by (reverse) induction on the set of integers $m_{2,q}(G)$ for $q \in \sigma(G)$. Thus (B) involves an assumption on those primes $q \in \sigma(G)$ for which $m_{2,q}(G) > m_{2,p}(G)$. Similar assumptions likewise appear in the hypotheses of (A).

We hope that this discussion satisfactorily motivates the precise form of our main results, which we shall now state.

THEOREM A. *Assume $e(G) \geqslant 4$ and the following conditions hold for some integer $m \geqslant 4$:*

(a) *for all $q \in \beta_{m+1}(G)$, $(\Gamma_2^0)_q$ holds;*
(b) *for all $q \in \beta_m(G)$, either $(\Gamma_2^0)_q$ holds or $(\Delta O)_q$ holds; and*
(c) *for some $q \in \beta_m(G)$, $(\Delta O)_q$ holds.*

Then there exists a prime $p \in \beta_m(G)$ with $m_{2,p}(G) = m$ such that $(\Delta O)_p$ holds and one of the following is satisfied:

(I) *$p = 3$ and G is of $GF(2)$-type for p; or*

(II) *G is of standard type for p.*

THEOREM B. *Assume $e(G) \geqslant 4$ and the following condition holds for some integer $m \geqslant 4$:*

$$(\beta) \text{ for all } q \in \beta_{m+1}(G), \quad (\Gamma_2^0)_q \text{ holds.}$$

Then, for any $p \in \beta_m(G)$, either $(\Gamma_2^0)_p$ holds or $(\Delta O)_p$ holds.

THEOREM C. *Assume $e(G) \geqslant 3$ and suppose the following condition is satisfied:*

$$(\gamma) \ (\Gamma_2^0)_p \text{ holds for every } p \in \sigma(G).$$

Then G possesses an almost strongly p-embedded maximal 2-local subgroup for every $p \in \sigma(G)$.

Finally we state the four results to be established in part (D). As remarked before, we shall not discuss them here. However, these results are available to Aschbacher at the outset of his analysis of the case in which G satisfies the conclusion of Theorem C.

THEOREM D_1. *Let the assumptions be as in Theorem C. Let $p \in \sigma(G)$, let $P \in Syl_p(G)$, and let M be a maximal 2-local subgroup of G containing $\Gamma_{P,2}^0(G)$. Then every 2-local subgroup H of G such that $m_p(H \cap M) \geqslant 3$ lies in M.*

THEOREM D_2. *Assume $e(G) \geqslant 3$ and suppose the following conditions are satisfied:*

(δ_1) *if $p \in \sigma(G)$ and $p \geqslant 5$, then $\Gamma_{P,2}(G)$ lies in a 2-local subgroup of G for any $P \in Syl_p(G)$; and*

(δ_2) *if $3 \in \sigma(G)$, then G possesses an almost strongly 3-embedded maximal 2-local subgroup.*

(In particular, G satisfies the assumptions of Theorem C.)

Let $p \in \sigma(G)$, let $P \in Syl_p(G)$ and let M be a maximal 2-local subgroup of G containing $\Gamma_{P,2}(G)$. Then, if Q is a p-subgroup of M of p-rank 2 such that $\Gamma_{Q,1}(G) \subseteq M$, every 2-local subgroup H of G containing Q lies in M.

Remark. Theorems D_1 and D_2 are used in the proof of Theorem C. Also, Theorem D_1 is used to verify the assumptions of Theorem D_2.

THEOREM D_3. *Assume $e(G) \geqslant 3$ and suppose the following condition is satisfied:*

(δ) *G possesses an almost strongly p-embedded maximal 2-local subgroup for every $p \in \sigma(G)$.*

Let $p \in \sigma(G)$, let $P \in Syl_p(G)$, and let M be a maximal 2-local subgroup of G containing P with M almost strongly p-embedded in G. Let Q be a noncyclic

q-subgroup of M for any odd prime q (including the possibility that $q \notin \sigma(G)$). Then, if $\Gamma_{Q,1}(G) \subseteq M$, every 2-local subgroup of G containing Q lies in M.

THEOREM D_4. *Let the assumptions and notation be as in Theorem D_3. Then if $T \in Syl_2(M)$, we have $N_G(T) \subseteq M$. In particular, $T \in Syl_2(G)$.*

3. General Results and Properties of $\mathscr{K}$-groups

We should like next to make some comments about the basic techniques underlying the entire analysis. Essentially our methods are the same as those used to pin down centralizers of involutions in groups of component type. They can be summarized as follows.

(A) General balance and generation properties of subgroups of G (in the present situation, relative to suitable p-subgroups of G for odd primes p).
(B) Signalizer functor techniques (trivial signalizer functors usually leading to quasisimple p-components, nontrivial signalizer functors leading to proper 2-generated p-cores).
(C) Verification of properties of the $\mathscr{K}$-subgroups of G needed to carry out the analysis under (A) and (B).

One of the first things that must be done is to show that all the general results of Gorenstein and Walter ("Balance and Generation in Finite Groups" [105]), which deal only with prime 2, carry over with no essential changes for odd primes. (This is discussed in some detail in our paper entitled "Non-solvable Signalizer Functors in Finite Groups" [101].) In particular, the *layer* $L(X)$ of any finite group X is again of crucial importance ($E(X)$ in the Bender notation). By definition $L(X)$ is the product of all quasisimple subnormal subgroups of X (the *components* of X), with $L(X) = 1$ if no such components exist. Likewise, for any prime p, a *p-component* of X is by definition a subgroup of X which is minimal subject to being subnormal and covering a component of $L(X/O_{p'}(X))$. The product of the p-components of X is called the *p-layer* of X and is denoted by $L_{p'}(X)$.

As in the case of the prime 2, "layer balance" is again of central importance.

PROPOSITION 3.1. *If X is a $\mathscr{K}$-group and P is a p-subgroup of X for any prime p, then*

$$L_{p'}(C_X(P)) \subseteq L_{p'}(X).$$

For $p = 2$, the assumption that X is a $\mathscr{K}$-group is unnecessary by virtue of Glauberman's result on the Schreier conjecture [78], which asserts that if Y is any corefree finite group, then the subgroup of $Aut(Y)$ which acts trivially on a fixed Sylow 2-subgroup of Y has a normal 2-complement (and, in particular, is solvable). No analogue of this result exists for odd primes p; however, using known properties of the outer automorphism group of every known simple group, the analogue can be verified when Y is a $\mathscr{K}$-group.

We refer to the conclusion of Proposition 3.1 as one form of *$L_{p'}$-balance.* There is a second form of $L_{p'}$-balance which applies to our simple group G under investigation. For brevity, we write C_x for $C_G(x)$ and L_x for $L_{p'}(C_x)$ for any p-element x of G, p any prime, and adopt this notational convention throughout this chapter.

PROPOSITION 3.2. *If x,y are commuting elements of G of order p, then we have*

(i) *$L_{p'}(L_x \cap C_y) \subseteq L_y$, and*

(ii) *if J is a p-component of $L_{p'}(L_x \cap C_y)$ and we set $K = \langle J^{L_y} \rangle$, then either K is a single x-invariant p-component of L_y or the product of p p-components of L_y cycled by x. Moreover, in either case, J is a p-component of $L_{p'}(C_K(y))$.*

Proposition 3.2 follows easily from Proposition 3.1 (which is applicable as every proper subgroup of G is a $\mathscr{K}$-group by hypothesis). Part (ii) of the proposition shows that the fundamental "pumping up" process for 2-components in the centralizers of involutions carries over with no essential change to centralizers of elements of odd prime order.

Of at least as great importance for the study of centralizers of involutions is the *B-property*, which at this writing has been nearly verified for all finite groups X. Our analysis actually yields a partial verification for G of the corresponding assertion for odd primes p (i.e., we argue that certain p-components in the centralizers of certain elements of order p are quasisimple). We term the corresponding general assertion the *B_p-property.* As with Glauberman's partial Schreier theorem, our arguments require the B_p-property only for proper sections of G and hence only for $\mathscr{K}$-groups. The B_p-property is indeed verifiable for such groups. Thus we have

PROPOSITION 3.3. *If X is a $\mathscr{K}$-group with $O_{p'}(X) = 1$ and x is an element of X of order p, p any prime, then $L_{p'}(C_X(x))$ is a semisimple group (i.e., all its p-components are quasisimple).*

Combining Propositions 3.2 and 3.3, one obtains a critical sharper form of $L_{p'}$-balance for G.

PROPOSITION 3.4. *Let x, y, J, and K be as in Proposition 3.2(ii) and set $\bar{C}_y = C_y/O_{p'}(C_y)$. Then $\bar{J}$ is quasisimple and $\bar{J}$ is a component of $L(C_{\bar{K}}(\bar{x}))$.*

The proposition is applied in the same way as in the case of involutions: given the isomorphism type of $J/O_{p'}(J)$ $(\cong \bar{J})$ as a $\mathscr{K}$-group, the possibilities for the pump-up $(\cong \bar{K})$ (which is also a $\mathscr{K}$-group) are determined from the known structure of the centralizers of inner and outer automorphisms of order p of the known simple groups. (This remark applies when K is a single x-invariant p-component. In the contrary case, it is immediate that each of the p-components of $\bar{K}$ (cycled by $\bar{x}$) is isomorphic to a covering group of $\bar{J}$, so the possible structures of $\bar{K}$ are determined in this case as well.)

McBride [144] has considerably improved the nonsolvable signalizer functor results obtained in [101]. Recall that for any element $A \in \mathscr{E}^p(G)$, p any prime, θ is called an *A-signalizer functor* on G provided for each $a \in A^\#$ there is associated an A-invariant p'-subgroup $\theta(C_a)$ of C_a with the property

$$\theta(C_a) \cap C_b = \theta(C_b) \cap C_a$$

for every $a,b \in A^\#$. The group

$$\theta(G;A) = \langle \theta(C_a) \mid a \in A^\# \rangle$$

is called the *closure* of θ. θ is said to be *closed* if its closure is a p'-group; and θ is said to be *complete* if θ is closed and, in addition,

$$C_{\theta(G;A)}(a) = \theta(C_a)$$

for each $a \in A^\#$. Likewise θ is *solvable* if each $\theta(C_a)$ is solvable; and θ is *nonsolvable* in the contrary case.

The Goldschmidt–Glauberman solvable signalizer functor theorem [83,85] asserts that, if $m_p(A) \geqslant 3$, every solvable A-signalizer functor on G is complete and hence closed. McBride's theorem (as well as ours) depends, of course, on this basic result. Furthermore, both of these nonsolvable signalizer functor theorems require some assumptions on the simple sections of the groups $\theta(C_a)$ for $a \in A^\#$, which can be shown to hold for simple $\mathscr{K}$-groups. Thus McBride's theorem is applicable to our group G.

THEOREM 3.5. *If $A \in \mathscr{E}^p(G)$ with $m_p(A) \geqslant 3$, p any prime, then every A-signalizer functor θ on G is complete. In particular, $\theta(G;A)$ is a p'-group.*

McBride's arguments also enable one to construct suitable A-signalizer functors on G (again as the proper simple sections of G are $\mathscr{K}$-groups) in the case that G is k-balanced with respect to A. Recall that G is said to be *k-balanced with respect to A* for any positive integer k provided, for any $E \in \mathscr{E}^p_k(A)$ and any $a \in A^\#$, we have

$$\Delta_G(E) \cap C_a \subseteq O_{p'}(C_a).$$

Thus we have

PROPOSITION 3.6. *Let $A \in \mathscr{E}^p(G)$ with $m_p(A) \geqslant k+2$ for some positive integer k, p any prime, and suppose that G is k-balanced with respect to A. If, for $a \in A^\#$, we set*

$$\theta(C_a) = \langle C_a \cap \Delta_G(E) \mid E \in \mathscr{E}^p_k(A) \rangle,$$

then θ is an A-signalizer functor on G.

This describes most of the general machinery which we use. However, there is one other general result, due to Thompson [212], which holds only for odd primes and which is also very useful to us. To state it, we introduce the following terminology: for any group X, we write $SC_p(X)$ for the set of p-subgroups $P \in \mathscr{E}^p(X)$ which contain every element of order p in $C_X(P)$. We then have

PROPOSITION 3.7. *Let X be a p-constrained group, p any odd prime, and let $P \in SC_p(X)$. Then every P-invariant p'-subgroup of X lies in $O_{p'}(X)$.*

Even though most of the preceding general results depend upon properties of the proper simple $\mathscr{K}$-groups involved in our group G, their statements turn out to have no exceptions. This is unfortunately not the case for many other properties which we need. We shall describe some of the key ones in the remainder of the section.

Consider first the property of k-balance. As in the case of involutions, questions of balance are closely related to properties of local k-balance of the components of the groups $C_a/O_{p'}(C_a)$ for appropriate elements a of order p in G. By definition, a quasisimple group X is said to be *locally k-balanced for the prime p, k a positive integer*, if for every subgroup Y with $Inn(X) \subseteq Y \subseteq Aut(X)$ and every $E \in \mathscr{E}_k^p(Y)$, we have $\Delta_Y(E) = 1$. (To prove k-balance with respect to A, one requires only a "relativized" form of local k-balance, similar to that for the prime 2, in which E is restricted to lie in F for some *fixed* $F \in \mathscr{E}^p(Y)$.) Note that as $Inn(X) \cong X/Z(X)$, it suffices to consider the simple factor $X/Z(X)$.

The result for simple $\mathscr{K}$-groups is the following.

PROPOSITION 3.8. *If X is a simple $\mathscr{K}$-group and p any odd prime, then one of the following holds:*

(i) *X is locally 2-balanced for the prime p;*
(ii) *$X \cong L_p(q)$, $p \mid q-1$, or $U_p(q)$, $p \mid q+1$;*
(iii) *$X \cong A_n$, $n = sp^2+r$, where either $2 \leqslant r \leqslant p-1$ or $p = 3$ and $r = 4$; or*
(iv) *$X \cong M(22)$ (the smallest Fischer group) and $p = 5$.*

Moreover, one of the following holds:

(v) *X is locally 3-balanced for the prime p; or*
(vi) *$X \cong A_n$, $n = sp^3+r$, $2 \leqslant r \leqslant p-1$, or $p = 3$ and $r = 4$.*

This result is very similar to the situation for $p = 2$, where the only non-locally 2-balanced simple groups are $L_2(q)$ for suitable odd q and A_n for suitable n, with the former groups being locally 3-balanced. This result indicates that the analysis of centralizers of elements of odd prime order will encounter difficulties which are very analogous to those which arose in the study of centralizers of involutions.

To illustrate, consider the case $X \cong A_n$ and suppose X is not locally 2-balanced, so that for suitable Y and $E \in \mathscr{E}_2^p(Y)$, we have $\Delta_Y(E) \neq 1$. Then

$E \subseteq X(\cong Inn(X))$ and, by checking the centralizers of elements of odd prime order in A_n, we see that in the natural representation of X on n letters, E must act semiregularly on sp^2 letters, fixing the remaining $r = n - sp$ letters. In this case, E centralizes the alternating group W on these r letters (the symmetric group W^* if $Y \cong S_n$). If $2 \leqslant r \leqslant p-1$, then W and W^* are both p'-groups and lie in $O_{p'}(C_Y(e))$ for each $e \in E^{\#}$, so they lie in $\Delta_Y(E)$. Hence $\Delta_Y(E) \neq 1$ for such values of r. Similarly if $r = 4$ and $p = 3$, $O_2(W) \subseteq \Delta_Y(E)$, so again $\Delta_Y(E) \neq 1$.

We see then that failure of local 2-balance for A_n occurs only with respect to E's of type (p,p) which are embedded in A_n in this very special way. This means that if a given $F \in \mathscr{E}^p(A_n)$ contains no semiregular subgroups of type (p,p), then X will be locally 2-balanced with respect to F.

PROPOSITION 3.9. *Let X be a $\mathscr{K}$-group with $O_{p'}(X) = 1$, p any odd prime, and let $B \in \mathscr{E}^p(X)$ with $m_p(B) \geqslant m_{2,p}(X)$. If L is a component of $L(X)$, then one of the following holds:*

(i) *B leaves L invariant; or*

(ii) *$p = 3$ and $L \cong L_2(2^n)$, n odd, $SL(3,q)$, $q \equiv 4,7 (mod\, 9)$, $SU(3,q)$, $q \equiv 2,5 (mod\, 9)$, $\hat{A}_6, \hat{A}_7$, or $\hat{M}_{22}$ (the $\hat{}$ indicating a nonsplit extension by Z_3). Moreover, $\langle L^B \rangle$ is a product of exactly three components of $L(X)$.*

The proposition applies directly in G to the groups $C_b/O_{p'}(C_b)$ for $b \in B^{\#}$ and $B \in \mathscr{B}_{\max}(G;p)$. It implies that B always leaves every p-component of C_b invariant when $p \geqslant 5$; the same holds for $p = 3$ unless C_b possesses 3-components of a very restricted type. (The same conclusions hold more generally for every proper subgroup H of G containing B.)

A result such as this should not be surprising, for if B_1 denotes the subgroup of B leaving L fixed and we set $K = \langle L^B \rangle$, then K is the product of $p^a = |B : B_1|$ isomorphic components of $L(X)$, each of order divisible by p (as $O_{p'}(X) = 1$) and each invariant under B_1. Here, $a = m_p(B) - m_p(B_1)$. Since a Sylow 2-subgroup of L centralizes the remaining components of K, it follows that

$$m_{2,p}(K) \geqslant p^a - 1.$$

In general, $p^a - 1$ is very much larger than $a+1$. Hence assuming that B does not leave L invariant, and using these inequalities, one would expect to be able to prove that

$$m_{2,p}(KB_1) \geqslant (a+1) + m_p(B_1) = (a+1) + (m_p(B) - a) = m_p(B) + 1,$$

thus contradicting our assumption that $m_p(B) \geqslant m_{2,p}(X)$. Using properties of the outer automorphism group of simple $\mathscr{K}$-groups, one can indeed obtain such a contradiction except in the cases listed in part (ii) of the proposition.

When dealing with elements of $\mathscr{B}_{\max}(G;p)$, we can also make stronger statements concerning relative local balance than those given by Proposition 3.8.

PROPOSITION 3.10. *Let X be a simple group and B an element of $\mathscr{E}^p(Aut(X))$, p an odd prime, such that $m_p(B) \geqslant m_{2,p}(Aut(X))$. Then X is locally 2-balanced with respect to B under either of the following conditions:*

(i) $X \cong A_n$;

(ii) $X \cong L_p(q)$ *or* $U_p(q)$ *and* $p \geqslant 5$.

Combining Propositions 3.8, 3.9 and 3.10, we obtain the following key result.

PROPOSITION 3.11. *If $B \in \mathscr{B}_{\max}(G;p)$ for $p \in \sigma(G)$, then we have*

(i) *if $p \geqslant 7$, then G is 2-balanced with respect to B; and*

(ii) *in any case, G is 3-balanced with respect to B.*

Note that for $p = 5$ the only obstructions to 2-balance are 5-components of C_b of type $M(22)$ for $b \in B^{\#}$. Because of the possibilities listed in Propositions 3.8(ii) and 3.9(ii), the construction of effective signalizer functors is severely hampered in the case $p = 3$ and $m_{2,3}(G) = 4$. To get around this difficulty, we introduce a notion of "weak 2-balance" (which we shall not define here) and show that G is weakly 2-balanced with respect to some *suitably chosen* element of $\mathscr{B}_{\max}(G;p)$. (Aschbacher also encounters serious difficulty in constructing appropriate signalizer functors in the case $\sigma(G) = \{3\}$ in the $e(G) = 3$ problem.)

We next describe a few of the important properties of $\mathscr{K}$-group generation. (For groups of Lie type, these reduce ultimately to Seitz's generational results [174].) Unfortunately, each of these statements, too, involves some exceptional cases, which also add considerably to the required analysis. The first two results deal with weak 2-generated p-cores covering, respectively, the p-constrained and non-p-constrained cases.

PROPOSITION 3.12. *Let X be a p-constrained group with $O_{p'}(X) = 1$ and $m_p(X) \geqslant 3$, p an odd prime. Then for $P \in Syl_p(X)$, we have*

(i) $\Gamma_{P,2}(X) = X$, *and*

(ii) *if $\Gamma^0_{P,2}(X) \subset X$, then $O_p(X)$ is a central product of an extra-special group of order p^3 and exponent p and a cyclic group and*

$$O^{p'}(X/O_p(X)) \cong SL(2,p).$$

PROPOSITION 3.13(?). *Let X be a non-p-constrained $\mathscr{K}$-group with $O_{p'}(X) = 1$ and $m_p(X) \geqslant 3$, p an odd prime. Then, for $P \in Syl_p(X)$, $\Gamma^0_{P,2}(X)$ covers $X/L(X)$. Moreover, if $\Gamma^0_{P,2}(X) \subset X$ and K is a component of $L(X)$ with $K \nsubseteq \Gamma^0_{P,2}(X)$, then one of the following holds:*

(i) *K is of Bender type for p of p-rank at least 2 (i.e., $K \cong L_2(p^n)$ or $U_3(p^n)$, or $p = 3$ and K is of Ree type of characteristic 3);*

(ii) $K \cong A_{2p}, A_{3p}$, *or* $PSp_4(p)$‡;
(iii) $p = 11$ *and* K *is of type* J_4;
(iv) $p = 5$ *and* $K \cong Sz(2^5), {}^2F_4(2)', {}^2F_4(2^5), Mc$, *or* $M(22)$; *or*
(v) $p = 3$ *and* $K \cong L_2(8), Sp_4(8), G_2(8), L_3(4), Sp_6(2), M_{11}$, *or* J_3.

The (?) here is meant to indicate that the assertion has not been completely checked for every sporadic group—in a few cases (in particular, for a group of type F_1) some additional arguments must be supplied.

As a corollary of these propositions, we have the following result on strong p-embedding.

PROPOSITION 3.14(?). *Let X be a $\mathcal{K}$-group with $O_{p'}(X) = 1$ and $m_p(X) \geqslant 2$, p an odd prime, and suppose that X possesses a strongly p-embedded subgroup. Then $K = L(X)$ is a single simple component, $X \subseteq Aut(K)$, and one of the following holds:*
(i) *K is of Bender type for p with $m_p(K) \geqslant 2$, or $K \cong A_{2p}$;*
(ii) $p = 11$ *and* K *is of type* J_4;
(iii) $p = 5$ *and* $K \cong Sz(2^5), {}^2F_4(2)', Mc$, *or* $M(22)$; *or*
(iv) $p = 3$ *and* $K \cong L_2(8), L_3(4)$, *or* M_{11}.

The complete analysis requires several variations of both local balance and these generational properties. We conclude this section with a result which links balance and generation. To state it, we need two definitions, which apply to a quasisimple normal subgroup X of a group Y, and an elementary abelian p-subgroup A of Y, p a prime.

We shall say that X is *strongly locally balanced* with respect to A in Y if every A-invariant p'-subgroup of Y centralizes X.

Remark. This notion is very close to Walter's concept of *local $\mathcal{E}$-balance with respect to* A, which is defined by the condition $\Delta_{Aut(X)}(E) = 1$ for every $E \in \mathcal{E}^p(A)$ [30]. Strong local balance for A in $Aut(X)$ implies local $\mathcal{E}$-balance for A. Moreover, in our analysis, except in the case $p = 3$ and $m_{2,3}(G) = 4$, the A-signalizers to which we apply the strong local balance condition are, in fact, generated by the corresponding $\Delta(E)$'s as E ranges over the hyperplanes of A. Hence, apart from this exception, we could have worked just as effectively with local $\mathcal{E}$-balance; and perhaps with a little more arguing, that concept would have sufficed in this case, too.

We mention the connection between these two notions because of the importance of local $\mathcal{E}$-balance for the prime 2. Indeed, Walter's approach to the B-conjecture and the unbalanced group conjecture is to attempt to derive both as corollaries of the classification of all non-locally $\mathcal{E}$-balanced simple groups. We wanted therefore to show that essentially the same con-

‡ Here, and in Proposition 3.14(i), it was formerly necessary to allow a group of Ree type over F_q with p dividing q^2-q+1; Landrock and Michler [138] have now shown that all such Sylow subgroups must be cyclic. Shortly thereafter, Bombieri (and Odlyzko) proved that the Ree groups are the only groups of Ree type [239].

siderations arise for odd primes in the study of simple groups of characteristic 2 type. Indeed, the content of the proposition we are about to state is a description of all non-strongly locally balanced quasisimple $\mathscr{K}$-groups.

Returning now to the main thread of the discussion, we make our second definition‡. If $m_p(A) \geqslant 4$, we say that X is *well-generated with respect to A* provided the following conditions hold:

(a) $X = \Gamma_{A,3}(X)$;
(b) $p \geqslant 5$ and $m_p(A) \geqslant 5$, then $X = \Gamma_{A,4}(X)$; and
(c) if $p = 3$ and $m_3(A) = 4$, then $X = \Gamma_{D,1}(X)$ for any $D \in \mathscr{E}_2^3(A)$.

Now we can state the desired proposition.

PROPOSITION 3.15. *Let X be a quasisimple normal $\mathscr{K}$-subgroup of a group Y, and let $B \in \mathscr{E}^p(Y)$, p an odd prime. Assume that $m_p(B) \geqslant 4$, $m_p(B) \geqslant m_{2,p}(BX)$, and $B \in SC_p(BX)$. Then one of the following holds:*
(i) *X is strongly locally balanced with respect to B in Y;*
(ii) *X is well-generated with respect to B; or*
(iii) *$p \leqslant 5$, $m_p(B) \leqslant 5$, and $X \cong L_3(4)$ $(p = 3)$, $\widehat{ON}$ $(p = 3)$, or $M(22)$ $(p = 5)$.*

Here $\widehat{ON}$ denotes the nonsplit extension of ON by Z_3. In essence (as in the case $p = 2$), the proposition expresses the fundamental dichotomy between the groups of Lie type of characteristic p (they are, in general, strongly locally balanced) and those of characteristic prime to p (they are, in general, well-generated).

The assumption that $B \in SC_p(BX)$ is actually a natural one; in the applications to our group G, it will hold whenever $B \in SC_p(G)$. Moreover, for elements of $\mathscr{B}_{\max}(G;p)$, it holds under quite general conditions, as the following elementary, basic lemma shows.

LEMMA 3.16. *If $B \in \mathscr{B}_{\max}(G;p)$, $p \in \sigma(G)$, and $O_{p'}(C_G(E))$ has even order for some $E \in \mathscr{E}^p(B)$, then $B \in SC_p(G)$.*

We hope that this will give some flavour of the general techniques and properties of $\mathscr{K}$-groups which underlie our arguments. We emphasize that many properties of $\mathscr{K}$-groups not listed here are also used. Several of these will arise in the ensuing discussion.

4. Theorem A

In this section we shall try to give some insights into the proof of Theorem A. We shall not attempt an outline of the entire proof, but instead shall focus on certain key points. In particular, we shall avoid discussion of the many

‡ Use of an Aschbacher-Goldschmidt signalizer functor has now enabled us to get by with a simpler definition. In particular, only condition (a) is needed.

exceptional configurations which arise throughout the argument. Suppose then that G satisfies the assumptions of Theorem A.

Let $p \in \beta_m(G)$ be such that $(\Delta O)_p$ holds. For $B \in \mathscr{B}_{\max}(G;p)$, we set

$$(1)\ B_e = \left\{ x \,\middle|\, \begin{array}{l} x \in B^{\#},\ B \text{ normalizes, but does not centralize, some 2-subgroup} \\ R \text{ of } C_x, \text{ with the proviso that } R \not\cong Q_8 \text{ if } p = 3 \end{array} \right\}$$

We call B_e the set of *essential* elements of B. For most of the analysis, we work only with essential elements. We remark that this set has already been used in our previous work on groups with solvable 2-local subgroups [100]. The restriction here for $p = 3$ is made for technical reasons, which we shall explain below.

The first point we wish to explain is the following:

A. How does the $GF(2)$-type alternative arise (with $p = 3$)?

Let $B \in \mathscr{B}_{\max}(G;p)$. Then there exists a maximal 2-local subgroup M of G containing B. Setting $T = O_2(M)$, B acts faithfully on T as G is of characteristic 2 type. Under certain circumstances, we can force T to be of symplectic type and p to be 3, in which case G will be of $GF(2)$-type with $p = 3$. This occurs first in the following easy lemma.

LEMMA 4.1. *One of the following holds:*
(i) $B_e \cup \{1\}$ *contains a hyperplane of* B; *or*
(ii) $p = 3$ *and* T *is extra-special.*

The second alternative arises because of our restriction $R \not\cong Q_8$ in the definition of B_e.

However, the principal occurrence of $GF(2)$-type is in "getting started" toward standard form, for clearly we cannot show that G is of standard type if the critical p-layers are all trivial. We prove

PROPOSITION 4.1. *One of the following holds:*
(i) T *is of symplectic type and* $p = 3$; *or*
(ii) $L_x \neq 1$ *for some* $x \in B_e$.

We shall outline the proof, for it depends crucially on the $(\Delta O)_p$ condition. Suppose the proposition is false. Then $L_x = 1$ for $x \in B_e$, so C_x is p-constrained. Using the Thompson $A \times B$ lemma, one easily concludes for any $b \in B^{\#}$ that

$$(2)\ O_{p'}(C_b) \cap C_x \subseteq O_{p'}(C_x).$$

We next prove that,

(3) for each $b \in B^{\#}$, B centralizes every B-invariant 2-subgroup of $O_{p'}(C_b)$.

Suppose false for some $b \in B^{\#}$ and some B-invariant 2-subgroup S of C_b. Then, for some hyperplane F of B, we have $R = [C_S(F), B] \neq 1$. On the other hand, as (i) is false, Lemma 4.1 implies that $B_e \cup \{1\}$ contains a hyperplane A of B and hence a hyperplane of F. Since $m_p(B) = m_{2,p}(G) \geqslant 4$, it follows that there is $D \in \mathscr{E}_2^p(F)$ with $D^\# \subseteq B_e$. Then as $R \subseteq O_{p'}(C_b)$ and $d \in B_e$ for each $d \in D^\#$, (2) implies that $R \subseteq O_{p'}(C_d)$ for each $d \in D^\#$. Hence

$$R \subseteq \bigcap_{d \in D^\#} O_{p'}(C_d) = \Delta_G(D).$$

Since B normalizes, but does not centralize R, the condition $(\Delta O)_p$ therefore fails, contrary to hypothesis. This proves (3).

Again let A be a hyperplane of B with $A^\# \subseteq B_e$ and set $T_a = [C_T(a), A]$ for $a \in A^\#$. We prove the following.

(4) If $a \in A^\#$ and $T_a \neq 1$, then $O_{p'}(C_a)$ is of odd order.

Indeed, as $a \in B^\#$, this follows from Lemma 3.16 if $B \notin SC_p(G)$; so we can assume $B \in SC_p(G)$. But C_a is p-constrained as $a \in B_e$, so $T_a \subseteq O_{p'}(C_a)$ by Proposition 3.7. However, this contradicts (3) as B normalizes but does not centralize T_a.

Finally set $\bar{C}_a = C_a/O_{p'}(C_a)$, so that $\bar{T}_a \cong T_a$ by (4). Furthermore, $\bar{T}_a$ acts faithfully on $O_p(\bar{C}_a)$ as C_a is p-constrained. Set $\bar{H}_a = O_p(\bar{C}_a)\bar{T}_a\bar{B}$ ($\bar{T}_a$ is $\bar{B}$-invariant). Then $m_p(\bar{B}) \geqslant m_{2,p}(\bar{H}_a)$ for any a such that $T_a \neq 1$. This puts a very strong restriction on the groups T_a, in view of the structure of $\bar{H}_a$. This situation has been completely analysed by Klinger and Mason in [136]. Since $T = \langle T_a \mid a \in A^\# \rangle$ and $m_p(A) \geqslant 3$, their results show under these conditions that $p = 3$ and T is of symplectic type. The proposition follows.

One can squeeze a little more out of the Klinger–Mason argument. Indeed, one can prove

PROPOSITION 4.3. *One of the following holds:*

(i) *T is of symplectic type and $p = 3$;*

(ii) *$B \in SC_p(G)$; or*

(iii) *for some $x \in B_e$, L_x possesses a p-component L such that $L/O_{p'}(L)$ is not of Lie type of characteristic p.*

This depends upon the fact that the p-locals in the groups of Lie type of characteristic p are p-constrained.

Henceforth we assume that G is not of $GF(2)$-type with $p = 3$, so that Lemma 4.1 (ii), Proposition 4.2(i), and Proposition 4.3(ii) or (iii) hold.

There is one further situation later in the proof of Theorem A, which leads to $GF(2)$-type. This arises in ruling out sporadic groups as possible standard components and will be discussed briefly below.

The next point we wish to consider is what we call

B. Partial standard form

Here if $B \in \mathscr{B}_{\max}(G;p)$, $x \in B_e$, and L is a p-component of C_x, we shall say that C_x is in *partial standard form with respect to L* if $C_G(L/O_{p'}(L))$ has cyclic

Sylow p-subgroups. (For C_x to be in standard form, L would have to be quasisimple.)

The fact that we can prove the existence of a partial standard form directly (under our assumption that G is not of $GF(2)$-type with $p = 3$) is in sharp contrast with the case of centralizers of involutions, where the initial analysis (under the assumption that G has the B-property and is of component type) only yields the existence of a standard component K whose centralizer $H = C_G(K)$ is tightly embedded in G (Aschbacher's component theorem [10]). However, the 2-rank of H may be arbitrary and it takes a separate analysis (Aschbacher–Seitz [21,22]) to reduce to the case that H has 2-rank 1. As we shall see, the assumption that G is of characteristic 2 type (together with Proposition 3.9 and the $(\Delta O)_p$ condition) allows us to establish the stronger conclusion of partial standard form.

As in the case of involutions, we must put an ordering on appropriate p-components and then establish partial standard form with respect to a *maximal* such p-component. In making the definition, it is essential, as we shall see, to consider *all* elements of $\mathscr{B}_{\max}(G;p)$ simultaneously. Thus we place our ordering on the set of all triples (B,x,L) with $B \in \mathscr{B}_{\max}(G;p)$, $x \in B_e$, and L a p-component of L_x. Given two such triples (B,x,L) and (B^*,x^*,L^*), we write $(B,x,L) \precsim (B^*,x^*,L^*)$ provided

(a) $x \in B^*$ and $B^* = (B \cap B^*)C_{B^*}(L/O_{p'}(L))$, and

(b) $x^* \in C_{B^*}(L/O_{p'}(L))$ and L^* is a p-component of the normal closure of $L_{p'}(C_L(x^*))$ in L_{x^*}.

Moreover, we write $(B,x,L) \prec (B^*,x^*,L^*)$ if, in addition,

(c) x does not centralize $L^*/O_{p'}(L^*)$.

(In the latter case, if x leaves L^* invariant, then L^* is a proper "pump-up" of $L/O_{p'}(L)$.)

A "maximal" triple (B,x,L) is essentially one which is maximal in this ordering. However, the actual situation is considerably more complicated; we must also place an ordering on the isomorphism types of the groups $L/O_{p'}(L)$ and then choose (B,x,L) not only so that (B,x,L) is maximal in the above ordering, but also so that $L/O_{p'}(L)$ is maximal among these isomorphism types. However, for simplicity, we completely ignore this second ordering here. A further complication arises if B does not leave L invariant (in which case $p = 3$ and $L/O_{p'}(L)$ is determined from Proposition 3.9). However, with some arguing, we can reduce to the case in which B leaves L invariant. We shall therefore also assume that this condition holds for maximal triples.

Our goal then is to prove the following result.

THEOREM 4.4. *If (B,x,L) is a maximal triple, then C_x is in partial standard form with respect to L.*

The theorem is proved in two stages.

(I) Proof that $C_B(L/O_{p'}(L)) = \langle x \rangle$.
(II) Shift from B to a second element of $\mathscr{B}_{\max}(G;p)$.

We shall focus here on (I), but first make a few comments on (II). Suppose (I) holds, but the theorem is false. Since B leaves L invariant, it follows that B leaves invariant a subgroup U of C_x of type (p,p) with $x \in U$ and U centralizing $L/O_{p'}(L)$. One then sets $B^* = C_B(U)U$, so that B^* is elementary and $m_p(B^*) \geqslant m_p(B)$.

Suppose that B^* normalizes a nontrivial 2-subgroup of G (in which case $B^* \in \mathscr{B}_{\max}(G;p)$ and $m_p(B^*) = m_p(B)$), and that $x \in B_e^*$ (even though $x \in B_e$, it is not automatic that x is an essential element of B^*). Then our ordering together with the maximality of (B,x,L) implies that (B^*,x,L) is also a maximal triple. Hence (I) applies equally well to B^* and yields that

$$C_{B^*}(L/O_{p'}(L)) = \langle x \rangle.$$

However, this contradicts the fact that $U \subseteq C_{B^*}(L/O_{p'}(L))$ and U is of type (p,p).

Thus the proof of (II) reduces to showing that, in fact, $B^* \in \mathscr{B}_{\max}(G;p)$ and $x \in B_e^*$. It is in the course of this analysis that we use our restriction that $R \not\cong Q_8$ (in the definition of B_e) and hence that $m_2(R) \geqslant 2$ (as B normalizes, but does not centralize R). Thus R contains a four subgroup, a fact which is essential to our argument.

Turning now to (I), we first prove a very general result concerning p-components which possess suitable nontrivial subcomponents.

PROPOSITION 4.5. *Let $B \in \mathscr{B}_{\max}(G;p)$ and $b \in B^\#$, let K be a B-invariant p-component of L_b, and assume that $J = L_{p'}(C_K(E)) \neq 1$ for some $E \in \mathscr{E}_3^p(B)$. Then $[O_{p'}(K),K]$ is of odd order.*

For simplicity of notation only, we assume here that $O_{p'}(K) = O_2(K)$. (Otherwise one uses a Frattini argument, which requires introduction of further notation.) Suppose the proposition is false, in which case J does not centralize $O_2(K)$ (as K does not centralize $O_2(K)$ and $K/O_2(K)$ is quasisimple). Thus there is $D \in \mathscr{E}_2^p(E)$ (as $m_p(E) \geqslant 3$) such that $S = [C_{O_2(K)}(D),J] \neq 1$. Furthermore, as $|O_{p'}(C_b)|$ is even, $B \in SC_p(G)$ by Lemma 3.16. Using a property of quasisimple groups, this implies that B does not centralize $J/O_{p'}(J)$ and consequently does not centralize S.

Now we argue that $S \subseteq O_{p'}(C_d)$ for each $d \in D^\#$, whence $S \subseteq \Delta_G(D)$. Since B normalizes, but does not centralize S, the $(\Delta O)_p$ assumption will again be contradicted.

Suppose false for some $d \in D^\#$ and set

$$\bar{C}_d = C_d/O_{p'}(C_d), \quad J_d = L_{p'}(C_K(d)),$$

and let K_d be the normal closure of J_d in L_d ($L_{p'}$-balance). Then $J \subseteq J_d$ and $\bar{J}_d$ is a product of *components of* $L(C_{\bar{K}_d}(\bar{b}))$ (B_p-property). Since the latter group clearly centralizes $O_{p'}(C_{\bar{K}_d}(\bar{b}))$, so does $\bar{J} \subseteq \bar{J}_d$. On the other hand, using that fact that J is perfect together with the definition of S, it follows that $S = [S,J]$, whence $\bar{S} = [\bar{S},\bar{J}]$. However, $\bar{S} \neq 1$ as $S \nsubseteq O_{p'}(C_d)$. Thus $\bar{J}$ does not centralize $\bar{S}$ and we conclude that $\bar{S} \nsubseteq O_{p'}(C_{\bar{K}_d}(\bar{b}))$.

We now contradict this last conclusion. Indeed,

$$S \subseteq C_d, \quad S = [S,J], \quad J \subseteq K_d, \quad \text{and} \quad K_d \lhd\lhd C_d,$$

so clearly $S \subseteq K_d$. Thus

$$S \subseteq O_{p'}(C_b) \cap K_d \subseteq O_{p'}(C_{K_d}(b)),$$

whence $\bar{S} \subseteq O_{p'}(C_{\bar{K}_d}(\bar{b}))$, giving the desired contradiction.

This general argument depended only on $L_{p'}$-balance, the B_p-property, the $(\Delta O)_p$ assumption, and the existence of the nontrivial subcomponent J. To establish (I), we require an extension of this argument, the proof of which is quite delicate. In particular, the isomorphism type ordering in the definition of maximal triples enters into the analysis. Moreover, this extension is a key result for the entire proof of Theorem A.

PROPOSITION 4.6. *If $B \in \mathscr{B}_{\max}(G;p)$, $x \in B_e$, L is a p-component of L_x (and some modest restriction is placed on the isomorphism type of $L/O_{p'}(L)$), then $[O_{p'}(L),L]$ is of odd order. In particular, this conclusion holds if (B,x,L) is a maximal triple.*

Finally let us see how (I) follows from Proposition 4.6. Indeed, let (B,x,L) be a maximal triple and suppose $C_B(L/O_{p'}(L))$ contains a subgroup D of type (p,p). A property of B_e (assuming, as we are, that G is not of $GF(2)$-type with $p = 3$) asserts now that for some element $d \in D \cap B_e$, C_d contains a nontrivial B-invariant 2-subgroup S such that $S = [S,D]$. (This depends on the fact that G is of characteristic 2 type.)

Choose such an element d, set $\bar{C}_d = C_d/O_{p'}(C_d)$, $J_d = L_{p'}(C_L(d))$ and $K_d = \langle J_d^{L_d} \rangle$. To simplify the discussion, suppose that B leaves every p-component of L_d invariant. Then, by $L_{p'}$-balance, K_d is a single such B-invariant p-component. Since B leaves every p-component of L_d invariant, so does $S = [S,D] = [S,B]$. Furthermore, as D centralizes $L/O_{p'}(L)$, $J_d \cong L/O_{p'}(L)$ and D centralizes J_d. However, $J_d \subseteq K_d$ and $d \in B_e$. Since (B,x,L) is a maximal triple, it follows that we must have $\bar{J}_d \cong \bar{K}_d$. Hence D centralizes $\bar{K}_d$ and, as $\bar{S} = [\bar{S},\bar{D}]$ with $\bar{S}$ leaving $\bar{K}_d$ invariant, we conclude that $\bar{S}$ centralizes $\bar{K}_d$.

Thus K_d normalizes $O_{p'}(K_d)S$. Again, for simplicity, assume that $O_{p'}(K_d) = O_2(K_d)$ (otherwise we would again use a Frattini argument). Since $J_d \subseteq K_d$, it follows that

$$J_d B \subseteq N = N_G(O_2(K_d)S).$$

Moreover, N is a 2-local subgroup of G as $S \neq 1$. Hence $J_d D$ acts faithfully on $O_2(N)$ and so, for some $d^* \in D^\#$,

$$S^* = [C_{O_2(N)}(d^*), J_d] \neq 1.$$

Here we consider only the case in which $d^* \in B_e$. Then, without loss of generality, we may suppose that $d^* = d$.

Since $J_d \subseteq K_d \lhd\lhd C_d$ and $S^* = [S^*, J_d]$, we have $S^* \subseteq K_d$. As $\bar{S}^* = [\bar{S}^*, \bar{J}_d]$ and $\bar{J}_d = \bar{K}_d$ is a component of $\bar{L}_d$, certainly $\bar{S}^* = 1$. Hence $S^* \subseteq O_{p'}(K_d)$. Again, as $1 \neq S^* = [S^*, J_d]$ and $J_d \subseteq K_d$, we conclude easily that $[O_{p'}(K_d), K_d]$ must have even order, contrary to Proposition 4.6.

Our remaining comments on Theorem A will be briefer. We next discuss

C. The characteristic of a maximal L

The next major step is to prove the following result.

THEOREM 4.7. *If (B,x,L) is a maximal triple with C_x in partial standard form with respect to L, then $L/O_{p'}(L)$ is of Lie type of characteristic* 2.

We must eliminate all other possibilities for the isomorphism type of $L/O_{p'}(L)$. Suppose first that $L/O_{p'}(L)$ is of Lie type of characteristic q, q odd. If $q = p$, then as $C_B(L/O_{p'}(L)) = \langle x \rangle$ it follows easily that $L_{p'}(C_L(E)) = 1$ for every $E \in \mathscr{E}_3(B)$. Our initial ordering of isomorphism types enables us now to conclude for any $b \in B_e$ and any p-component J of L_b that $J/O_{p'}(J)$ is of Lie type of characteristic p, and moreover, with the aid of Proposition 4.3, that $B \in SC_p(G)$. The first condition implies that

(5) $O_{p'}(C_A) \cap C_b \subseteq O_{p'}(C_b)$

for all $A \subseteq B$ and $b \in B_e$. (Thus G is very nearly "$\mathscr{E}$-balanced with respect to B" in the sense of Walter.) Now, using (5) and the fact that $B \in SC_p(G)$, we conclude, as usual, that $(\Delta O)_p$ fails, a contradiction. Hence $q \neq p$.

Since $L/O_{p'}(L)$ is of Lie type of characteristic $q \neq p$, it follows, except in some low dimensional cases, that there is $D \in \mathscr{E}_2^p(B)$ and $E \in \mathscr{E}_3^p(B)$ with $D \subseteq E$ such that

(6) (a) $L_{p'}(C_L(E)) \neq 1$, and
 (b) if $J = L_{p'}(C_L(D))$, then $C_B(J/O_{p'}(J)) = D$ and $J/O_{p'}(J)$ is unambiguously of Lie type of characteristic q.

In this case, the idea of the proof is essentially to force $[O_{p'}(L), L]$ to have even order, contrary to Proposition 4.6. The proof is similar in spirit to that of assertion (I) of the preceding subsection, but is more complicated because the elements of $D^\#$ we must work with need not lie in B_e. It also hinges on a delicate property of the groups of Lie type of characteristic $q \neq p$. Moreover, considerable additional analysis is required to treat the cases in which there are no D and E satisfying these conditions.

If $L/O_{p'}(L) \cong A_n$, we again derive a contradiction using Proposition 4.6. This time we choose $D \in \mathscr{E}_2^p(B)$ so that $D \cap L$ ($\cong Z_p$, as B induces inner automorphisms of L and $C_B(L/O_{p'}(L)) = \langle x \rangle$) corresponds to a p-cycle in $L/O_{p'}(L)$. Then, if $J = L_{p'}(C_L(D))$, it follows that $J/O_{p'}(J) \cong A_{n-p}$. Special arguments are required to handle the cases $n-p \leqslant 9$ (whence $p = 3$ and $n \leqslant 12$); in all orther cases, the only pump-ups of J are of type A_{n-p+rp}, $r \geqslant 0$, and the argument is straightforward.

Finally suppose $L/O_{p'}(L)$ is sporadic. The groups F_5 and ON (which can occur for $p = 3$) are handled by special arguments. In the first case, we use the fact that $A_{12} \subseteq F_5$ and follow the corresponding argument of the alternating case. On the other hand, in ON, no element y of order 3 has a centralizer in partial standard form; however, for a suitable choice of y,

$$C_{ON}(y) \cong Z_3 \times Z_3 \times A_6.$$

This fact is exploited to derive a 3-fusion contradiction using information obtained from the pumping up process.

The remaining possibilities for L are handled in a uniform way by forcing G to be of $GF(2)$-type with $p = 3$, contrary to our assumption on G. Note, first of all, that as $m_p(B) \geqslant 4$ and $C_B(L/O_{p'}(L)) = \langle x \rangle$, only certain possibilities for L can arise. Moreover, except in the case that L is of type F_1 (where we can also have $p = 5$), we must have $p = 3$. The idea of the proof is the following. We argue first that L contains a B-invariant extraspecial 2-subgroup R with

$$R \in Syl_2(O_{p'}(N_L(R)))$$

and then study a maximal 2-local subgroup M of G containing $N_G(R)$. The aim is to show that $O_2(M)$ is of symplectic type. To accomplish this, we again use a suitable $D \in \mathscr{E}_2^p(B)$ with $x \in D$ and $J = L_{p'}(C_L(D))$ and study the groups $C_{O_2(M)}(d)$ for $d \in D^\#$. It turns out in all cases that $m_p(JB) > m_p(B)$. Since $m_p(B) = m_{2,p}(G)$, this forces $|O_{p'}(C_d)|$ to be odd for all $d \in D^\#$, a fact which greatly simplifies the analysis. The possible pump-ups of J are also important for the analysis. Ultimately each $C_{O_2(M)}(d)$ is shown to be extraspecial (or of order 2), from which we easily deduce that $O_2(M)$ is extraspecial. The action of $J \cap M$ on $O_2(M)$ is important in this analysis.

It therefore remains only to prove that $p = 3$ and hence to eliminate the case $p = 5$, L of type F_1. However, in this case, Timmesfeld's results [221] imply, on the one hand, that $O^{5'}(M/O_{5'}(M))$ is simple (as $O_2(M)$ has width at least 12); on the other hand, it follows from the structure of $L_{5'}(M \cap J)$ (of type Co_1) and the fact that $m_5(M) = m_5(B)$ that no such simple $\mathscr{K}$-group exists.

Finally we make a few comments on

D. Quasisimplicity of L and the order of a Cartan subgroup

On the basis of Theorem 4.7 and the full hypotheses of Theorem A, we can now prove

THEOREM 4.8. *If (B,x,L) is a maximal triple with C_x in partial standard form with respect to L, then L is quasisimple.*

It is here that we use the $(\Gamma_2^0)_q$ assumption for $q \in \beta_{m+1}(G)$ for the first time. Set $X = [L, O_{p'}(L)]$, so that X is of odd order by Proposition 4.6. If L is not quasisimple, then L does not centralize X, whence L does not centralize $F(X)$ and so L does not centralize $Q = O_q(X)$ for some prime q. In particular, q is odd.

On the other hand, $L/O_{p'}(L)$ is a homomorphic image of $\tilde{L} = L/X$ and $\tilde{L}$ is quasisimple, so by Theorem 4.7, $\tilde{L}$ is a covering group of a group of Lie type of characteristic 2. Also $C_B(\tilde{L}) = \langle x \rangle$, so $B/\langle x \rangle$ acts faithfully on $\tilde{L}$. It follows, in general, from these conditions that the 2-rank of $\tilde{L}$ must be much larger than that of $B/\langle x \rangle$ and hence of B. Choose $\tilde{V} \in \mathscr{E}^2(\tilde{L})$ with $m_2(\tilde{V}) = r$ as large as possible subject to $\tilde{V} \cap Z(\tilde{L}) = 1$. Since $\tilde{L}$ is quasisimple, $\tilde{V}$ must then act faithfully on Q, so by the Thompson dihedral lemma [214,I: Lemma 5.38] $\tilde{V}Q$ contains the direct product of r dihedral groups of order $2q$. Hence $m_{2,q}(\tilde{V}Q) \geqslant r-1$, whence $m_{2,q}(L) \geqslant r-1$. Now, if $r \geqslant m_p(B)+2$ (which will be true, in general), we have

(7) $m_{2,q}(G) \geqslant m_{2,q}(L) \geqslant m_p(B)+1 > m_{2,p}(G) = m.$

Thus $q \in \beta_{m+1}(G)$ and consequently $(\Gamma_2^0)_q$ holds. Expand Q to $Q^* \in Syl_q(G)$, so that $\Gamma_{Q^*,2}^0 \subseteq M$ for some maximal 2-local subgroup M of G. Since certainly $m_q(Q) \geqslant 3$, we have $C_x \subseteq N_G(Q) \subseteq M$.

Again, in general, there is $E \in \mathscr{E}_3^p(B)$ such that $I = L_{p'}(C_L(E))$ is unambiguously of Lie type of characteristic 2. Then I acts nontrivially on Q and so for some $D \in \mathscr{E}_2^p(E)$, $Q_0 = [C_Q(E), I] \neq 1$. Given the possible isomorphism types of $I/O_{p'}(I)$, we have $m_q(Q_0) \geqslant 3$. Using the uniqueness properties of M, we now argue that $C_d \subseteq M$ for every $d \in D^\#$.

These conditions lead at once to a contradiction. Indeed, as ID acts faithfully on $O_2(M)$, there is $d \in D^\#$ such that $S = [C_{O_2(M)}(d), I] \neq 1$. By $L_{p'}$-balance, $I \subseteq K$ for some p-component K of L_d (possibly K is a product of three p-components when $p = 3$). However, $S \subseteq O_2(C_d)$ as $C_d \subseteq M$, so certainly $[O_2(C_d), I] \neq 1$, whence also $[O_2(C_d), K] \neq 1$. Thus $[O_{p'}(C_d), K]$ has even order which, in general, conflicts with Proposition 4.6.

Again there are many small cases which require separate treatment.

The proof that either p can be chosen to divide the order of a Cartan subgroup of L or $p = 3$ and L is defined over $GF(2)$ is very similar. Indeed, in the contrary case, it follows in general, from the structure of the groups of Lie type of characteristic 2, that $m_{2,q}(L) \geqslant m_p(B)$ for some odd prime q. (Recall that $B/\langle x\rangle$ acts faithfully on L.) Moreover, either the equality is strict or q can be taken to divide the order of a Cartan subgroup of L.

If $(\Gamma_2^0)_q$ holds, we choose $Q \in Syl_q(L)$ and define Q^*, M, E, I and D as before. Using generational properties of groups of Lie type of characteristic 2 (relative to q-subgroups) together with the uniqueness of M, we again obtain, in general, that $C_d \subseteq M$ for every $d \in D^{\#}$, which leads to the same contradiction as in the proof of Theorem 4.8.

Hence by the hypothesis of Theorem A, we must have

(8) $m_{2,q}(G) = m_{2,q}(L) = m_p(B) = m_{2,p}(G)$,

and

(9) $(\Delta O)_q$ holds.

In particular, q divides the order of a Cartan subgroup of L. Furthermore, by (8) and (9), all our results for p hold equally well for q. Now choosing an appropriate element x^* of order q in L, it is not difficult to show that C_{x^*} has a component L^* with all the properties of L such that, in addition, q divides the order of a Cartan subgroup of L^*. Hence replacing p by q, if necessary, we see that p can be chosen to have the required properties.

The remaining conclusions of Theorem A are proved by very similar arguments, which again ultimately rely on Proposition 4.6. In particular, there is one further place in the analysis in which a prime shift may occur. If L is a twisted group defined over $GF(2^n)$, then $p \mid 2^{kn}-1$, where $k = 2$ or 3, by what we have already shown. However, the exact conclusion of Theorem A involves the stronger assertion that $p \nmid 2^n-1$ in certain of these cases; to obtain this conclusion, it may again be necessary to replace p by an appropriate prime q in the same way as above.

5. Theorem B

We shall discuss the proof of Theorem B in the same way as we have just done with Theorem A. Suppose then that G satisfies the assumptions of Theorem B and let $p \in \beta_m(G)$ be such that

(1) $(\Delta O)_p$ fails.

To establish the theorem, we must prove in this case that $(\Gamma_2^0)_p$ holds.

It is in this proof that we employ signalizer functor methods. The use of these methods (for any prime p, odd or even) breaks up into two distinct parts: the first can be suggestively called "getting started" and the second, "moving around". The aim of the first part is to produce a proper subgroup M of G which contains the normalizers in G of certain p-subgroups of M, while that of the second is to show that, in fact, M contains the normalizers in G of most p-subgroups of M (specifically that $\Gamma^0_{P,2}(G) \subseteq M$ for $P \in Syl_p(G)$).

In view of assumption (1), there are $B \in \mathscr{B}_{\max}(G;p)$ and $D \in \mathscr{E}^p_2(B)$ such that $\Delta_G(D)$ has even order and, in addition, if $p = 3$, such that $[\Delta_G(D),B]$ has even order. We denote by $\mathscr{B}^0_{\max}(G;p)$ the subset of those $B \in \mathscr{B}_{\max}(G;p)$ with this property. The set $\mathscr{B}^0_{\max}(G;p)$ is very important for the analysis.

A. Getting started

We shall discuss this part of the analysis only in the case in which G is 2-balanced with respect to some $B \in \mathscr{B}^0_{\max}(G;p)$, limiting ourselves to a few comments in the remaining cases. Note that by Proposition 3.11, this condition holds for every element of $\mathscr{B}^0_{\max}(G;p)$ when $p \geqslant 7$.

The goal is to prove the following result.

THEOREM 5.1. *If $B \in \mathscr{B}^0_{\max}(G;p)$ and G is 2-balanced with respect to B, then $\Gamma_{B,3}(G)$ is contained in a 2-local subgroup of G.*

In the cases when no such B exists, we establish slightly weaker forms of Theorem 5.1. For example, if $p \leqslant 5$ and $m_{2,p}(G) \geqslant 5$, then G is 3-balanced with respect to any $B \in \mathscr{B}^0_{\max}(G;p)$ by Proposition 3.11 and we show that $\Gamma_{B,4}(G)$ is contained in a 2-local subgroup of G by essentially the same argument that proves Theorem 5.1 but with the 3-balanced functor in place of the 2-balanced functor. Likewise, when $p = 3$, we are able to prove that G satisfies a weakened form of 2-balance with respect to some $B \in \mathscr{B}^0_{\max}(G;p)$ and then use the corresponding "weak" 2-balanced functor for B. On the other hand, the cases $p \leqslant 5$, $m_{2,p}(G) = 4$, are exceptional, and special arguments are required to obtain an analogue of Theorem 5.1‡.

Assume now that B satisfies the conditions of Theorem 5.1. We then obtain at once

PROPOSITION 5.2. *If θ denotes the 2-balanced B-signalizer functor on G (Proposition 3.6), then the closure $\theta(G;B)$ of θ is a p'-group of even order.*

That $\theta(G;B)$ is a p'-group follows from the signalizer functor theorem (Theorem 3.5). Moreover, as $B \in \mathscr{B}^0_{\max}(G;p)$, $\Delta_G(D)$ has even order for some $D \in \mathscr{E}^p_2(G)$. Now

(2) $\theta(G;B) = \langle \Delta_G(D) \mid D \in \mathscr{E}^p_2(G) \rangle$

by definition of the 2-balanced functor θ, so clearly $\theta(G;B)$ must have even order.

We now set $M_0 = N_G(\theta(G;B))$. By the proposition, $\theta(G;B)$ is a nontrivial

‡ These arguments have now been streamlined by use of the Aschbacher-Goldschmidt type signalizer functor.

proper subgroup of G and as G is simple, it follows that M_0 is a proper subgroup of G. Moreover, we have

PROPOSITION 5.3. $\Gamma_{B,3}(G) \subseteq M_0$.

Indeed, let $E \in \mathscr{E}_k^p(B)$ with $k \geqslant 3$. For any $D \in \mathscr{E}_2^p(B)$, E leaves $\Delta_G(D)$ invariant and, as $m_p(E) \geqslant 3$,

(3) $\Delta_G(D) = \langle \Delta_G(D) \cap C_G(F) | F \in \mathscr{E}_2^p(E) \rangle$.

Now using the condition of 2-balance with respect to B, it follows that

(4) $\Delta_G(D) \subseteq \langle \Delta_G(F) \mid F \in \mathscr{E}_2^p(E) \rangle$.

Denote the right side of (4) by $\theta(G;E)$. Since (4) holds for each $D \in \mathscr{E}_2^p(B)$, (2) now yields that $\theta(G;B) \subseteq \theta(G;E)$. The reverse inclusion being clear, we conclude that

(5) $\theta(G;B) = \theta(G;E)$.

However, by its definition, $\theta(G;E)$ is invariant under $N_G(E)$, so $\theta(G;B)$ is as well. Thus $N_G(E) \subseteq M_0$. Since this holds for every $E \in \mathscr{E}_k^p(B)$ with $k \geqslant 3$, we conclude from the definition that $\Gamma_{B,3}(G) \subseteq M_0$, as asserted.

Hence to establish Theorem 5.1, it will suffice to prove that $O_2(M_0) \neq 1$, for then $N_G(O_2(M_0))$ will be a 2-local subgroup of G containing $\Gamma_{B,3}(G)$, as required.

The proof of this assertion reduces ultimately to an elementary lemma of Thompson [214, I: Lemma 6.1].

LEMMA 5.4. *If X is a group of characteristic 2 type and $U \in \mathscr{U}(2)$, then U centralizes every U-invariant subgroup of X of odd order.*

By definition, if S is a 2-group, $\mathscr{U}(S)$ is the set of normal four subgroups U of S with the restriction that $U \subseteq Z(S)$ if $Z(S)$ is noncyclic. Moreover, for any group X, $\mathscr{U}(2)$ denotes the set of four subgroups U of X such that $U \in \mathscr{U}(S)$ for some $S \in Syl_2(X)$.

Now set $W = O_{p'}(M_0)$. Thus $\theta(G;B) \subseteq W$ and so W has even order. Clearly $O_2(M_0) \neq 1$ if and only if $O_2(W) \neq 1$. We shall derive a contradiction from the assumption $O_2(W) = 1$. It will suffice to show that there is $U \in \mathscr{U}(2)$ in G with the following two properties:

(6) (a) $U \subseteq M_0$; and
(b) $U_0 = U \cap O_{2'2}(W) \neq 1$.

Indeed, suppose (6) holds. By Lemma 4, U, and hence also U_0, centralizes $O_{2'}(W) = O(W)$. Setting $Y = O_{2'2}(W)$, it follows that $O(Y) = O(W)$ and $C_Y(O(Y)) \supseteq U_0$. As $O(Y)$ is a normal 2-complement in Y, this immediately implies that $U_0 \subseteq O_2(Y) = O_2(W)$, so $O_2(W) \neq 1$, contrary to assumption.

To describe the proof that (6) holds for some $U \in \mathscr{U}(2)$, we make a few preliminary observations. Let H be a 2-local subgroup of G such that $H = N_G(O_2(H))$ (in particular, any maximal 2-local). Let $T \in Syl_2(G)$ with $T \cap H \in Syl_2(H)$ and set $R = O_2(H)$. Since $H = N_G(R)$, $Z(T) \subseteq H$, so $Z(T) \subseteq C_H(R) \subseteq R$ (as H is 2-constrained with trivial core). Now, if U is any element of $\mathscr{U}(T)$, $U \cap R \neq 1$ as $U \cap Z(T) \neq 1$. Since $U \lhd T$, it follows that U normalizes R, whence $U \subseteq H$. Thus we have

(7) if $U \in \mathscr{U}(T)$, then $U \subseteq H$ and $U \cap R \neq 1$.

Suppose some $U \in \mathscr{U}(T)$ is not contained in R. Setting $\bar{H} = H/R$, it follows that $\bar{U} \cong Z_2$ (as $U \cap R \neq 1$ and $|U| = 4$). Also U does not centralize R (as $C_H(R) \subseteq R$). Now $|T : C_T(U)| \leqslant 2$ as $U \lhd T$, so, in fact, $|R : C_R(U)| = 2$. Hence, if we set $\bar{R} = R/\Phi(R)$, we conclude that $C_{\bar{R}}(\bar{U})$ is a hyperplane of $\bar{R}$. Thus a generator of $\bar{U}$ acts like a transvection on $\bar{R}$. Also $\bar{H}$ acts faithfully on $\bar{R}$ and $O_2(\bar{H}) = 1$. Under these conditions, a theorem of McLaughlin [243] describes the normal closure of $\bar{U}$ in $\bar{H}$. Indeed, we have

(8) $\langle \bar{U}^{\bar{H}} \rangle = \bar{K}_1 \times \bar{K}_2 \times \ldots \times \bar{K}_r$, where $\bar{K}_i \cong L_n(2)$, $Sp_{2n}(2)$, $O_{2n}(2)$, or S_n for some n, $1 \leqslant i \leqslant r$, and $\bar{U} \subseteq \bar{K}_i$ for some i, $1 \leqslant i \leqslant r$.

Now let R be a B-invariant Sylow 2-subgroup of $O_{2',2}(W)$. Using the preceding remarks, we shall show next that (6) is a consequence of the following conditions:

(9) $R \neq 1$ and $R = O_2(N_G(R))$.

Assume (9) and set $H = N_G(R)$. Let $T \in Syl_2(G)$ with $T \cap H \in Syl_2(H)$ and let $U \in \mathscr{U}(T)$. If $U \subseteq R$, then $U \subseteq O_{2',2}(W)$, in which case (6) clearly holds; so we can assume that $U \nsubseteq R$. Then $U_0 = U \cap R \cong Z_2$ by (7) and $\bar{K} = \langle \bar{U}^{\bar{H}} \rangle$ has the form specified in (8).

Let J be the preimage of $\Gamma_{\bar{B},3}(\bar{K})$ in H ($B \subseteq H$ as R is B-invariant). Then $R \subseteq J \subseteq M_0$ by Proposition 5.3. Now, using generational properties of the groups $L_n(2)$, $Sp_{2n}(2)$, $O^{\pm}_{2n}(2)$, and S_n, it follows that $\bar{U}^{\bar{y}} \subseteq \bar{J}$ for some $\bar{y} \in \bar{K}$, whence $U^y \subseteq M_0$ for some $y \in H$. Moreover, $U_0^y \subseteq R \subseteq O_{2',2}(W)$. Since $U \in \mathscr{U}(2)$ in G, (6) therefore holds with U^y in the role of U.

To establish (9), we first prove

LEMMA 5.5. *W is 2-constrained.*

Suppose false and set $\bar{M}_0 = M_0/O(W)$, so that $L(\bar{W}) \neq 1$. Let $\bar{L}$ be a component of $L(\bar{W})$. We claim first that

(10) if N is a 2-local subgroup of G containing B, then N does not cover $\bar{L}$.

If not, then $S = O_2(N) \subseteq \Gamma_{B,3}(G) \subseteq M_0$, so $\bar{S} \subseteq \bar{M}_0$ and $\bar{S} \cong S$. However, as N/S acts faithfully on $S/\Phi(S)$ and N covers $\bar{L}$, we see that $[\bar{S},\bar{L}] \neq 1$, which is impossible as $\bar{L}$ is a component of $L(\bar{W})$ and hence of $L(\bar{M}_0)$.

We conclude at once from (10) that $O_2(\bar{W}) = 1$ and that B permutes the components of $L(\bar{W})$ transitively. Indeed, in the contrary case, $\bar{L}$ centralizes a nontrivial $\bar{B}$-invariant 2-subgroup of $\bar{W}$ and the existence of a 2-local N violating (10) follows directly by a Frattini argument.

We want to show that B also acts *faithfully* on $L(\bar{W})$. To do so, we need

(11) if S is a B-invariant Sylow 2-subgroup of W, then $S = O_2(N_G(S))$.

Indeed, as $|W|$ is even, $S \neq 1$, so $N = N_G(S)$ is a 2-local subgroup of G. Furthermore, N covers $\tilde{M}_0 = M_0/W$ by the Frattini argument. On the other hand,

$$S \subseteq S^* = O_2(N) \subseteq \Gamma_{B,3}(G) \subseteq M_0.$$

Since $S^* \lhd N$, it follows that $\tilde{S}^* \lhd \tilde{M}_0$, whence $\tilde{S}^* \subseteq O_2(\tilde{M}_0)$. However, $O_2(\tilde{M}_0) = 1$ as $W = O_{p'}(M_0)$. Hence $\tilde{S}^* = 1$, so $S^* \subseteq W$. Since $S \in Syl_2(W)$, this forces $S^* = S$, proving (11).

Now we can establish the faithful action of B:

(12) $C_B(L(\bar{W})) = 1$.

Indeed, set $B_0 = C_B(L(\bar{W}))$. Since $O_2(\bar{W}) = 1$, $C_{\bar{W}}(L(\bar{W})) = 1$ and it follows at once that B_0 centralizes $\bar{W}$. Hence B_0 centralizes a B-invariant Sylow 2-subgroup S of W. But $S = O_2(N)$, where $N = N_G(S)$, by (11). However, N is a 2-local subgroup of G, so B_0 must act faithfully on S. Thus $B_0 = 1$, as asserted.

Finally we use our hypothesis on $\beta_{m+1}(G)$. Since $\bar{L}$ is a p'-group (and a $\mathscr{K}$-group), $Out(\bar{L})$ has cyclic Sylow p-subgroups. Since B permutes the components $\bar{L} = \bar{L}_1, \bar{L}_2, \ldots, \bar{L}_r$ of $L(\bar{W})$ transitively and B is faithful on $L(\bar{W})$ by the preceding arguments, we have

(13) $r \geqslant |B|/p > m_p(B)$.

Hence there is an odd prime divisor q of $|\bar{L}|$ such that $m_{2,q}(G) > m_p(B) = m$. Thus $(\Gamma_2^0)_q$ holds and we conclude by the usual argument that $M_0 \subseteq M^*$ for some maximal 2-local subgroup M^* of G. (Take $M^* \supseteq \Gamma_{Q,2}^0(G)$, where $Q \in Syl_q(G)$ and $Q \cap M_0 \in Syl_q(M_0)$.) However, this violates (10) with M^* in the role of N. Thus W is 2-constrained, as asserted.

Since W has even order, it follows from the 2-constraint of W that a B-invariant Sylow 2-subgroup R of $O_{2',2}(W)$ is nontrivial. Now repeating the argument of (11) with R and $M_0/O_{2',2}(W)$ in the roles of S and M_0/W, we conclude in the same way that $R = O_2(N_G(R))$. Thus (9) holds and Theorem 5.1 is proved.

B. Prelude to moving around: the embedding of $\Gamma_{B,2}(G)$

Again we assume that G is 2-balanced with respect to the element $B \in \mathscr{B}_{\max}^0(G;p)$, so that by Theorem 5.1, $\Gamma_{B,3}(G) \subseteq M$ for some maximal 2-local subgroup M of G. The next major step is to establish the stronger inclusion $\Gamma_{B,2}(G) \subseteq M$. The proof of this assertion involves ideas which will form the basis for the subsequent moving-around analysis.

We shall need two important consequences of Theorem 5.1. First of all, using the fact that $\Gamma_{B,3}(G) \subseteq M$ in conjunction with Proposition 3.15, we easily obtain

PROPOSITION 5.6. *For any $b \in B^{\#}$, we have*

(i) $O_{p'}(C_b) \subseteq M$, *and*

(ii) *for any p-component L of L_b, one of the following holds:*

(1) $L \subseteq M$;

(2) *B leaves L invariant and $L/O_{p'}(L)$ is strongly locally balanced in $C_b/O_{p'}(C_b)$ with respect to B; or*

(3) $p \leqslant 5$ *and* $L/O_{p'}(L) \cong L_3(4), \widehat{ON}$ *or* $M(22)$.

The exceptions listed in (ii)(3) are neither well generated nor well balanced. The existence of such p-components, especially those of type $L_3(4)$, adds considerably to the entire proof of Theorem B. For simplicity, we assume here that (ii) (1) or (ii) (2) holds for any b and L.

Using this assumption, we can easily prove

PROPOSITION 5.7. *For any $b \in B^{\#}$, we have $C_{O_2(M)}(b) \subseteq O_2(C_b)$.*

Indeed, set $X_b = C_{O_2(M)}(b)$ and $\bar{C}_b = C_b/O_{p'}(C_b)$. Since $X_b \lhd C_M(b)$ and $O_{p'}(C_b) \subseteq C_M(b)$ by Proposition 5.6, we need only show that $X_b \subseteq O_{p'}(C_b)$ to obtain the desired conclusion,

$$X_b \subseteq O_2(O_{p'}(C_b)) = O_2(C_b).$$

Using Thompson's Proposition 3.7, noting that $B \in SC_p(G)$, it follows at once that $\bar{X}_b$ centralizes $O_p(\bar{C}_b)$. Moreover, if L is a p-component of L_b with $L \subseteq M$, then L normalizes X_b and consequently $\bar{X}_b$ centralizes $\bar{L}$. Hence, by the preceding proposition, $\bar{X}_b$ centralizes the product of those components of $\bar{L}_b$ which are not strongly locally balanced with respect to B. Furthermore, by definition of the term, $\bar{X}_b$ centralizes any component of $\bar{L}_b$ which is strongly locally balanced with respect to B and which is left invariant by $\bar{X}_b$. However, as

$$B \in \mathscr{B}_{\max}(G;p) \cap SC_p(G),$$

it is not difficult to show that, in fact, X_b leaves every p-component of L_b invariant. Our argument thus yields that $\bar{X}_b$ centralizes $\bar{L}_b O_p(\bar{C}_b)$. It follows now from general principles, as $O_{p'}(\bar{C}_b) = 1$, that

$$C_{\bar{C}_b}(\bar{L}_b O_p(\bar{C}_b)) \subseteq O_p(\bar{C}_b).$$

Thus $\bar{X}_b \subseteq O_p(\bar{C}_b)$ and consequently $\bar{X}_b = 1$, whence $X_b \subseteq O_{p'}(C_b)$, as required.

These two results indicate a strong connection between the groups $O_2(C_b)$ and $C_{O_2(M)}(b)$ for $b \in B^\#$. Indeed, they imply directly, for any $D \in \mathscr{E}_2(B)$, that

(14) $O_2(M) \subseteq \langle O_2(C_d) \mid d \in D^\# \rangle \subseteq M.$

It is therefore natural to define the groups

(15) $\Sigma^*(D) = \langle O_2(C_d) \mid d \in D^\# \rangle$ and $\Sigma(D) = O_2(\Sigma^*(D))$;

then, as an immediate consequence of (14), we have

(16) $O_2(M) \subseteq \Sigma(D) \subseteq M$ for every $D \in \mathscr{E}_2^p(B)$, and $\Sigma(D)$ is a 2-group.

The groups $\Sigma^*(D)$ and $\Sigma(D)$ are certainly $N_G(D)$-invariant. Hence, if each $\Sigma(D) = O_2(M)$, it would follow that each $N_G(D) \subseteq M$ ($M = N_G(O_2(M))$ as M is a maximal 2-local). Thus, combined with Theorem 5.1, we would obtain our objective that $\Gamma_{B,2}(G) \subseteq M$. Therefore the embedding of the groups $\Sigma(D)$ and $\Sigma^*(D)$ in M is clearly very important for us. Unfortunately, there is no *a priori* reason to have $\Sigma(D) = O_2(M)$. In fact, we cannot obviously assert this even when $\Sigma(D) \subseteq O_{p'}(M)$. Furthermore, if we set $\bar{M} = M/O_{p'}(M)$, $\Sigma^*(D)$ centralizes $O_p(\bar{M})$ (by the $A \times B$ lemma) and, as a consequence of what is called $L^*_{p'}$-balance (not defined here), $\overline{\Sigma^*(D)}$ leaves each component of $L(\bar{M})$ invariant. However, we cannot prove that $\overline{\Sigma^*(D)}$ (or $\overline{\Sigma(D)}$) must centralize every such component.

Indeed, for $d \in D^\#$, we do have $\overline{O_2(C_d)} \subseteq O_2(C_{\bar{M}}(\bar{d}))$, and the latter group also leaves every component of $L(\bar{M})$ invariant by $L^*_{p'}$-balance. Clearly then, if $\bar{K}$ is a component of $L(\bar{M})$ not centralized by $\overline{O_2(C_d)}$ and hence not by $O_2(C_{\bar{M}}(\bar{d}))$, $\bar{K}$ will not be locally balanced (with respect to B). Because we are dealing here only with O_2's, this is a special kind of local balance, which we refer to as "local O_2-balance". Thus the non-locally O_2-balanced components of $L(\bar{M})$ represent the obstruction to the inclusion $\Sigma(D) \subseteq O_{p'}(M)$.

We would like to reduce the possible non-locally O_2-balanced components which we are forced to consider. For the analysis, any "functorially" defined subgroup of $\Sigma(D)$ containing $O_2(M)$ will work essentially as well as $\Sigma(D)$ itself. In fact, there exists a rather natural such subgroup to take, which will at least allow us to ignore those components $\bar{K}$ of $L(\bar{M})$ for which the Thompson (Z,J)-factorization holds; i.e., for which either $C_{\bar{M}}(Z(\Sigma(D)))$ or $N_{\bar{M}}(J(\Sigma(D)))$ covers $\bar{K}$. To define it, we first set for any nontrivial 2-subgroup T of G,

$$\gamma(T) = T \cap \Bigg(\bigcap_{1 \neq T_0 \operatorname{char} T} O_2(N_G(T_0)) \Bigg) = \gamma^1(T),$$

(17) $\gamma^{i+1}(T) = \gamma(\gamma^i(T))$ for all $i \geqslant 1$, and

$$\gamma^\infty(T) = \bigcap_{i=1}^{\infty} \gamma^i(T).$$

Moreover, for any $D \in \mathscr{E}_2^p(G)$, we define $\Sigma^*(D)$ and $\Sigma(D)$ in accordance with (15). Now, if $D \in \mathscr{E}_2^p(G)$ and $\Sigma(D) \neq 1$ (in particular, if $D \subseteq B$), we set

(18) $\Gamma(D) = \gamma^{\infty}(\Sigma(D))$.

Note that

(19) $\Gamma(D) \subseteq \Sigma(D)$, $\Gamma(D)$ is $N_G(D)$-invariant, and $(\Gamma(D))^g = \Gamma(D^g)$ for $g \in G$.

The reason for iterating γ is that with the present definition of $\Gamma(D)$ we can prove the following result, which does not seem easy to verify if one works instead with the group $\gamma(\Sigma(D))$: namely,

(20) $\Gamma(D) \subseteq O_2(N_G(T_0))$ for any $1 \neq T_0$ char $\Gamma(D)$.

We can prove

Proposition 5.8. *For any $D \in \mathscr{E}_2^p(B)$, we have $O_2(M) \subseteq \Gamma(D)$.*

The next result pinpoints the obstruction to the equality $\Gamma(D) = O_2(M)$. It vividly displays the close connection with Aschbacher's theory of blocks [17b]. In fact, it was this connection which motivated our use of the γ operator.

In Aschbacher's terminology, a group X is said to be *short* if $F^*(X) = O_2(X)$, $X/O_2(X)$ is quasisimple (or of prime order), X centralizes $O_2(X)/\Omega_1(Z(O_2(X)))$, and, if

$$V = [\Omega_1(Z(O_2(X))), X],$$

then X acts irreducibly (and nontrivially) on $V/C_V(X)$. Moreover, a subnormal short subgroup of a group Y is called a *block* of Y. For convenience, we refer to the group $V/C_V(X)$ above as the *associated $GF(2)$-module* of X.

Because our result is also used in the moving-around part of the analysis, we state it in greater generality that we need in this subsection (its assumptions hold for $D \subseteq B$ by Propositions 5.6, 5.7, and 5.8).

Proposition 5.9. *Let $D \in \mathscr{E}_2^p(M)$ and suppose that the following conditions hold:*

(a) *for each $d \in D^{\#}$, $O_2(C_d) \subseteq M$ (whence $\Gamma(D) \subseteq \Sigma(D) \subseteq M$);*
(b) *for each $d \in D^{\#}$, $C_{O_2(M)}(d) \subseteq O_2(C_d)$; and*
(c) *$O_2(M) \subseteq \Gamma(D)$.*

Let L be a p-component of $L_{p'}(M)$ and set $\bar{M} = M/O_{p'}(M)$. Then we have the following:

(i) *$\Gamma(D) \cap O_{p'}(M) = O_2(M)$;*
(ii) *$\overline{\Gamma(D)}$ centralizes $O_p(\bar{M})$;*
(iii) *$\overline{\Gamma(D)}$ normalizes $\bar{L}$; and*

(iv) *either* $\overline{\Gamma(D)}$ *centralizes* $\bar{L}$ *or the following conditions hold:*

(1) L *is a block of* M *with* $L/O_2(L) \cong A_{kp+2}$, $k \geqslant 1$, $\Omega^{\pm}_{2n+2}(2)$, *or* A_7 (*with* $p = 3$);

(2) *the associated* $GF(2)$*-module of* L *is the natural module, except for* A_7, *where it is the 4-dimensional* $GF(2)$*-module; and*

(3) $\overline{\Gamma(D)}/C_{\overline{\Gamma(D)}}(\bar{L}) \cong Z_2, Z_2$, *or* $Z_2 \times Z_2$, *respectively.*

(*Actually when* $p = 3$, *assumption* (*b*) *must be sharpened in order to handle certain difficulties related to components* $\bar{K}$ *of* $L(\bar{M})$ *with* $\bar{K}/Z(\bar{K}) \cong A_7$).

We shall not prove Propositions 5.8 or 5.9, nor any other results needed to complete the proof of Theorem B, but will confine ourselves to a few relevant comments.

Aschbacher's theory of blocks deals primarily with simple groups G of characteristic 2 type in which some maximal 2-local subgroup possesses a block X. The aim is then to determine G (i.e., to show that G is a $\mathscr{K}$-group) for various choices of the block X. This objective has nearly been attained when $X/O_2(X) \cong L_2(2^n)$, $n \geqslant 2$, or A_n, n odd, $n \geqslant 5$, and the associated $GF(2)$-modules of X are the natural ones (these are the Aschbacher χ-blocks). Work has also begun to extend this analysis to blocks of the type which occur in the conclusion of Proposition 5.9.

If such a block theorem could be established, the possibilities of Proposition 5.9 (iv) would be immediately eliminated, since there exists no simple $\mathscr{K}$-group G of characteristic 2 type which possesses a maximal 2-local subgroup with the property of M (namely, with $\Gamma_{B,3}(G) \subseteq M$). Thus we would obtain the stronger conclusion $\Gamma(D) = O_2(M)$ and so would directly obtain our present objective, $\Gamma_{B,2}(G) \subseteq M$ (using Theorem 5.1 and (19)). Moreover, this stronger form of the proposition would also greatly simplify the subsequent moving around portion of the analysis.

Since, unfortunately, no such block theorem exists at present‡, we must take into account the possible presence of these exceptional components $\bar{L}$ in $L(\bar{M})$. To deal with them, we make yet a further refinement in the definition of $\Sigma(D)$.

We let $\pi(p)$ be the set of quasisimple groups L such that $O_{p'}(L) = 1$ and $L/Z(L)$ is isomorphic to one of the groups listed in Proposition 5.9(iv). For any subgroup H of G, we then define $\pi_{p'}(H)$ to be the product of all p-components L of $L_{p'}(H)$ such that $L/O_{p'}(L) \in \pi(p)$ (with $\pi_{p'}(H) = 1$ if no such p-components exist). Moreover, we write π_x for $\pi_{p'}(C_x)$, when $x \in G$ and $|x| = p$.

With this terminology, we set, for any $D \in \mathscr{E}^{p}_{2}(G)$,

(21) $\pi^*(D) = \langle \pi_d O_{p'}(C_d) \mid d \in D^{\#} \rangle$,

‡ The work of Aschbacher, Foote, Harada, F. Smith, Solomon, and S. K. Wong has now established this block theorem.

and

(22) $\Gamma^*(D) = \Gamma(D) \cap O_2(\pi^*(D))$.

Thus we have

(23) $\Gamma^*(D) \subseteq \Gamma(D)$, $\Gamma^*(D)$ is $N_G(D)$-invariant, and $(\Gamma^*(D))^g = \Gamma^*(D^g)$ for $g \in G$.

Using the fact that $B \in \mathscr{B}_{\max}(G;p)$, we can prove

PROPOSITION 5.10. *For any $D \in \mathscr{E}_2^p(B)$, we have $\Gamma^*(D) = O_2(M)$.*

The point here is that if L is a p-component of $L_{p'}(M)$ such that $\bar{L}$ is not centralized by $\overline{\Gamma(D)}$, Proposition 5.9(iv) gives the possible isomorphism types of $\bar{L}$ and, as $B \in \mathscr{B}_{\max}(G;p)$, we can determine the embedding of $\bar{B}$ in $\bar{L}\bar{B}$. In particular, we can establish a suitable generational statement for $\bar{L}$ relative to the elements of $D^{\#}$, which enables us ultimately to show that $\overline{\pi^*(D)}$ covers $\bar{L}$. Then, as $\Gamma^*(D) \subseteq O_2(\pi^*(D))$, we see that $[\bar{L}, \overline{\Gamma^*(D)}]$ must be a 2-group, so $\bar{L}$ centralizes $\overline{\Gamma^*(D)}$. It follows that $\Gamma^*(D) \subseteq O_2(M)$. Furthermore, an argument similar to that of Proposition 5.8 establishes the reverse inclusion.

We remark that if D is an arbitrary element of $\mathscr{E}_2^p(M)$ which satisfies the hypothesis of Proposition 5.9, we cannot establish the same generational statement for $\bar{L}$ as in the case $D \subseteq B$. As a consequence, we are unable to prove Proposition 5.10 for arbitrary such D. This point is central for understanding the moving around part of the analysis.

Using Proposition 5.10, Theorem 5.1, and (23), we immediately obtain the goal of this subsection.

THEOREM 5.11. *We have $\Gamma_{B,2}(G) \subseteq M$.*

C. Moving around

Now we are in a position to establish Theorem B. As expected, the argument involves the notion of connectivity (for the prime p). Recall that if A, $A^* \in \mathscr{E}^p(G)$ with A, A^* noncyclic, then A is said to be *connected* to A^* in G provided there is a chain of noncyclic elementary abelian p-subgroups $A = A_1, A_2, \ldots, A_n = A^*$ in G such that A_i centralizes A_{i+1}, $1 \leqslant i \leqslant n-1$. In particular, if P is any p-subgroup of G containing our element $B \in \mathscr{B}^0_{\max}(G;p)$, then any element $A \in \mathscr{E}_3^p(P)$ is connected to B, since each is clearly connected to any normal subgroup U of P of type (p,p).

The goal of the moving around analysis is to prove the following result.

THEOREM 5.12. *If $D \in \mathscr{E}_2^p(M)$ and D is connected to B in G, then the following conditions hold:*

(i) *D satisfies the hypotheses of Proposition 5.9 and hence its conclusions;*
(ii) *$O_2(M) \subseteq \Gamma^*(D)$; and*
(iii) *if $O_2(M) \subseteq T \subseteq \Gamma^*(D)$, then $N_G(T) \subseteq M$.*

Theorem B is a corollary of Theorem 5.12. Indeed, let $P \in Syl_p(M)$ with $B \subseteq P$ and let $Q \subseteq P$ with $m_p(Q) \geqslant 2$ and $m_p(QC_P(Q)) \geqslant 3$. We must show that $N_G(Q) \subseteq M$ for any such Q, for this is equivalent to the assertion $\Gamma^0_{P,2}(G) \subseteq M$. We need only show that any element x of $N_G(Q)$ lies in M. It is easy to see from our conditions on Q (and the fact that $m_p(P) \geqslant m_p(B) \geqslant 4$) that there is $D \in \mathscr{E}^p_2(Q)$ such that for all integers $j \geqslant 0$, D^{x^j} $(\subseteq Q)$ lies in an element A_j of $\mathscr{E}^p_3(P)$. Since each A_j is connected to B, so therefore is each D^{x^j}, and consequently (ii) holds for each D^{x^j}.

Hence, if we set

$$T = \bigcap_{j=1}^{\infty} \Gamma^*(D^{x^j}),$$

we conclude that

(24) $O_2(M) \subseteq T \subseteq \Gamma^*(D)$.

Now $N_G(T) \subseteq M$ by (iii). On the other hand, x permutes the groups $\Gamma^*(D^{x^j})$ by (23) and so normalizes their intersection T. Therefore $x \in N_G(T) \subseteq M$, as required.

Of course, if the desired block theorem were proved‡, Theorem 5.12(i) would suffice for Theorem B. Indeed, we would then know that $\Gamma(D) = O_2(M)$ for any $D \in \mathscr{E}^p_2(M)$ such that D is connected to B. Then an even easier argument than the one just outlined would yield the conclusion $\Gamma^0_{P,2}(G) \subseteq M$.

The main point to emphasize about the proof of Theorem 5.12 is that the bulk of the argument has nothing to do with the subgroup B. Indeed, if the theorem fails for some D and one considers a connected chain from D to B, one quickly arrives at a pair of elements A_1, A_2 of $SC_p(G)$ with $A_1, A_2 \subseteq M$, having the following properties:

(25) (a) Theorem 5.12 holds for every element of $\mathscr{E}^p_2(A_1)$;
(b) Theorem 5.12 fails for some element of $\mathscr{E}^p_2(A_2)$;
(c) $A_0 = A_1 \cap A_2$ is noncyclic; and
(d) $\langle A_1, A_2 \rangle$ is a p-group.

One establishes Theorem 5.12 by contradicting condition (*b*)—in other words, by showing that the theorem, in fact, holds for every element of $\mathscr{E}^p_2(A_2)$. Properties of the centralizers of elements of $A_2^{\#}$ and ultimately of the normalizers of 2-subgroups T of $\Gamma^*(D_2)$ for $D_2 \in \mathscr{E}^p_2(A_2)$ (with $O_2(M) \subseteq T$) are established by using the fact that A_1 is "well behaved". The subgroup A_0 provides the link between A_2 and A_1, and its action on the p'-layers and p'-cores of these centralizers and normalizers is central to the analysis.

6. Theorem C

Finally we discuss Theorem C in the same way as Theorems A and B. We assume then that G satisfies the assumptions of the theorem, we let $p \in \sigma(G)$,

‡ See footnote on p. 120.

and we let M be a maximal 2-local subgroup of G containing $\Gamma^0_{P,2}(G)$, where $P \in Syl_p(G)$. Note that our hypotheses allow $m_{2,p}(G) = 3$ here. We should also point out that the proof involves no further use of signalizer functors—that method is confined to the proof of Theorem B. On the other hand, as remarked earlier, Theorem C does depend upon Theorems D_1 and D_2.

A. Proper 2-generated p-cores

The first major step in the proof is to establish the following result.

THEOREM 6.1. *We have* $\Gamma_{P,2}(G) \subseteq M$.

We argue by contradiction. Since $\Gamma_{P,3}(G) \subseteq \Gamma^0_{P,2}(G) \subseteq M$, we quickly produce a subgroup D of P with the following properties:

(1) (a) $m_p(D) = m_p(DC_P(D)) = 2$;
(b) $N = N_G(D)$ is not contained in M;
(c) $Q = P \cap N \in Syl_p(N)$ and $D = O_p(N)$; and
(d) N is p-constrained.

The p-constraint of N comes from the fact that otherwise $m_p(DC_P(D)) \geqslant 3$, contrary to (a). We choose D so that $|Q|$ is maximal and, subject to this, so that $|D|$ is maximal.

We set $Z = \Omega_1(Z(P))$ and $N_0 = N_G(Z)$, so that $P \subseteq N_0$. Since $m_p(DC_P(D)) = 2$, we have $Z \subseteq D$. Furthermore, our conditions directly yield that $|Z| = p$. We can prove the following result concerning N_0.

LEMMA 6.2. *If N_0 is not p-constrained, then* $N_0 \subseteq M$.

The proof depends upon the fact that the quasisimple groups listed in the conclusion of Proposition 3.13 are all simple, in conjunction with the fact that $Z \subseteq O_p(N_0)$.

The proof of Theorem 6.1 subdivides into two cases, according as $Q = P$ or $Q \subset P$. Moreover, the preceding lemma is important in the $Q \subset P$ case. We first prove

PROPOSITION 6.3. *We have* $Q \subset P$.

Suppose false, in which case $Q = P$ and so $m_p(N) = m_p(P) \geqslant 3$. For any $A \in \mathscr{E}^p_3(P)$, we have

$$\Gamma_{A,2}(G) \subseteq \Gamma^0_{P,2}(G) \subseteq M.$$

It follows at once that $O_{p'}(N) \subseteq M$. Maximality of D now yields that $D \in Syl_p(O_{p',p}(N))$, whence

(2) $O_{p',p}(N) = O_{p'}(N) \times D$.

Since N is p-constrained and $\Gamma^0_{P,2}(N) \subset N$ as $N \nsubseteq M$, the structure of $N/O_{p'}(N)$ is given by Proposition 3.12. In particular, D is a central product of an extraspecial group of order p^3 and exponent p and a cyclic group,

$m_p(P) = 3$, and $|P : D| = p$. Thus we are necessarily in the $e(G) = 3$ case here.

It is immediate from the structure of P that P has a unique normal subgroup U of type (p,p). The key step in the proof is to show that

(3) $\langle \mathsf{И}_G(U;p') \rangle \subseteq M$.

Since $U \subseteq D$, this yields at once that

(4) $\langle \mathsf{И}_G(D;p') \rangle \subseteq M$.

Denote the left side of (4) by X. Since $O_2(M)$ is a D-invariant 2-group, $O_2(M) \subseteq X$. As $X \subseteq M$ by (4), $O_2(M) \subseteq O_2(X)$. Thus $O_2(X) \neq 1$. On the other hand, clearly $N_G(D)$ permutes by conjugation the set of all D-invariant p'-subgroups of G and so normalizes their join, which is X. We therefore conclude that

(5) $N = N_G(D) \subseteq N_G(O_2(X))$.

Now we invoke Theorem D_1 to derive a contradiction. Indeed, $H = N_G(O_2(X))$ is a 2-local subgroup of G containing P and so $m_p(H \cap M) = 3$. Therefore $H \subseteq M$ by Theorem D_1, whence $N \subseteq M$ by (5), which is not the case.

We establish Theorem 6.1 by contradicting Proposition 6.3. Thus we prove

PROPOSITION 6.4. *We have* $Q = P$.

Suppose false. If N_0 is p-constrained, then $m_p(P \cap O_{p',p}(N_0)) \geqslant 2$ and it follows at once from the maximality of D that $N_0 \subseteq M$. We conclude therefore from Lemma 6.2 that

(6) $N_0 \subseteq M$.

However, $Z \subseteq D$ and D centralizes $O_{p'}(N)$, so $O_{p'}(N) \subseteq M$ and now (2) holds in this case as well, again by the maximality of D.

Now a straightforward argument yields

(7) $D \cong Z_p \times Z_p$ and $O^{p'}(N/O_{p',p}(N)) \cong SL(2,p)$.

In particular, Q is nonabelian of order p^3 and exponent p, so $D \in Syl_p(C_G(D))$. Furthermore, by the structure of $SL(2,p)$ in its action on D, we see that $N_G(Q)/QC_G(Q)$ contains a cyclic subgroup of order $p-1$ which acts Frobeniusly on D. Also $N_G(Q) \subseteq M$ by the maximality of D. Likewise, as $N \nsubseteq M$, we see that M does not cover $O^{p'}(N/O_{p',p}(N))$, so $N_M(D)/C_M(D)$ does not involve $SL(2,p)$. Using all these properties of D together with the fact that M is a $\mathscr{K}$-group, we deduce the following structure for M.

LEMMA 6.5. *Either M is of "p-constrained type", or else the following conditions hold:*

(i) *$L_{p'}(M)$ is the product of p p-components of p-rank 1 cycled by an element of P;*

(ii) $|P:P\cap L_{p'}(M)| = p$ *and* $O_{p',p}(M) = O_{p'}(M)$; *and*
(iii) $m_p(P) = p$.

By saying that M is of p-constrained type, we mean that every p-local subgroup of $M/O_{p'}(M)$ is p-constrained with trivial p'-core. Note that, if $p = 3$, then this must be the case. Indeed, in the contrary case,

$$m_{2,3}(G) = m_3(M) = m_3(P) = 3$$

by (iii), whence $\sigma(G) = \{3\}$ by definition of $\sigma(G)$. Thus $m_{2,q}(G) \leqslant 2$ for all $q \geqslant 5$. But (i) and (ii) easily imply that $m_q(L_{p'}(M)) \geqslant 3$ for some prime $q \geqslant 5$. Since M is a 2-local, it follows that $m_{2,q}(G) \geqslant 3$, a contradiction.

Using this structure of M, we now prove

LEMMA 6.6. *$\langle И_M(D;p')\rangle$ is a p'-group and is invariant under a Sylow p-subgroup of M.*

This is immediate from Proposition 3.7 if M is of p-constrained type since $D \in SC_p(G)$; however, the assertion may seem surprising in the non- p-constrained case. But as an element of D cycles the p p-components of $L_{p'}(M)$ in this case, we can show that $L/O_{p'}(L)$ is isomorphic to a subgroup of the alternating group A_p for any such p-component L; the desired conclusion essentially follows from the fact that subgroups of order p in A_p normalize no nontrivial p'-groups.

The main step in the proof of Proposition 6.4 is to establish the following properties of the maximal D-invariant 2-subgroups of G and M.

LEMMA 6.7.
(i) *If $T \in И^*_M(D;2)$, then $T \in И^*_G(D;2)$.*
(ii) *$C_G(D)$ acts transitively by conjugation on the elements of $И^*_G(D;2)$.*

Note that, by Lemma 6.6 and a Frattini argument, $m_p(N_M(T)) \geqslant 3$. Now Theorem D_1 implies that $N_G(T) \subseteq M$, which immediately yields (i). The proof of (ii) is more involved; it is similar to the corresponding argument in the solvable 2-local case [100].

Lemma 6.7 gives a final contradiction at once, for by (ii)

(8) $N = N_G(D) = N_N(T)C_G(D)$

for any $T \in И^*_M(D;2)$ (as $T \in И^*_G(D;2)$ by (i)). On the other hand, $C_G(D) \subseteq C_G(Z) \subseteq N_0 \subseteq M$ by (6) and $N_G(T) \subseteq M$ by the previous paragraph, so $N \subseteq M$ by (8), which is not the case. Thus Proposition 6.4 and hence Theorem 6.1 is proved.

B. Centralizers not in M

Clearly the existence of elements x of order p in M with $C_x \nsubseteq M$ will prevent M from being strongly p-embedded. Such elements x fall into two categories, according as $m_p(C_M(x)) \geqslant 3$ or $m_p(C_M(x)) = 2$. We now consider the first of these cases; for convenience, we set

(9) $\mathscr{P}(M) = \{x \mid x \in M, |x| = p, m_p(C_M(x)) \geqslant 3, C_x \nsubseteq M\}$.

Using our generational results (Propositions 3.12 and 3.13) together with Theorem D_1, we easily obtain

PROPOSITION 6.8. *Let $x \in \mathscr{P}(M)$ and set $\bar{C}_x = C_x/O_{p'}(C_x)$. Then the following conditions hold:*

(i) *$O_{p'}(C_x)$ is of odd order;*
(ii) *$C_{\bar{C}_x}(\bar{L}_x) = O_p(\bar{C}_x)$ and is cyclic;*
(iii) *M covers C_x/L_x; and*
(iv) *one of the following holds:*
   (1) *$\bar{L}_x \cong L_2(p^n), n \geqslant 2, U_3(p^n), n \geqslant 1$, or A_{2p};*
   (2) *$p = 11$ and $\bar{L}_x$ is of type J_4;*
   (3) *$p = 5$ and $\bar{L}_x \cong Sz(32), {}^2F_4(2)', Mc$, or $M(22)$; or*
   (4) *$p = 3$ and $\bar{L}_x \cong L_2(8), L_3(4), M_{11}$, or a group of Ree type of characteristic 3.*

We assume $\mathscr{P}(M) \neq \emptyset$ and we fix $x \in \mathscr{P}(M)$. The fact that $|O_{p'}(C_x)|$ is odd has important fusion consequences. Indeed, we can prove

LEMMA 6.9. *If $D \in \mathscr{E}_2^p(C_x)$ with $x \in D$, then $\langle x \rangle$ is not G-conjugate to p distinct subgroups of D of order p.*

Suppose false. Since M contains a Sylow p-subgroup Q of C_x (as $m_p(C_x \cap M) \geqslant 3$ and $\Gamma_{P,2}(G) \subseteq M$), we can assume without loss that $D \subseteq Q$. Setting $R = C_Q(D)$, it follows easily from the possible structures of C_x that $m_p(R) \geqslant 3$.

Since D acts faithfully on $O_2(M)$, D does not centralize $C_{O_2(M)}(E)$ for at least *two* subgroups E of D of order p. Since the lemma is assumed false, we can therefore choose E so that

(10) $T = [C_{O_2(M)}(E), D] \neq 1$ and $E^g = \langle x \rangle$ for some $g \in G$.

Since T is clearly R-invariant, we have $T^g R^g \subseteq C_x$ with T^g invariant under R^g. As $m_p(R^g) \geqslant 3$, it follows now from the structure of C_x that T^g must lie in $O_{p'}(C_x)$, so $|O_{p'}(C_x)|$ is even, a contradiction.

For example, if L_x is of type $L_2(p^n)$ with $p \geqslant 5$, the lemma implies that $\langle x \rangle$ must be weakly closed in the Sylow p-subgroup Q of C_x. Hence $Q \in Syl_p(G)$ and $\langle x \rangle$ is weakly closed in a Sylow p-subgroup of G. We remark that, because $L_x \nsubseteq M$, no element of Q of order p induces a nontrivial field automorphism on $L_x/O_{p'}(L_x)$ (otherwise $L_x \subseteq \Gamma_{Q,2}(C_x) \subseteq M$); hence the G-conjugates of x in C_x all induce inner automorphisms on $L_x/O_{p'}(L_x)$.

The oddness of $|O_{p'}(C_x)|$ also restricts the action of x on an x-invariant Sylow 2-subgroup S of $O_{p'}(M)$. Without loss we can assume that S is Q-

invariant. Thus $C_S(x)$ is Q-invariant. Since $C_S(x) \cap O_{p'}(C_x) = 1$, the structure of C_x immediately yields

LEMMA 6.10. *We have* $|C_S(x)| \leqslant 2$ *and, if equality holds, then* $L_x/O_{p'}(L_x) \cong M(22)$ *(with* $p = 5$*) or* $L_3(4)$ *(with* $p = 3$*).*

At a later stage of the analysis, we prove that, in fact, $O_{p'}(M)\langle x\rangle$ is a Frobenius group.

The main result concerning elements of $\mathscr{P}(M)$ is the following.

THEOREM 6.11. *If* $x \in \mathscr{P}(M)$, *then* $p \geqslant 5$ *and* $L_x \cong L_2(p^n)$, $n \geqslant 2$. *In particular,* L_x *is simple.*

We must eliminate all other possibilities for $L_x/O_{p'}(L_x)$ and then prove that L_x is quasisimple. We use several distinct lines of argument. If L_x is of type $U_3(p^n)$, $n \geqslant 2$, or Ree type over $\boldsymbol{F}_{3^n}$, then as $|O_{p'}(C_x)|$ is odd and $L \nsubseteq M$, we can find an involution t in L_x such that M does not contain $J = L_{p'}(C_{L_x}(t))$ (with, respectively, $J/O_{p'}(J) \cong L_2(p^n)$ or $L_2(3^n)$) and such that $m_p(C_Q(t)) \geqslant 3$ ($Q \in Syl_p(C_x)$ with $Q \subseteq M$). However, as $m_p(C_t \cap M) \geqslant 3$, Theorem D_1 implies that $C_t \subseteq M$, so $J \subseteq M$, a contradiction.

On the other hand, if L_x is of type $M(22)$ (with $p = 5$) or A_{2p}, $p \geqslant 7$, we use the fact that then L_x has 2-local 3-rank at least 4, whence $3 \in \beta_4(G) = \sigma(G)$, so $\Gamma_{P^*,2}(G) \subseteq M^*$ for some maximal 2-local subgroup M^* of G, where $P^* \in Syl_3(G)$, by the hypothesis of Theorem C and Theorem 6.1. Taking a suitable choice of P^* and using Theorem D_1, we easily conclude that $C_x \subseteq M^*$ and $M^* = M$, a contradiction.

These are the easy cases. Apart now from the case L_x of type $U_3(p)$, $p \geqslant 5$, which requires a special argument, all remaining cases are treated in essentially the following way. The critical fact in these cases is that there is $D \in \mathscr{E}_2^p(Q)$ with $x \in D$ and a D-invariant 2-subgroup S of L_x with $S \nsubseteq M$. Thus $\langle И_G(D;2)\rangle \nsubseteq M$. The goal of the analysis is then to show, to the contrary, that

(11) $\langle И_G(D;2)\rangle \subseteq M$.

The proof of (11) involves a careful study of the 2-local subgroups H of G containing D with $H \nsubseteq M$. If $m_p(H) \geqslant 3$, then $m_p(H \cap M) \geqslant 3$ (as $D \subseteq H$ and $\Gamma_{P,2}(G) \subseteq M$), so $H \subseteq M$ by Theorem D_1, a contradiction. Hence $m_p(H) = 2$. Moreover, the G-fusion of x, which we have previously obtained (Lemma 6.9 and its consequences), places severe further restriction on the structure of H. Eventually we argue that $D \cap O_{p',p}(H) \neq 1$ and this is critical in verifying (11).

Note that this argument includes the case

$$L_x/O_{3'}(L_x) \cong L_2(3^n)$$

since a subgroup of order 3 in $L_2(3^n)$ normalizes a four subgroup. On the other hand, a subgroup of order p in $L_2(p^n)$ for $p \geqslant 5$ normalizes no nontrivial p'-subgroups, so the argument cannot be applied when L_x is of the latter type. This should help explain the precise statement of Theorem 6.11.

Finally if L_x is not quasisimple, it follows easily that $m_{2,q}(L_x) \geqslant 4$ for some (odd) prime divisor q of $|O_{p'}(C_x)|$. Thus $q \in \beta_4(G)$ and we conclude, as usual, from this that $C_x \subseteq M$, a contradiction; so L_x must be quasisimple.

C. Almost strong p-embedding when $\mathscr{P}(M) \neq \emptyset$

As a direct consequence of Theorem 6.11, we have the following basic result (continuing the above notation).

THEOREM 6.12. *Assume that either $\mathscr{P}(M) \neq \emptyset$ and $3 \in \sigma(G)$ or that $3 \in \beta_4(G)$. Then G possesses a strongly 3-embedded maximal 2-local subgroup.*

Indeed, as $p \geqslant 5$ by Theorem 6.11, it follows in the first case from the definition of $\sigma(G)$ (and the assumption $3 \in \sigma(G)$) that $\sigma(G) = \beta_4(G)$. Thus $m_{2,3}(G) \geqslant 4$ in either case. But again by the hypothesis of Theorem C and Theorem 6.1, $\Gamma_{P^*,2}(G) \subseteq M^*$ for some maximal 2-local subgroup M^* of G, where $P^* \in Syl_3(G)$. Furthermore, applying Theorem 6.11 to M^* for the prime 3, it follows that $C_{x^*} \subseteq M^*$ whenever x^* has order 3 in M^* and $m_3(C_{M^*}(x^*)) \geqslant 3$.

We claim that $C_{x^*} \subseteq M^*$ for *every* element x^* of order 3 in M^*. We can suppose that $x^* \in P^*$. Now $m_3(P^*) = m_{2,3}(P^*) \geqslant 4$ and so, by a result of [137], P^* possesses an elementary normal subgroup A of rank 4. Considering the action of x^* on A, we see that $m_3(C_A(x^*)) \geqslant 2$ and that $m_3(C_{A\langle x^*\rangle}(x^*)) \geqslant 3$, so the desired conclusion follows from the preceding paragraph.

Now, if P_0 is any cyclic subgroup of P^* and we set $P_1 = \Omega_1(P_0)$, our argument yields that $C_G(P_0) \subseteq C_G(P_1) \subseteq M^*$. Since $m_3(C_{M^*}(P_1)) \geqslant 3$, it follows by a Frattini argument that $N_G(P_1) \subseteq M^*$, so $N_G(P_0) \subseteq M^*$. Since $\Gamma_{P^*,2}(G) \subseteq M^*$, we conclude at once from the definition that M^* is strongly 3-embedded in G.

Note now that if $\mathscr{P}(M) \neq \emptyset$ the hypothesis of Theorem C together with Theorem 6.12 show that the assumptions of Theorem D_2 are satisfied. We use that theorem in proving the main result of this subsection.

THEOREM 6.13. *If $\mathscr{P}(M) \neq \emptyset$, then M is almost strongly p-embedded in G.*

Let $x \in \mathscr{P}(M)$, so that $p \geqslant 5$ and $L_x \cong L_2(p^n)$ by Theorem 6.11. We have noted above that $\langle x\rangle$ is weakly closed in a Sylow p-subgroup Q of C_x, which can be taken to lie in M. Then $Q \in Syl_p(G)$ and, without loss, we can assume that $Q = P$. By Theorem 6.1, $O^p(M) = O^p(G) \cap M = M$, which implies that P induces inner automorphisms on L_x. Thus $P = (P \cap L_x) \times C_P(L_x)$. The first factor is elementary and the second is cyclic, so P is abelian.

Set $R = P \cap L_x$. We can argue quite easily that $\Gamma_{R,1}(G) \subseteq M$ and now Theorems D_1 and D_2 (according as $m_p(R) \geqslant 3$ or $m_p(R) = 2$) imply that every 2-local subgroup of G containing R lies in M. In particular, any 2-local subgroup of G containing L_x lies in M. Thus we have

(12) L_x is not contained in a 2-local subgroup of G.

Using (12), together with the fact that x acts fixed-point-freely on any x-invariant Sylow 2-subgroup of $O_{p'}(M)$, it is not difficult to show that $O_{p'}(M)\langle x\rangle$ is a Frobenius group with kernel $O_2(M)$ and complement $O_{p'}(C_x)\langle x\rangle$.

Finally to prove that M is solvable, one has only to show that $O_{p'}(M)\langle x\rangle \lhd M$ and the result will follow from the structure of $C_M(x) = C_x \cap M$, which is solvable. Since P is abelian, it suffices, in view of the weak closure of $\langle x\rangle$, to prove that M is p-constrained. If false, we conclude easily from this weak closure and the solvability of $C_M(x)$ that $K = L_{p'}(M)$ is a single p-component with cyclic Sylow p-subgroups, $x \in K$, and R centralizes $K/O_{p'}(K)$.

Observe now that any p'-element y of $N_K(\langle x\rangle) \cap C_K(R)$ acts on L_x and centralizes its Sylow p-subgroup R, so y must centralize L_x. Then, likewise, $N_K(\langle y\rangle) \cap C_K(R)$ centralizes L_x. With a little more arguing, this forces L_x to lie in a 2-local subgroup, contrary to (12). Thus M is solvable (of p-length 1).

These are the main points in showing that M is almost strongly p-embedded in G when $\mathscr{P}(M) \neq \emptyset$.

D. Almost strong p-embedding when $\mathscr{P}(M) = \emptyset$

Assume now that $\mathscr{P}(M)$ is empty. In this case, only one question remains to be settled in order to have M strongly p-embedded in G, namely the following.

(13) If $x \in M$ with $|x| = p$ and $m_p(C_M(x)) = 2$, must C_x be contained in M?

The complete answer to (13) requires a rather lengthy argument. Ultimately we obtain an affirmative answer when $p \geqslant 5$. Furthermore, the proof of Theorem 6.12 shows that there are no such x when $p = 3$ and $m_{2,3}(G) \geqslant 4$, so (13) holds vacuously in this case. In the remaining case ($e(G) = 3$ and $\sigma(G) = \{3\}$), certain configurations can arise for which we are unable to verify (13); these are incorporated into the definition of almost strong 3-embedding.

Let x satisfy the assumptions of (13) and suppose $C_x \nsubseteq M$. Note first that if $y \in Z(P)$ with $|y| = p$, then $C_y \subseteq M$ as $\mathscr{P}(M) = \emptyset$. Furthermore, as $\Gamma_{P,2}(G) \subseteq M$, it follows, with the aid of the Alperin–Goldschmidt conjugation family [87] that M controls the G-fusion of subsets of P. Using these two facts, we easily obtain the following result.

LEMMA 6.14.
(i) *There exists* $D \in \mathscr{E}_2^p(P)$ *such that* $D \in SC_p(G)$ *and* $x \in D$; *and*
(ii) $\langle x \rangle$ *is G-conjugate to every element of* $\mathscr{E}_1(D)$ *except* $\Omega_1(Z(P))$.

Using these properties of D and the fact that M is a $\mathscr{K}$-group, it is possible to determine the structure of M.

LEMMA 6.15. *One of the following holds:*
(i) M *satisfies one of the alternatives of Lemma* 6.5; *or*
(ii) *if* $\bar{M} = M/O_{p'}(M)$, *then*
$L(\bar{M}) \cong SL(p,q)$ *or* $L_p(q)$, *where* $p \mid q-1$
$SU(p,q)$ *or* $U_p(q)$, *where* $p \mid q+1$,
$L_2(p^p)$ *or* Co_1 (*with* $p = 5$).
Moreover, $C_{\bar{M}}(L(\bar{M})) = O_p(\bar{M})$ *and is trivial unless* $Z(L(\bar{M})) \neq 1$.

The exceptional possibilities of (ii) require some special analysis. We exclude them here, for brevity. Thus if M is not of p-constrained type we assume that x cycles the p p-components of $L_{p'}(M)$.

As a corollary of the lemma and our assumption, one can easily obtain the following properties of maximal D-signalizers in M.

LEMMA 6.16. *If* $F \in И_M^*(D;f)$ *for some prime* $f \neq p$, *then we have*
(i) $m_p(N_M(F)) = m_p(P)$; *and*
(ii) *if* $f = 2$, *then* $F \in И_G^*(D;2)$.

Note that (ii) follows at once from (i) and Theorem D_1.

We now list the main moves of the analysis, limiting ourselves to a few comments. Our results depend upon the preceding lemmas as well as a detailed description (which we omit here) of the structure of proper subgroups H of G containing D with $H \nsubseteq M$.

PROPOSITION 6.17. *If* X *is a nontrivial D-invariant p'-subgroup of* M *of odd order, then* $N_G(X) \subseteq M$.

PROPOSITION 6.18. *If* H *is a 2-local subgroup of* G *containing* D *such that* $H \cap M$ *contains a nontrivial D-invariant 2-subgroup, then* $H \subseteq M$.

Note that we cannot apply Theorem D_2 to obtain Proposition 6.18 since the crucial assumption $\Gamma_{D,1}(G) \subseteq M$ fails here as $C_x \nsubseteq M$ and $x \in D$.

Using Proposition 6.18, we prove

PROPOSITION 6.19. *One of the following holds:*
(i) M *covers* $C_x/O(O_{p'}(C_x))$; *or*
(ii) $p = 3$, $C_x \cong Z_3 \times L_3(2)$ *or* $Z_3 \times Aut(L_3(2))$, *and if* T *is an x-invariant Sylow 2-subgroup of* $O_{p'}(M)$, *then* $C_T(x) \cong Z_2 \times Z_2$.

We next use Propositions 6.17 and 6.18 to study the case in which M covers C_x/Y, where $Y = O(O_{p'}(C_x))$. Since $C_x \nsubseteq M$, $Y \nsubseteq M$ and so certainly

$Y \neq 1$. We let F be a minimal normal subgroup of C_x with $F \subseteq Y$. Then Proposition 6.17 implies that $F \cap M = 1$. Therefore $Z = \Omega_1(Z(P))$ acts Frobeniusly on F (as $Z \subseteq C_x$ and $N_G(Z) \subseteq M$). Moreover, as $D \in SC_p(G)$, Lemma 6.14(ii) implies that, if T denotes a D-invariant Sylow 2-subgroup of $O_{p'}(M)$, then Z does not centralize $S = C_T(x)$. By Proposition 6.18, S ($\subseteq C_x$) acts faithfully on F (as $F \cap M = 1$), so SZ acts faithfully on F. However, F is an elementary abelian f-group for some odd prime $f \neq p$. If $m_2(C_F(u)) \geqslant 4$ for some $u \in S^{\#}$, then $f \in \beta_4(G)$ and this leads to a contradiction. By a careful analysis of the action of SZ on F, we eventually prove

PROPOSITION 6.20. *If M covers $C_x/O(O_{p'}(C_x))$, then $p = 3$, and if T is an x-invariant Sylow p-subgroup of $O_{p'}(M)$, then $C_T(x) \cong Q_8 \circ Z_{2^n}$ for some n.*

As an immediate corollary of Propositions 6.19 and 6.20, we obtain

THEOREM 6.21. *If $p \geqslant 5$ and $\mathscr{P}(M) = \emptyset$, then M is strongly p-embedded in G.*

It remains therefore to analyse the exceptional cases of Propositions 6.19 and 6.20. We prove

PROPOSITION 6.22. *If $p = 3$ and $\mathscr{P}(M) = \emptyset$, then*
(i) *M is solvable,*
(ii) *$P \cong Z_3 \wr Z_3$, and*
(iii) *Proposition 6.19(ii) does not hold.*

We note that the proof of (iii) depends upon Theorem D_2. The conditions that we have reached are precisely those of almost strong 3-embedding. Thus we have

THEOREM 6.23. *If $p = 3$ and $\mathscr{P}(M) = \emptyset$, then M is almost strongly 3-embedded in G.*

This completes our discussion of Theorem C.

5

The Uniqueness Case for Groups of Characteristic 2 Type

MICHAEL ASCHBACHER

1. Introduction

In earlier chapters we have seen that the classification of the finite simple groups may be reduced to the solution of certain subproblems. In particular one such subproblem involves the analysis of groups of the following sort.

Uniqueness Case. Let G be a finite group of characteristic 2 type with $e(G) \geqslant 3$ in which all proper subgroups are $\mathscr{K}$-groups. Define $\sigma(G)$ to be the following set of primes:

if

$$\sigma_1(G) = \{p \mid p \text{ odd}, m_{2,p}(G) \geqslant 4\},$$
$$\sigma_2(G) = \{p \mid p > 5, m_{2,p}(G) = 3\},$$
$$\sigma_3(G) = \{p \mid p > 3, m_{2,p}(G) = 3\},$$
$$\sigma_4(G) = \{3\},$$

and

$$i_0 = \min\{i \mid \sigma_i(G) \neq \emptyset\},$$

then

$$\sigma(G) = \sigma_{i_0}(G).$$

Then, for each $p \in \sigma(G)$, G possesses an almost strongly p-embedded 2-local subgroup.

To complete the classification of the finite simple groups in the manner envisaged, we must show that the Uniqueness Case leads to a contradiction.

If G is a generic group in the Uniqueness Case, then $m_{2,3}(G) \geqslant 4$. The author has shown that under this extra restriction the Uniqueness Case leads to a contradiction. It seems likely that the same techniques will extend to obtain a contradiction in general.

The first six sections of this chapter discuss some of the techniques used in this analysis. These techniques can be applied outside the Uniqueness Case, particularly to groups G with $e(G) \leqslant 2$. The final section in this chapter outlines how the techniques can be used to derive a contradiction to the Uniqueness Case when $m_{2,3}(G) \geqslant 4$.

Let G be a group of characteristic 2 type and let $\mathscr{M}$ be the set of maximal 2-local subgroups of G. For any subset X of G, let $\mathscr{M}(X)$ consist of those members of $\mathscr{M}$ containing X. If G is in the Uniqueness Case and $p \in \sigma(G)$ then $|\mathscr{M}(X)| = 1$ for each noncyclic p-subgroup X of G. More generally, the techniques of this chapter depend upon the existence of a large collection of subgroups contained in unique maximal 2-locals. The larger the collection, the more successful the techniques.

Let $M \in \mathscr{M}$ and let $T \in Syl_2(M)$. We seek to obtain factorizations of the form

$$M = N_M(T_1)N_M(T_2)$$

for suitable subgroups T_i of T, or sometimes results like

$$M = \langle N_M(S) \mid S \in \mathscr{S} \rangle$$

for suitable collections $\mathscr{S}$ of subgroups of T. We also refer to statements of the latter sort as factorizations.

In order to establish such factorizations we must often consider the action of M on certain elementary abelian sections V in $O_2(M)$. Thus $M/C_M(V)$ is faithfully represented on the $GF(2)$-module V, and we are led to various questions about $GF(2)$-representations.

In Section 13 of the N-group paper [214,IV], Thompson showed that an analogue of the Uniqueness Case led to a contradiction for N-groups. The principles discussed above appeared already in his analysis.

The following notational convention will be adopted in this chapter. If $\mathscr{C}$ is a collection of groups, given either abstractly or as subgroups of a group G, and $H \subseteq G$, then

$$\mathscr{C} \cap H = \{K \subseteq H \mid K \in \mathscr{C}\}.$$

2. The Characteristic Core

Let G be a finite group of characteristic 2 type and let $T \in Syl_2(G)$. For $S \subseteq T$ define $C(G,S)$ to be the subgroup of G generated by the normalizers of all nontrivial characteristic subgroups of S. $C(G,T)$ is called the *characteristic core* of G and is of course determined up to conjugacy in G. We are interested in the case where the characteristic core is proper.

For example, assume that $O_2(G) = 1$ and $\mathscr{M}(T) = \{M\}$. Then $C(G,T) \subseteq M \subset G$. So G has a proper characteristic core in the case where T is in a unique maximal 2-local.

In analysing this problem, we introduce some terminology and notation. A group X is *short* if the following conditions are satisfied:

(1) $X = O^2(X)$ and $F^*(X) = O_2(X)$;
(2) $[X,O_2(X)] = U(X) \subseteq \Omega_1(Z(O_2(X))$;
(3) $X/O_2(X)$ is quasisimple or of prime order; and
(4) X acts irreducibly on $\tilde{U}(X) = U(X)/C_{U(X)}(X)$.

A *block* of G is a subnormal short subgroup.

Blocks behave much like components. For example, distinct blocks of a group commute. Short subgroups and blocks seem to be very interesting objects which deserve more study.

Let $\mathscr{X}$ be the collection of short groups X such that $X/O_2(X) \cong L_2(2^n)$ or A_{2n+1} with $\tilde{U}(X)$ the natural module for $X/O_2(X)$. Here the natural module for $L_2(2^n)$ is the 2-dimensional $GF(2^n)$-module considered as a $GF(2)$-module, while the natural module for A_n is the unique noncentral chief factor in the permutation module of degree n. Notice that $L_2(4) \cong A_5$, but the groups have different natural modules. Let $\mathscr{Y}$ be the set of $X \in \mathscr{X}$ with $X/O_2(X) \cong L_2(2^n)$ or A_{2^n+1}.

THEOREM 2.1. *Let G be of characteristic 2 type with $O_2(G) \neq 1$, $T \in Syl_2(G)$, and $M = C(G,T) \neq G$. Then*

$$G = ML_1L_2 \ldots L_r$$

where $L_i \in \mathscr{Y}$ and L_i is a block of G for each i.

Thus the 2-constrained groups of characteristic 2 type with proper characteristic cores are determined. There are also some partial results for the non-2-constrained case. For this case we need still more terminology and notation.

Let $\mathscr{X}(G)$ be the set of $X \in \mathscr{X} \cap G$ such that, for $R \in Syl_2(C_G(X/O_2(X)))$, $X \trianglelefteq\trianglelefteq N_G(R)$ and either $X = [J_e(N_G(X)),X]$ or $X \trianglelefteq\trianglelefteq N_G(J_e(R))$. Here $\mathscr{A}_e(X)$ is the set of elementary abelian 2-subgroups of X of maximal order and $J_e(X) = \langle \mathscr{A}_e(X) \rangle$. A partial ordering $X \to Y$ is defined on $\mathscr{X} \cap G$ if either

(1) $X = Y$, or
(2) $X/O_2(X) \cong A_n$, $Y/O_2(Y) \cong A_m$, $U(X) \subseteq U(Y)$, $X \subseteq Y$, and $XO_2(Y)/O_2(Y)$ is the stabilizer of $m-n$ points in the permutation representation of degree m.

We will see how this order arises later. Let $\mathscr{X}^*(G)$ be the maximal members of $\mathscr{X}(G)$ under this partial order.

THEOREM 2.2. *Let G be of characteristic 2 type and suppose that there exists $X \in \mathscr{X} \cap G$ with $X \trianglelefteq\trianglelefteq M \in \mathscr{M}$. Then $\mathscr{X}(G)$ is nonempty.*

THEOREM 2.3. *Let G be of characteristic 2 type and assume that $M = C(G,T) \neq G$. Then either*

(1) *M is strongly embedded in G, or*

(2) *$\mathscr{X}(G)$ is nonempty.*

Actually Theorem 2.3 is a rather easy corollary to Theorems 2.1 and 2.2.

THEOREM 2.4. *Let G be of characteristic 2 type, and suppose that $X \in \mathscr{X}^*(G)$. Then $\mathscr{M}(X) = \{M\}$, and $X \trianglelefteq\trianglelefteq M$.*

These results appear in [17b]. Theorems 2.2 and 2.3 reduce the problem of determining groups G of characteristic 2 type with a proper characteristic core to the problem of determining the groups G of characteristic 2 type with a subgroup $X \in \mathscr{X}$ such that $\mathscr{M}(X) = \{M\}$ and $X \trianglelefteq\trianglelefteq M$. The analogous situation, with short groups replaced by quasisimple groups, was considered in Theorem 5 of the component paper [10] where it was shown that, with known exceptions, $X \trianglelefteq M$. Presumably one would attempt to establish the same result for short groups, or at least for groups in $\mathscr{X}$. Next, if $X \trianglelefteq M$, then $C_G(X)$ is tightly embedded in G, and the theory of tightly embedded subgroups may presumably be employed to show that $C_G(X)$ is small, much as in the analysis of groups with a standard subgroup. At this point the 2-local subgroup M is more or less determined and one should presumably be able to use this information to determine G. For example, Solomon has determined the groups G of characteristic 2 type with $X \in \mathscr{X} \cap G$, $\mathscr{M}(X) = \{N_G(X)\}$, and $X/O_2(X) \cong A_n$ in [191]. In any case it seems probable that this type of analysis can be used to complete the determination of groups with a proper characteristic core.

The proofs of these results are also reminiscent of analogous results in the theory of groups of component type. Here is a brief sketch of the proof of Theorem 2.1. Set $Z = \Omega_1(Z(T))$, $V = \langle Z^G \rangle$ and $\bar{G} = G/C_G(V)$. Then V is a $GF(2)$-module for $\bar{G}$ and $O_2(\bar{G}) = 1$. We consider two cases.

Case I. $\mathscr{M}(\bar{T}) \subseteq \bar{M}$.

Case II. There exists $\bar{H} \in \mathscr{M}(\bar{T})$ with $\bar{H} \nsubseteq \bar{M}$.

Case I will be discussed later. Consider Case II. Observe that $C(H,T) \subseteq M \cap H \neq H$. Hence, proceeding by induction on the order of G, there is a block L of H with $L \in \mathscr{Y}$ and $[V,L] \neq 1$. Since $L \trianglelefteq\trianglelefteq H$ and $V \trianglelefteq H$, $[V,L] \subseteq L$ and hence

$$[V,L] \subseteq [O_2(L),L] = U(L).$$

But L is short so L has just one noncentral chief factor in $O_2(L)$ and that factor is $\tilde{U}(L)$. Thus $[V,L] = U(L)$. Also $L/U(L)$ is quasisimple or Z_3, and

$$\bar{L} = LC(V)/C(V) \cong L/C_L(V)$$

is an image of $L/U(L)$, so that $\bar{L}$ is quasisimple or Z_3. Finally, in its action on V, $\bar{L}$ has a unique noncentral chief factor $\tilde{U}(L)$.

Indeed we know a little more. Let $\bar{S} = O_2(\bar{H})$. As $\bar{H} \in \mathcal{M}$, $\bar{H} = N_{\bar{G}}(\bar{S})$. Also $\bar{L} \trianglelefteq\trianglelefteq \bar{H}$, so $\bar{L}$ is a component of $C_{\bar{G}}(\bar{S})$ (ignoring the case $\bar{L} \cong Z_3$). Hence it suffices to determine all groups $\bar{G}$ for which $O_2(\bar{G}) = 1$, having a nontrivial 2-group $\bar{S}$ such that $C(\bar{S})$ has a component $\bar{L}$ of type $L_2(2^n)$ or A_{2n+1}, and such that there is a faithful $GF(2)$-module for $\bar{G}$ on which $\bar{L}$ has a unique noncentral chief factor which is natural. My approach to this problem was to use the theory of groups of component type, although I now believe there may be better proofs.

In any case we are led to a certain partial order on short groups (or possibly on some slightly larger class of groups). Namely, if $L \subseteq K$, we write $L \prec K$ if there exists a group K_1 containing K and a 2-subgroup T of K_1 normalizing K such that

(1) T contains $O_2(K)$ as a subgroup of index 2,
(2) $L \trianglelefteq\trianglelefteq N_K(T)$, and
(3) $LO_2(K)/O_2(K)$ is a standard subgroup of $TK/O_2(K)$,

and extend this relation by transitivity to a partial order $\rightarrow$. When restricted to $\mathcal{X}$, this is the partial order defined earlier. If one analyses $\bar{G}$ using component theory, this order is evidently relevant.

Notice that condition (3) is about embedding one quasisimple group in another, while condition (2) ensures the compatibility of the associated modules inside the short groups. With the relation $\prec$ extended to cover the quasisimple factors, there are some indications that $L/O_2(L) \prec K/O_2(K)$ only in the following cases‡.

(1) $A_n \prec A_{n+2}$ on natural modules.
(2) $Sp_{2n}(2^m) \prec \Omega^{\pm}_{2n+2}(2^m)$ on natural modules.
(3) $U_4(2) \prec \hat{U}_4(3)$ where $U_4(2)$ acts on its natural module, and $\hat{U}_4(3)$ is the 3-fold covering group of $U_4(3)$ acting on a 12-dimensional module when considered as a subgroup of $SU_6(2)$.

If this proves to be the case, then it should be possible to develop a reasonably elegant theory of short groups.

Let us now return to Case I, where $\mathcal{M}(\bar{T}) \subseteq \bar{M}$. There we prove the following.

THEOREM 2.5 [17a]. *Let G be a group and suppose that $F^*(G) = O_2(G)$. Let $T \in Syl_2(G)$ and set $Z = \Omega_1(Z(T))$, $V = \langle Z^G \rangle$ and $\bar{G} = G/C_G(V)$. Assume that there is a G-invariant subset D of $\mathcal{A}_e(G)$ such that, whenever $A \in D$, $B \in \mathcal{A}_e(G)$ and $\bar{B} \subseteq \bar{A}$, then either $\bar{B} = \bar{A}$ or $\bar{B} = 1$. Assume also that*

$$M = \langle N_G(T \cap D), C_G(Z) \rangle \neq G$$

‡ R. Foote has now established this

and that D is minimal subject to these conditions. Then G has a normal subgroup H such that

(1) $G = HM$,

(2) $\bar{H}$ *is the direct product of the conjugates of a subgroup* $\bar{L}$,

(3) $U = [V,H]/C_{[V,H]}(H)$ *is the direct sum of the conjugates of* $W = [U,L]$, *and*

(4) *W is the natural module for* $\bar{L}$ *and either*

(a) $\bar{L} \cong SL_m(2^n)$ *and* $\bar{D} \cap \bar{L}$ *is the collection of root groups of transvections, or*

(b) $\bar{L} \cong S_{2n+1}$ *and* $\bar{D} \cap \bar{L}$ *is the collection of transpositions.*

This result says something rather weak about the situation where the Thompson factorization $G = C_G(Z)N_G(J_e(T))$ fails. Namely it is applicable to the case where $\langle \mathcal{M}(\bar{T}) \rangle \neq \bar{G}$.

In particular, in Case I we can pick $A \in \mathcal{A}_e(G)$ such that A acts nontrivially on V with $|\bar{A}|$ minimal, and let D be the set of G-conjugates of A. If W is the weak closure of A in T, then $N(W) \subseteq M$ since $N(W) \supseteq T$. Now Theorem 2.5 applies and we conclude that $G = TL_1 \ldots L_r$ where T permutes the conjugates of $L_1 = L$ transitively, and $\bar{L} \cong L_2(2^n)$ or A_{2^n+1}. At this point we need several lemmas. For example we need to know that $J_e(T)$ acts on L. Actually $J_e(T)$ fixes each component of $\bar{G}$, as will be established in Theorem 5.4. We next consider $S = C_T(\Omega_1(Z(J_e(T))))$. As $J_e(T)$ fixes L, so does S. The interesting case occurs when $\bar{L} \cong L_2(2^n)$. There a lemma of Baumann shows that $S \in Syl_2(\langle S^L \rangle)$. But then by induction on the order of G, $T = S$ and $G = L$. To complete the proof we require an important theorem established independently by Baumann [24] and Niles [155].

THEOREM 2.6. *Let G be a group of characteristic 2 type with* $O_2(G) \neq 1$, $T \in Syl_2(G)$ *and* $C(G,T) \neq G$. *Assume that T is contained in a unique maximal subgroup of G, and that* $G/K \cong L_2(2^n)$ *for some normal subgroup K of G. Then G has a unique noncentral chief factor on* $O_2(G)$.

Actually Glauberman and Niles have extended Theorem 2.6 to the following result [86], discussed in Chapter 9.

THEOREM 2.7. *Let T be a 2-group. Then there exist nontrivial characteristic subgroups* T_i *of T,* $i = 1,2$, *with* $T_1 \subseteq Z(T)$, *such that whenever G is a group of characteristic 2 type with* $O_2(G) \neq 1$, $T \in Syl_2(G)$, $G/K \cong L_2(2^n)$ *for some normal subgroup K of G, and T contained in a unique maximal subgroup of G, then* T_1 *or* T_2 *is normal in G, or G has a unique noncentral chief factor in* $O_2(G)$.

Using Theorem 2.7 and the argument sketched above, it is possible to extend Theorem 2.1 to the following.

THEOREM 2.8. *Let T be a 2-group. Then there exist nontrivial characteristic subgroups* T_1 *of T and* T_2 *of* $C_T(\Omega_1(Z(J_e(T))))$ *with* $T_1 \subseteq Z(T)$ *such that, whenever G is a group of characteristic 2 type with* $O_2(G) \neq 1$ *and* $T \in Syl_2(G)$, *then* $G = HL_1 \ldots L_r$ *where* $H = \langle C_G(T_1), N_G(T_2)\rangle$ *and* $L_i \in \mathscr{X}$ *is a block of G.*

3. Weak Closures

Assume that G is of characteristic 2 type, $M \in \mathscr{M}$, and there is a set of subgroups X of G with $\mathscr{M}(X) = \{M\}$. If this set of *uniqueness subgroups* is sufficiently large, we would expect that $G = M$, so that $O_2(G) \neq 1$. For example, in the Uniqueness Case we have $|\mathscr{M}(X)| = 1$ for each noncyclic p-subgroup X with $p \in \sigma(G)$. The most effective technique known to the author in this situation comes from Section 13 of the N-group paper and involves the study of weak closures of normal elementary abelian 2-subgroups of M.

Let $\mathscr{V} = \mathscr{V}(M)$ be the set of normal elementary abelian 2-subgroups of M. If $V \in \mathscr{V}$ and $X \subseteq G$, for each nonnegative integer i define

$$\Gamma_i(X) = \Gamma_i(X,V) = \{U^g \subseteq X \mid U \subseteq V, \quad m(V/U) = i, \quad g \in G\},$$

$$W_i(X) = W_i(X,V) = \langle \Gamma_i(X)\rangle,$$

and

$$C_i(X) = C_i(X,V) = C_X(W_i(X)).$$

We wish to establish statements like

$$H = C_H(C_j(S))N_H(W_i(S)),$$

or

$$H = \langle C_H(C_j(S)), \quad N_H(W_i(S))\rangle,$$

when $F^*(H) = O_2(H)$, $S \in Syl_2(H)$, and $j \geqslant i$. Such statements will be referred to as *factorizations.*

For example, assume that G is in the Uniqueness Case, $p \in \sigma(G)$ and that X is a p-subgroup of G of exponent p and order at least p^3 permuting with a Sylow 2-subgroup $T$of G. Then $\mathscr{M}(Y) = \{M\}$ for each subgroup Y of X of order at least p and $O_2(TX) \neq 1$, so that $TX \subseteq M$. We assume that, whenever $T \subseteq H \subseteq G$ with $F^*(H) = O_2(H)$, then $H = C_H(C_{i+1}(T))N_H(W_i(T))$ for $i = 0$ and 1, with respect to some $V \in \mathscr{V}$.

Define

$$G_1 = N_G(W_0(T)), \quad G_2 = N_G(Z(W_1(T)) \quad \text{and} \quad G_3 = C_G(C_2(T)).$$

Then, from our factorizations,

$$H = (G_i \cap H)(G_j \cap H)$$

for distinct $i,j \in \{1,2,3\}$. In particular, either $|X \cap G_i| > p$ or $|X \cap G_j| > p$ for $i \neq j$, so there are distinct i,j in $\{1,2,3\}$ with $\mathscr{M}(G_k \cap X) = \{M\}$ for $k = i$ and j. Hence $G_k \subseteq M$. Now if groups of characteristic 2 type with a proper characteristic core are classified, then we may assume that $H \in \mathscr{M}(T)-\{M\}$. But $H = (H \cap G_i)(H \cap G_j) \subseteq M$, yielding a contradiction.

This example illustrates how weak closure arguments can be used in the Uniqueness Case. Let $V \in \mathscr{V}$. If our set of uniqueness subgroups is sufficiently large, then either

(3.1) V is a TI-set in G, or
(3.2) if $Q \in Syl_2(C_M(V))$, then $\mathscr{M}(N_M(Q)) = \{M\}$.

For example, in the Uniqueness Case let $p \in \sigma(G)$ and suppose that $Syl_p(M) \subseteq Syl_p(G)$. Then either $m_p(C_M(V)) > 1$ and (3.1) holds or $m_p(M/C_M(V)) > 1$ and (3.2) holds.

Both conditions are advantageous in applying weak closure arguments. In order to establish factorizations, one must show that $C_G(U) \subseteq M$ for large subgroups U of V. If V is a TI-set this holds automatically. On the other hand, one must also show that if $T \in Syl_2(M)$, then $N_G(W_i(T)) \subseteq M$ for suitable i. The easiest way to establish such statements is to show that $W_i(T)$ centralizes V. Then, if (3.2) holds, a Frattini argument shows that $N_G(W_i(T)) \subseteq M$.

Let $\mathscr{V}^*$ be the maximal members of $\mathscr{V}$ under inclusion. One bad case occurs when (3.2) fails for $V \in \mathscr{V}^*$. The problem becomes worse when V is small. At the extreme we have the case where $\mathscr{V}^*$ consists of a subgroup of order 2. In this case $O_2(M)$ is of *symplectic type*; that is, $O_2(M)$ has no noncyclic characteristic abelian subgroups. However groups with a 2-local of this type have been classified, and these are discussed in Chapter 6.

More generally, if either (3.1) or (3.2) holds for each $V \in \mathscr{V}$ and (3.2) fails for some $V \in \mathscr{V}^*$, then V is a TI-set. This case should lead to the situation where $O_2(M)$ is a special group with $V = Z(O_2(M))$, and then to a classification of G as a group of Lie type over the field of order $|V|$. Timmesfeld and others have results which are close to such a classification and these are also discussed in Chapter 6. In any event some of their more elementary results, together with a little extra analysis, are sufficient to deal with this situation in the Uniqueness Case, hence ensuring that $\mathscr{V}$ is large enough.

In order to factorize a subgroup H of G with respect to some $V \in \mathscr{V}$, it seems to be necessary that either H is "nearly solvable" or that very precise information about M and its representation on V is available. In what sense should H be "nearly solvable"? Let $S \in Syl_2(H)$ and let Ω be an H-invariant set of elementary abelian 2-subgroups of H. For $A \in \Omega$ define $\mathscr{E}(H,S,A)$ to be the set of subgroups K of H minimal subject to

$$C_S(AO_2(H)/O_2(H)) \subseteq S \cap K \in Syl_2(K)$$

and $A \nsubseteq O_2(K)$. The correct definition in more technical but this will capture the spirit. Let $\mathscr{E}_i(H,S,A)$ consist of those $K \in \mathscr{E}(H,S,A)$ with

$$m(A/A \cap O_2(K)) \leqslant i.$$

Set
$$E_i(H,S,\Omega) = \langle \mathscr{E}_i(H,S,A) \mid A \in S \cap \Omega \rangle$$

and say that $H \in \mathscr{E}_i\Omega$ if $H = \langle N_H(S \cap \Omega), E_i(H,S,\Omega) \rangle$, and $H \in \mathscr{E}_i$ if $H \in \mathscr{E}_i\Omega$ for each choice of Ω.

If H is solvable then $H \in \mathscr{E}_1$. Thus if $H \in \mathscr{E}_i$ for small i, for example $i = 1$ or 2, then H is nearly solvable. It seems to be the case that if $H \notin \mathscr{E}_2$ then H has a composition factor which is a group of Lie type over $GF(2^n)$, $n > 2$, a twisted group over $GF(4)$, ${}^3D_4(2)$, or an alternating group of high degree.

In considering weak closures in the Uniqueness Case, it seems necessary to first show that if $T \in Syl_2(M)$, $T \subseteq H \subseteq G$, $T \in Syl_2(G)$ and $H \in \mathscr{E}_2$, then $H \subseteq M$. This information can then be used to get precise information about M so that further weak closure arguments are possible.

I have found it convenient to break weak closures up into three cases:

(3.3) V is a noncyclic TI-set;
(3.4) $O_2(M/C_M(V)) = 1$ and (3.2) holds; and
(3.5) $1 \neq V_0 \in \mathscr{V}$, $V_0 \subseteq V$, V_0 is a TI-set, (3.2) holds, and $V = \langle v^M \rangle$ for each $v \in V - V_0$. V is not a TI-set.

Case 3.5 is the most difficult and technical. I shall not discuss this case. Cases 3.3 and 3.4 are discussed in Sections 4 and 6, respectively.

4. *TI*-Sets

In this section, G is a finite group and V is a noncyclic elementary abelian 2-subgroup of G which is a TI-set in G.

The following situation is of interest. Let $H \subseteq G$, and let E be an abelian normal subgroup of H. Set

$$\Omega = \{V^g \subseteq H \mid V^g \cap E \neq 1 \quad \text{and} \quad V^g \nsubseteq O_2(H)\}.$$

Assume that $V \in \Omega$, and set

$$K = \langle \Omega \rangle \quad \text{and} \quad I = \langle U \cap O_2(H) \mid U \in \Omega \rangle.$$

Let $J \trianglelefteq H$ with $I \subseteq J \subseteq O_2(H)$ and set $\bar{H} = H/J$. Finally let

$$D = \{u \mid u \in U - O_2(H),\ U \in \Omega\}.$$

LEMMA 4.1.
(1) *$\bar{D}$ is a set of root involutions of $\bar{H}$ and I is abelian.*
(2) *$\bar{V}$ is a TI-set in $\bar{H}$ with $N_{\bar{H}}(\bar{V}) = N_H(V)J/J$.*
(3) *If $\langle \bar{V}, \bar{U} \rangle$ is a 2-group for some $U \in \Omega$, then $[V,U] = 1$.*
(4) *If $\bar{V} = \langle \bar{v} \rangle$ is of order 2, then $\bar{v}^H$ is a set of odd transpositions of $\langle \bar{v}^H \rangle$.*

Here a *set of root involutions* of a group X is an X-invariant set $\tilde{D}$ of involutions of X which generate X and such that for each choice of a and b in $\tilde{D}$ the order of ab is odd, 2 or 4, and, in the final case, $ab \in \tilde{D}$. A set $\tilde{D}$ of root involutions is a set of *odd transpositions* when ab is never of order 4.

Timmesfeld has classified the groups generated by root involutions in [218]. This makes 4.1(1) quite useful. The lemma is an easy consequence of the following elementary observation about TI-sets first made by Timmesfeld. (cf. Proposition 4.5 of Chapter 6.)

LEMMA 4.2. *Let $g \in G - N(V)$ and assume that $U = V \cap N(V^g) \neq 1$. Set $X = \langle V, V^g \rangle$, $Y = O_2(X)$ and $\bar{X} = X/Y$. Assume that $[V, V^g] \neq 1$. Then the following hold.*

(1) $Y = U \times U^g$.
(2) $\bar{X} \cong L_2(2^m)$, $Sz(2^m)$ *or* D_{2n}, *n odd.*
(3) $m(\bar{V}) = m(\bar{X})$.
(4) $U = C_Y(v)$ *for each* $v \in V - U$.
(5) *Y is the sum of natural modules for* $\bar{X}$.
(6) *vx has odd order for each* $v \in V - U$, $x \in V^g - U^g$.

If $\tilde{D}$ is a set of root involutions of X and $\tilde{D} = A \cup B$ is an X-invariant partition of $\tilde{D}$, then $[A,B] \subseteq O_2(X)$. Hence with little loss of generality one may take X transitive on $\tilde{D}$.

THEOREM 4.3. *Suppose that K is transitive on Ω, $J = O_2(K)$, and $O(K) = 1$. Then K is transitive on D and one of the following holds.*

(1) $\bar{K}/Z(\bar{K}) \cong Sp_m(2^n)$, $U_m(2^n)$, *or* $Sz(2^n)$, *$\bar{V}$ is a root subgroup of $\bar{K}$, and $\bar{D}$ is a set of odd transpositions of $\bar{K}$.*
(2) $\bar{K}/Z(\bar{K}) \cong L_m(2^n)$ *and $\bar{V}$ is the subgroup generated by all transvections with a fixed centre.*
(3) $\bar{K} \cong \hat{A}_6$ *and $\bar{V}$ is a 4-group.*
(4) $\bar{K}/Z(\bar{K}) \cong Sp_m(2^n)$, $U_m(2^n)$, $O_m^{\pm}(2^n)$, *or* $L_2(2^n) \wr S_m$ *with $\bar{V}$ of order 2.*
(5) *$\bar{V}$ has order 2 and* $\bar{K}/O(\bar{K}) \cong S_n$.
(6) *$\bar{V}$ has order 2 and* $\bar{K} \cong P\hat{O}_6^-(3)$.

(In (3) and (6), $\hat{X}$ denotes a 3-fold central extension of a group X.)

When $J/I \subseteq Z(K/I)$ the action of $K/C_K(I)$ on $I/C_I(K)$ seems to be the sum of natural modules, although I have not attempted to establish this in all cases. Hence the group K is more or less determined.

Let us now see how this result may be used to produce factorizations. Suppose that $V \subseteq H \subseteq G$ with $F^*(H) = O_2(H)$ and $m(H/O_2(H)) = k < m(V)$. Let $T \in Syl_2(H)$. We wish to show that $H = N_H(W)C_H(Z)$, $W = W_0(T,V)$ and $Z = C_k(T,V)$. Assume otherwise. Then evidently $W \nsubseteq O_2(H)$ so we may take $V \nsubseteq O_2(H)$. Also, without loss, $H = \langle W^H \rangle T$ and the largest normal subgroup H_0 of H contained in $N(W)$ is 2-closed. Let $P = W_k(O_2(H),V)$ and $H_1 = C_H(P)$. $m(V/V \cap O_2(H)) \leqslant k$, so $V \cap O_2(H) \subseteq P$

and hence $H_1 \subseteq N(V)$. Thus $Z \subseteq H_1 \subseteq H_0$, so $Z \subseteq C_k(O_2(H)) = Q$. Thus we may assume $[V,Q] \neq 1$. But $Q \subseteq C(V \cap O_2(H)) \subseteq M$, so $[Q,V] \subseteq Q \cap V \subseteq E = P \cap Q$. Now E is an abelian normal subgroup of H with $1 \neq V \cap E$ and $V \nsubseteq O_2(H)$, so we are in a position to apply Theorem 4.3.

Here is a different application. Suppose that $V \subseteq H \subseteq G$ with $F^*(H) = O_2(H)$ and $T \in Syl_2(N_H(V)) \subseteq Syl_2(H)$. Let $\mathscr{H}_0$ be the set of subgroups H_0 of H containing T with $V \subseteq O_2(H_0)$. As $F^*(H) = O_2(H)$ and $V \trianglelefteq T \in Syl_2(H)$, V intersects $E = Z(O_2(H))$ nontrivially. So if $H \notin \mathscr{H}_0$ then we are in a position to apply Theorem 4.3. Then we can show, under some additional weak hypotheses on H (e.g., $C_H(z) \subseteq N(V)$ for $z \in Z(T)^\#$ or $N_H(W_0(T)) \subseteq N(V)$), that $H = \langle N_H(V), N_H(W_0(T)) \rangle$. Thus with little loss we take $H \in \mathscr{H}_0$. Now if $U = V^g \subseteq H$ and $U \cap O_2(H) \neq 1$ then, as $V \subseteq O_2(H) \subseteq M$, Lemma 4.2 implies that $[U,V] = 1$. Thus unless $V^g \cap O_2(H) = 1$ for some $V^g \subseteq T$, we have $H = C_H(V)N_H(W_0(T))$. On the other hand, if $U \cap O_2(H) = 1$ then, as U is a TI-set, $C_E(U) = C_E(u)$ for each $u \in U^\#$, a strong restriction on the action of H on E, as we shall see in the next section. For example H cannot be solvable.

Theorem 4.3 can also be used to investigate questions concerning $GF(2)$-representations. We will see some of these applications in the next section.

5. *GF*(2)-Representations

As we have seen a number of times already, the analysis of groups of characteristic 2 type involves the study of various types of $GF(2)$-representations.

Let G be a group with $O_2(G) = 1$ and V a faithful $GF(2)$-module for G. Certain parameters associated with the representation are of interest. $m(G,V)$ is the minimum codimension $m(V/C_V(t))$ for the centralizer of an involution t on V. If G has odd order, set $m(G,V) = m(V)$. $\mathscr{A}_k(G,V)$ is the set of nonidentity elementary abelian 2-subgroups A of G for which $C_V(A) = C_V(B)$ for each $B \subseteq A$ with $m(A/B) < k$. $a(G,V)$ is the maximum k for which $\mathscr{A}_k(G,V)$ is nonempty. If G has odd order set $a(G,V) = 0$.

Representations in which $m(G,V)$ is small or $a(G,V)$ large cause trouble in much characteristic 2 type analysis, particularly in weak closure theory. Also of interest are representations corresponding to the failure of the Thompson factorization. There one produces a nonidentity elementary abelian 2-subgroup A of G with $|V : C_V(A)| \leqslant |A|$. In this case and in others, one also encounters 2-groups A with *quadratic action* on V; that is, $[V,A,A] = 0$.

This section discusses representations in which one or more of these unusual situations occur. We begin with some applications of Theorem 4.3. The following observation, whose proof is trivial, puts us in a position to apply Theorem 4.3 to the semidirect product GV.

LEMMA 5.1. *Suppose that* $C_V(G) = 0$, *let a be an involution in G*, $W = [V,a]$ *and* $Y \subseteq C_G(\langle a,W \rangle)$ *with* $G = \langle Y, Y^g \rangle$ *for* $g \in G - C(a)$. *Let H be the semi-direct product of G and V*, $A = C_G(W) \cap C_G(V/W)$, *and* $U = AW$. *Then U is an elementary abelian TI-set in H.*

LEMMA 5.2. *Let a be an involution in G, L a component of G and A a 4-subgroup of G containing a. Suppose that* $[V,a,A] = 0$ *and* $C_A(L) = 1$. *Then either*

(1) $a \in N(L)$, *or*

(2) $\langle A,L \rangle / O_2(\langle A,L \rangle) \cong O_4^+(2^n)$ *and* $[L^A, V]$ *is a sum of natural modules for* $\langle L^A \rangle \cong \Omega_4^+(2^n)$.

We shall sketch a proof of this lemma. Take $G = \langle L,A \rangle$, to be a counterexample, set $X = \langle L^A \rangle$, $Z = Z(X)$, $W = [V,a]$ and $K = C_X(a)^{(\infty)}$. Then $K = [K,A] \subseteq C(W)$. If a inverts $z \in Z^{\#}$, then $U = [V,z] = [U,a] \oplus [U,a]^z$ and, as K centralizes W and W^z, $[V,z,K] = 0$. But now $z \in X = \langle K^X \rangle \subseteq C([V,z])$, which is a contradiction. So $[Z,a] = 1$. Set $Y = AK$. Then the hypothesis of Lemma 5.1 is satisfied and we apply Theorem 4.3 to finish the proof.

The lemma says that quadratic 4-groups fix most components of G. It can also be used to establish an important property of groups in which the Thompson factorization fails.

Define $\mathscr{P}(G,V)$ to be the set of nonidentity elementary abelian 2-subgroups A of G with

$$|A| \,.\, |C_V(A)| \geqslant |B| \,.\, |C_V(B)|$$

for each subgroup B of A. If H is a group with $F^*(H) = O_2(H)$, $T \in Syl_2(H)$, and $Z = \Omega_1(Z(T))$, and the Thompson factorization $H = N_H(J_e(T))C_H(Z)$ *fails*, then there is $A \in \mathscr{A}_e(T)$ with $\bar{A} \neq 1$, where $V = \langle Z^H \rangle$ and $\bar{H} = H/C_H(V)$. Moreover, for each $A \in \mathscr{A}_e(T)$ with $\bar{A} \neq 1$, $\bar{A} \in \mathscr{P}(\bar{H}, V)$.

LEMMA 5.3. $\mathscr{P}(G,V)$ *fixes each component of G which is a* $\mathscr{K}$*-group.*

COROLLARY 5.4. *Let H be a* $\mathscr{K}$*-group with* $F^*(H) = O_2(H)$. *Let* $T \in Syl_2(H)$, $Z = \Omega_1(Z(T))$ *and* $V = \langle Z^H \rangle$. *Then* $J_e(T)$ *fixes each component of* $H/C_H(V)$.

We use Lemma 5.2 to prove that each minimal A of $\mathscr{P}(G,V)$ fixes each component L of G. For suppose L is not fixed by some $a \in A$. By the Thompson replacement theorem (e.g., 8.2.4 in [G]), and minimality of A, $[a,V,A] = 0$. Let $X = \langle A,L \rangle$ and $U = C_V(O_2(X))$. Then X/U is faithful on U and $A/C_A(U) \in \mathscr{P}(X/U, U)$, so we may take $U = V$ and $X = G$. Now by Lemma 5.2, either $G \cong O_4^+(2^n)$ and V is the sum of natural modules for G or $A = \langle a \rangle$. In either case it is easy to check that $A \notin \mathscr{P}(G,V)$.

To determine representations in which $m(G,V)$ is small, one must first consider the situation where $F^*(G) = L$ is quasisimple and irreducible on V.

The worst cases arise when L is a group of Lie type of even characteristic or an alternating group. Mason and Cooperstein have complete results in the situation where G is the homomorphic image of a universal group of Lie type of even characteristic and V is irreducible [45]. Using their results it is possible to determine the modules V for $\mathscr{K}$-groups G with $m(G,V) \leqslant 3$.

THEOREM 5.5. *Assume that* $m(G,V) \leqslant 3$, *G is a $\mathscr{K}$-group, and*

(a) *G is faithful on no proper submodule of V, and*

(b) *if $V = V_1 + V_2$ is a G-invariant decomposition then $V = V_j$ for $j = 1$ or 2.*

Then there is a subgroup L of G such that, if $W = [L,V]$, then $V = \langle W^G \rangle$, $[LW, L^g] = 1$ for $g \in G - N(L)$, and one of the following holds.

(1) *$L \cong Z_3$ and $m(W) = 2, 4$, or 6.*

(2) *$L \cong Z_5, 3^{1+2}, E_9$, or Z_7 and $m(W) = 4, 6, 8$, or 3, respectively.*

(3) *$L \cong E_{p^r}$, $m(W) = s(r+1)$, and W is the direct sum of $r+1$ subspaces of rank s permuted by $N(W)$. Either $p = 3$ and $s = 2$ or $p = 7$ and $s = 3$.*

(4) *$L \cong KK^g$, $K \cong L_3(2)$ is a component of G, and W is the tensor product of natural modules for K and K^g.*

(5) *$L \cong SL_n(2), Sp_n(2), \Omega_n^{\pm}(2)$, or A_n is a component of G and $W/C_W(L)$ is the direct sum of at most 3 natural modules conjugate under $N_G(L)$.*

(6) *$W \in Irr(L,V)$ for some component L of G and $m(Aut_G(W),W) \leqslant 3$.*

Here $Irr(X,V)$ is the collection of subspaces $I = [I,X]$ of V with X irreducible on $I/C_I(X)$. The possibilities in (6) are enumerated in the next result. Also, E_{p^r} denotes an elementary abelian group of order p^r and 3^{1+2} a nonabelian group of order 27 and exponent 3.

THEOREM 5.6. *Let $F^*(G) = L$ be quasisimple and irreducible on V with $L/Z(L) \in \mathscr{K}$ and $m(G,V) \leqslant 3$. Then one of the following holds.*

(1) *$L \cong SL_n(q), Sp_n(q), \Omega_n^{\pm}(q)$, $q = 2^m$, $m \leqslant 3$; V is the natural module.*

(2) *$L \cong SU_n(2)$, A_n or $G_2(2)$; V is natural.*

(3) *$L \cong \hat{A}_6$, $\hat{U}_4(3)$, A_7, $Sp_6(2)$, $L_5(2)$, M_{22}, $L_3(4)$, or $L_2(7)$ with $m(V) = 6$, 12, 4, 8, 10, 10, 9, or 8, respectively. ($\hat{}$ denotes a central extension by Z_3.)*

Representations in which $a(G,V)$ is large are related to those discussed above. I will mention only one result which gives the flavour of the situation.

LEMMA 5.7. *Let G be solvable. Then $a(G,V) \leqslant 1$.*

For suppose $A \in \mathscr{A}_2(G,V)$ and set $U = C_V(A)$. Then $U = C_V(B)$ for each hyperplane B of A. Now $X = [F(G),A] \neq 1$ and

$$X = \langle [C_X(B),A] \,|\, |A:B| = 2 \rangle$$

while $[C_X(B),A] \subseteq [N(U),A] \subseteq C(U)$. So $[X,U] = 0$, whereas $[X,V] \neq 0$ so that

$$0 \neq [X,V] \cap U \subseteq [X,V] \cap C_V(X) = 0.$$

To close this section I record a lemma which will be used in the next section.

LEMMA 5.8. *Let X be a nontrivial cyclic subgroup of G regular on $[V,X]^{\#}$ and normal in $N_G(C_V(X))$, with $m(G,V) > m([V,X])$. Then $L = \langle X^H \rangle$ is the direct product of the conjugates of a subgroup Y and $[L,V]$ is the direct sum of the conjugates of $[Y,V]$. Either $Y = X$ or $X \subseteq Y$, X has order 3 and Y has order 21.*

6. 2-Reduced Weak Closures

Let G be a group of characteristic 2 type, $M \in \mathcal{M}$, and $V \in \mathcal{V}(M)$. Define

$$r(M,V) = \min \{r \mid U \subseteq V, C_G(U) \nsubseteq M, m(V/U) = r\}$$

and

$$s(M,V) = \min \{r(M,V), \quad m(M/C_M(V), V)\}.$$

V is said to be *2-reduced* if $O_2(M/C_M(V)) = 1$. In this situation the following observation makes weak closure relatively easy.

LEMMA 6.1. *Suppose that $B \subseteq A \subseteq V$, E is an A-invariant 2-subgroup of G, and $m(V/B) < s(M,V)$. Then $C_E(A) = C_E(B)$.*

For, as $m(V/B) < s(M,V)$, $O^{2'}(C_G(B)) \subseteq C(V)$, and the lemma holds.

Let us consider an example. Suppose that $H \subseteq G$ with $F^*(H) = O_2(H)$, $T \in Syl_2(H)$, $Z = \Omega_1(Z(T))$, $E = \langle Z^H \rangle$, and $\bar{H} = H/C_H(E)$. Assume that $a(\bar{H},E) < s(M,V)$. We will show $H = N_H(W_0(T,V))C_H(Z)$. If not, $W_0(T,V)$ does not centralize E so we may take $V \subseteq T$, $\bar{V} \neq 1$. However if $B \subseteq V$ with $m(V/B) < k = s(M,V)$ then, by Lemma 6.1, $C_E(V) = C_E(B)$, so $\bar{V} \in \mathcal{A}_k(\bar{H},E)$. However, by hypothesis, $k > a(\bar{H},E)$, so $\mathcal{A}_k(\bar{H},E)$ is empty, a contradiction. In particular with Lemma 5.7 we see that if H is solvable and $s(M,V) > 1$, then $H = N_H(W_0(T))C_H(Z)$.

The following lemma is a more definitive result along the same lines.

LEMMA 6.2. *Suppose that $H \subseteq G$ with $F^*(H) = O_2(H)$, $T \in Syl_2(H)$, $H \in \mathcal{E}_i$, and $s(M,V) > i+j$. Then*

$$H = \langle N_H(W_j(T)), C_H(C_{i+j}(T))\rangle.$$

To be able to apply results of this type we must be able to show that $s(M,V)$ is not too small. This seems to be feasible in the following situation.

Hypothesis 6.3. V is 2-reduced and $\mathcal{M}(N_M(Q)) = \{M\}$ for $Q \in Syl_2(C_M(V))$.

As V is 2-reduced, Theorems 5.5 and 5.6 give good control where $m(M/C_M(V),V)$ is small. Thus it remains to show that $r(M,V)$ is large.

THEOREM 6.4. *Assume Hypothesis* 6.3 *with* $m(M/C_M(V),V) = m > 2$. *Then* $r(M,V) \geqslant m$.

Theorem 6.4 is proved using Lemma 5.8. Here is an idea of that proof. Let $U \subseteq V$ with $H = C_G(U) \nsubseteq M$ and $m(V/U) = r(M,V) < m(M/C_M(V),V)$. As $m(V/U) < m(M/C_M(V),V)$, a Sylow 2-group Q of $C_M(V)$ is a Sylow subgroup of $C_M(U)$. By Hypothesis 6.3, $C(G,Q) \subseteq M$, so that $Q \in Syl_2(H)$ and $C(H,Q) \subseteq H \cap M \neq H$. Hence, by Theorem 2.1, H has a block L with $L \in \mathscr{Y}$. An easy reduction shows that $L \trianglelefteq H$ and $L/O_2(L) \cong L_2(2^n)$, $n > 1$. Moreover there is a subgroup X of order $2^n - 1$ in $L \cap M$ with $V = U \oplus [V,X]$ and X regular on $[V,X]^\#$. As $L \trianglelefteq H$, $X(Q \cap L) \trianglelefteq C_M(U)$. Thus we may apply Lemma 5.8 to the action of $M/C_M(V)$ on V to determine the embedding of X in M. From here on the proof is relatively easy.

As a corollary to Lemmas 5.7 and 6.2, and Theorem 6.4 we have the following.

LEMMA 6.5. *Assume Hypothesis* 6.3 *with* $m(M/C_M(V),V) > 2$. *Let* $H \subseteq G$ *with* H *solvable and* $F^*(H) = O_2(H)$. *Then* $H = N_H(W_i(T))C_H(C_{i+1}(T))$ *for* $i = 0$ *and* 1.

In particular, if all 2-local subgroups of G are solvable, Hypothesis 6.3 holds, $m(M/C_M(V),V) > 2$, and we are in the Uniqueness Case, then the example discussed in Section 3 shows how to derive a contradiction.

7. The Uniqueness Case

I will now outline how to derive a contradiction in the generic subcase of the Uniqueness Case, where the 2-local 3-rank is high.

Assume then that G is in the Uniqueness Case with $m_{2,3}(G) > 3$. Let $M \in \mathscr{M}$ be strongly 3-embedded in G and $T \in Syl_2(M)$. I recall some facts from Gorenstein's discussion in Chapter 4.

(7.1) $N_G(T) \subseteq M$.

(7.2) If p is an odd prime and X is a noncyclic p-subgroup of G with $\Gamma_{1,X}(G) \subseteq M$, then $\mathscr{M}(X) = \{M\}$ and any maximal subgroup N of G with $X \subseteq N$ and $O_2(M \cap N) \neq 1$ is contained in M.

In particular (7.2) says that each noncyclic 3-subgroup of G is contained in a unique maximal 2-local. The first step is to extend this collection to an even larger collection of uniqueness subgroups which is large enough to carry out the necessary weak closure arguments. I will not mention details here.

We assume that groups with a proper characteristic core are classified so that T is contained in subgroups H of G with $F^*(H) = O_2(H)$ and H not contained in M. By (7.1) and (7.2), H has cyclic Sylow 3-groups so that $F^*(H/O_{3'}(H))$ is isomorphic to $L_2(q)$, $L_3(q)$, $U_3(q)$, J_1, or Z_{3^n}.

We next show that $H \notin \mathscr{E}_1$, hence forcing $F^*(H/O_2(H)) \cong L_2(2^n)$ or $L_3(2^n)$, $n > 1$ odd. This is done using weak closures. We first show that $\mathscr{V}(M)$ is sufficiently large as outlined in Section 3. We then pick $V \in \mathscr{V}(M)$ so that one of the following holds.

(7.3) V is a TI-set with $m(V) > 2$.

(7.4) $O_2(M/C_M(V)) = 1$ and $\mathscr{M}(N_M(Q)) = \{M\}$ for $Q \in Syl_2(C_M(V))$.

(7.5) $1 \neq V_0 \subseteq V$, V_0 is a TI-set, $m(V_0) \leqslant 2$, $V = \langle v^M \rangle$ for each $v \in V - V_0$, and V is not a TI-set.

This procedure was also discussed in Section 3. In (7.4) set $\tilde{V} = V$ and in (7.5) set $\tilde{V} = V/V_0$. To apply weak closure methods in these cases we must show that $m(M/C_M(\tilde{V}), \tilde{V}) > 2$. The necessity for this condition in (7.4) was discussed in Section 6. If $m(M/C_M(\tilde{V})) \leqslant 2$ then Theorems 5.5 and 5.6 give us very strong control over M and its action on V. Indeed, usually $m_3(M/C_M(v)) > 1$ for each $v \in V^\#$, so by (7.2) V is a TI-set in G. Hence $W = W_0(T,V)$ centralizes V by Lemma 4.2, so by (7.4) and a Frattini argument, $N_G(W) \subseteq M$. But then several of the observations in Section 4 complete the proof that $H \subseteq M$ when $H \in \mathscr{E}_1$. The interested reader can search out the pertinent observations.

This difficult case aside, we are able to establish the factorizations, for $i = 0$ and 1,

(7.6) $H = \langle C_H(C_{i+1}(T)), N_H(W_i(T)) \rangle$ when $H \in \mathscr{E}_1$,

and even

(7.7) $H = C_H(C_{i+1}(T)) N_H(W_i(T))$ when H is solvable.

As the example in Section 3 indicates, it now remains to show that two of the subgroups

$$G_1 = N_G(W_0(T)), \quad G_2 = N_G(Z(W_1(T)), \quad \text{and} \quad G_3 = C_G(C_2(T))$$

are contained in M. This is usually accomplished by showing that T permutes with a p-subgroup X of order at least p^3 and exponent p with $\Gamma_{1,X}(G) \subseteq M$. Then (7.2) and (7.7) complete the proof, again as indicated in Section 3.

We now have $T \subseteq H \subseteq G$ with $H \nsubseteq M$ and $H/O_{3'}(H) \cong L_2(2^n)$ or $L_3(2^n)$, $n > 1$ odd. In particular, if p is a prime divisor of $2^n - 1$ there is a subgroup B of $N_H(T)$ and a 2-element of $H - M$ inverting B. Hence $N_G(B) \nsubseteq M$ while, by (7.1), $B \subseteq N_G(T) = N_M(T)$.

We now analyse the possible embeddings of B in M. As $N_G(B) \nsubseteq M$, $N_M(B)$ contains no subgroups contained in a unique maximal subgroup of M. If $m_3(N_M(B)) > 1$ then (7.2) says that $O_2(M) \cap C(B) = 1$, and indeed there are many more restrictions which tend to force $m_3(N_M(B)) \leqslant 1$. The condition that B acts on a Sylow 2-group T of M is quite restrictive in itself. The condition that p divides $2^n - 1$ for n *odd* is very useful.

In any case the end result is that M and the embedding of B in M are extremely restricted. We have already restricted the structure of 2-locals containing T but not contained in M to an even greater degree. With this much information it is now possible to study weak closures further and complete the proof. In particular it is not hard to show that $C_G(Z) \subseteq M$ where $Z = \Omega_1(Z(T))$. Now choose $H \in \mathcal{M}(T) - \{M\}$ and let $L = O^{3'}(H)$. As all solvable subgroups of L permuting with T are in M, it is easy to show that $L/O_2(L)$ is quasisimple and then, if L is maximal in a suitable sense, that $\mathcal{M}(LT) = \{H\}$. We now consider $V = \langle Z^H \rangle$ and observe that it satisfies

(7.8) $O_2(H/C_H(V)) = 1$ and $\mathcal{M}(N_H(P)) = \{H\}$ for $P \in Syl_2(C_H(V))$.

Thus Hypothesis 6.3 is satisfied with an appropriate change of notation. The results in Section 6 are then applicable and allow us to carry out the necessary weak closure arguments.

This then is a rather brief outline of the generic subcase of the Uniqueness Case. Necessarily many details have been omitted. Indeed a preliminary manuscript dealing with this subcase is some 200 typewritten pages. This seems an excessive length to devote to the empty set. Still, no easier approach yet suggests itself.

6

Groups of $GF(2)$-type and Related Problems

FRANZ TIMMESFELD

1. Introduction

In general classification problems, as for example Thompson's work on N-groups or Aschbacher's work on thin groups, one argues, after the suitable uniqueness theorems are established, on abelian normal subgroups of some maximal 2-local subgroup M. The same type of argument, the so called weak closure arguments, seems to be essential for the classification of groups with a strongly p-embedded 2-local subgroup or with $e(G) = 2$, which is in progress at the moment and which is an important step in the classification of finite simple groups. (See Chapters 5 and 7.) One always has the special case in which all abelian normal subgroups of M are cyclic, that is $F^*(M)$ is a 2-group of symplectic type. Moreover, the weak closure arguments seem to be more difficult if such a uniqueness subgroup of M acts trivially on a chosen abelian normal subgroup V of M. In this case, as is easily seen, V is a TI-set in G; that is, $V \cap V^g = V$ or 1 for all $g \in G$.

In this chapter we consider the following hypotheses which describe the above mentioned problems.

Hypothesis A. G is a finite group which contains an involution z such that $Q = F^*(C(z))$ is a 2-group of symplectic type.

Hypothesis B. G is a finite group with $O_2(G) = 1$, M is a maximal 2-local subgroup satisfying $F^*(M) = O_2(M)$ and some abelian normal subgroup B of M is a TI-set in G.

On several occasions we will consider TI-sets in general. But for the major theorems under Hypothesis B one has to assume that B is a maximal abelian normal subgroup of M.

A special case of Hypothesis A occurs as an important subcase in the work of Gorenstein and Lyons, namely the so called groups of $GF(2)$-type. (See Chapter 4.) It should also be mentioned that most of the sporadic groups satisfy Hypothesis A, so that the classification of groups satisfying this hypothesis gives some insight into how these sporadic groups arise. Actually several of the sporadic groups have been found by considering a special centralizer which satisfies Hypothesis A.

The organization of this chapter is as follows: Section 2 contains notation and preliminary results, Section 3 the statement of the theorems, Section 4 some lemmas on TI-subgroups and Section 5 some lemmas on Hypothesis A. In Section 6 the case where Q is of symplectic type but not extraspecial is treated, and in Section 7 the case in which Q is extraspecial and z is weakly closed in Q. In Section 8 we introduce an important tool which makes it possible to reduce the general Hypothesis A to several specific centralizer problems. The final identification of groups satisfying Hypothesis A is explained in Section 9. Finally in Section 10 we sketch some of the proofs of the theorems on TI-subgroups.

2. Notation and Preliminary Results

A 2-group Q is of *symplectic type* if all abelian characteristic subgroups of Q are cyclic. Q is *extraspecial* if $\Phi(Q) = Z(Q)$ is cyclic of order 2.

The connection between a 2-group of symplectic type and an extraspecial 2-group is given by a theorem of P. Hall.

THEOREM 2.1. *Suppose that G is a 2-group of symplectic type. Then $G = P \circ Q$ (i.e., G is the central product of P and Q), where Q is extraspecial and P is cyclic, dihedral, quasidihedral or quaternion, or one or both factors is trivial.*

See [H: III(13.10)]. For the definition of *central product* see [H: I(9.10)].

A subgroup A of a group G is a *TI-subgroup* of G if $A \cap A^g = A$ or 1 for all $g \in G$. This notation is closely connected with Aschbacher's notation of *tightly embedded* subgroups. Subgroups of prime order are of course always TI-subgroups. So in this case the notation is redundant.

The following trivial lemma gives a connection between TI-subgroups and Hypothesis A.

LEMMA 2.2. *Suppose Hypothesis A holds. If B is an abelian characteristic subgroup of Q, then B is a TI-subgroup of G.*

Proof. Let $g \in G$ such that $B \cap B^g \neq 1$. Then $\langle z \rangle = \Omega_1(B) = \Omega_1(B^g)$, so $g \in C(z)$ and $B = B^g$.

Actually this is the main tool of Aschbacher's treatment of the part of Hypothesis A where Q is not extraspecial. In the generic case, where Q is extraspecial, (2.2) is of course useless, since in this case $|B| = 2$ for every such B.

Before we prove a further general lemma, which is useful when one considers Hypothesis A, we state a well known lemma about the action of an involution on an elementary abelian 2-group.

LEMMA 2.3. *Suppose A is an elementary abelian 2-subgroup of G and $t \in N(A)$ such that $t^2 \in C(A)$. Then the following hold.*

(1) *The map $a \to [t,a]$, $a \in A$, is $\mathbf{F}_2$-linear.*
(2) *$[A,t] \subseteq C_A(t)$ and $|A:C_A(t)| \leqslant |C_A(t)|$.*
(3) *$|[A,t]| = |A:C_A(t)|$.*
(4) *$t^A = t[t,A]$.*
(5) *$[A,t] = C_A(t)$ if and only if $|A:C_A(t)| = |C_A(t)|$.*

Proof. (1) and (3) are trivial. (2) follows from the fact that the map $a \to [t,a]$ is $(t-1)$ in $End(A)$. (4) is trivial and (5) a consequence of (2) and (3).

LEMMA 2.4. *Suppose Hypothesis A holds and Q is of exponent 4. If $z \neq z^g \in Q$, then $z \in Q^g$ and $\langle Q,Q^g \rangle$ induces $L_2(2)$ on $\langle z,z^g \rangle$.*

Proof. Let $t = z^g$ and $W = \langle z,t \rangle$. By (2.1) the hypothesis implies that $[W,Q] = \langle z \rangle$ and $[W,Q^g] = \langle t \rangle$ if $W \subseteq Q^g$. So it suffices to show that $z \sim t$ in $N(W)$.

Let $S \in Syl_2(C(t) \cap C(z))$. Since $C_Q(t) \trianglelefteq C(t) \cap C(z)$, $C_Q(t) \subseteq S$. Hence $|SQ:S| = 2$ and so, if $Z = \Omega_1(Z(S))$, (2.3)(2) implies that $|C_Z(Q)| \geqslant 4$ if $|Z| \geqslant 8$. This is a contradiction since $\Omega_1(Z(C(Q))) = \langle z \rangle$ as $Q = F^*(C(z))$.

So $Z = W$. Since $N_T(S) \supset S$ for $T \in Syl_2(C(t))$ containing S, we have $z \sim zt$ in $N_T(W)$. Since $t \sim zt$ in $N_Q(W)$ this proves (2.4).

(2.5) A 2-group Q of symplectic type and exponent 4 is equipped naturally with the structure of a symplectic space over $GF(2)$ by the following definition:

If $\tilde{Q} = Q/Z(Q)$ and $\langle z \rangle = \Omega_1(Z(Q))$, then

$$(\tilde{a},\tilde{b}) = 1 \quad \text{if and only if} \quad [a,b] = z, \quad a,b \in Q, \quad \text{and}$$
$$(\tilde{a},\tilde{b}) = 0 \quad \text{if and only if} \quad [a,b] = 1, \quad a,b \in Q.$$

If Q is extraspecial, then $Z(Q) = \langle z \rangle$. Hence one may define in addition a quadratic form on $\tilde{Q}$ by

$$q(\tilde{a}) = 1 \quad \text{if and only if} \quad a^2 = z, \quad a \in Q, \quad \text{and}$$
$$q(\tilde{a}) = 0 \quad \text{if and only if} \quad a^2 = 1, \quad a \in Q.$$

By [H: III(13.7), (13.8)] (,) is a symmetric scalar product and q is a quadratic form over $GF(2)$. Furthermore, $\tilde{Q}$ is a nondegenerate symplectic or orthogonal space over $GF(2)$ respectively with this notation. It is obvious that automorphisms of Q act as symplectic or orthogonal transformations on $\tilde{Q}$. Hence, if Q is of exponent 4 under Hypothesis A, one may identify $C(z)/Q$ with a subgroup of the symplectic or orthogonal groups on $\tilde{Q}$ respectively.

Let V be a nondegenerate symplectic space over $GF(2)$, t an involution in $Sp(V)$ and set

$$V(t) = \{v \in V \mid (v,v^t) = 0\}$$

and

$$m = \dim([V,t]).$$

We say that t is of *type* a_m if $V(t) = V$, and that t is of *type* b_m or *type* c_m if *codim* $V(t) = 1$ and m is odd or m is even respectively.

By Sections 7 and 8 of [21] we have the following.

LEMMA 2.6. *The following hold.*

(1) *Each involution of $Sp(V)$ is of type a_m, b_m or c_m for some $m \leqslant \frac{1}{2}\dim(V)$.*
(2) *Involutions of the same type are conjugate in $Sp(V)$.*
(3) $[V,t]^{\perp} = C_V(t)$ *for each involution $t \in Sp(V)$.*

LEMMA 2.7. *Suppose V is a nondegenerate orthogonal space over $GF(2)$. Then the following hold.*

(1) *An involution $t \in O(V)$ is of type a_m if and only if $[V,t]$ is totally singular. Further $m \equiv 0(2)$ in this case.*
(2) *If t is of type b_m or c_m then $[V,t] = \langle v\rangle \perp W$, $q(v) = 1$ and W is totally singular.*
(3) *Involutions of the same type are already conjugate in $O(V)$.*

We shall require one further important lemma about the action of involutions on an extraspecial 2-group.

LEMMA 2.8. *Suppose that Q is an extraspecial 2-subgroup of G and $t \in N(Q)$ an involution satisfying $|Q : C_Q(t)| \leqslant 2$. Then $t \in QC(Q)$.*

Proof. Assume false. Let $\langle z\rangle = Z(Q)$ and $Z = Z(C_Q(t))$. Then $[Q,t] \neq 1$ and if $[Q,t] = \langle z\rangle$ then there exists some $x \in Z$ such that $[Q,xt] = 1$, a contradiction.

So $[\tilde{Q},t] = \langle\tilde{v}\rangle \neq 1$, where $\tilde{Q} = Q/\langle z\rangle$. Use the notation of (2.5). By (2.6) (3) we have

$$\widetilde{C_Q(v)} = \tilde{v}^{\perp} = C_{\tilde{Q}}(t) = \widetilde{C_Q(t)},$$

since $|Q : C_Q(t)| = 2$. In particular, $v \in C_Q(t)$. Choose $w \in Q$ such that $v = [w,t]$. Then $v^2 = [w,t]^2 = [w,t^2] = 1$. So $q(\tilde{v}) = 0$. But on the other hand we have

$$q(\tilde{w}) = q(\tilde{w}^t) = q(\tilde{v}+\tilde{w}) = q(\tilde{v})+q(\tilde{w})+(\tilde{v},\tilde{w}) = q(\tilde{v})+q(\tilde{w})+1.$$

Hence $q(\tilde{v}) = 1$, a contradiction. This proves (2.8).

(2.9) Suppose that D is a set of involutions generating a group G such that the following hold:

(1) $D = D^G$ (D is closed under conjugation),
(2) if $d,e \in D$ then $o(de)$ is 2,4 or odd, and
(3) if $o(de) = 4$ for $d,e \in D$, then $(de)^2 \in D$.

Then D is a *set of root involutions* of G. Groups generated by root involutions were classified in [218]. They are essentially, if $O_2(G) = 1$, direct products of groups of Lie type in characteristic 2. If possibility (3) never occurs, then D is a set of *odd transpositions* of G or a set of *degenerate* root involutions. Groups generated by odd transpositions were classified in [8].

There is an important connection between root involutions and TI-subgroups, which will be explained in Section 4. (See also Chapter 5.)

A set D of root involutions of G is a set of $\{3,4\}^+$-*transpositions*, if $o(de) \leqslant 4$ for all $d,e \in D$. It is a set of 3-*transpositions*, if $o(de) \leqslant 3$. Groups generated by $\{3,4\}^+$-transpositions were classified in [67] and [217]. They will be useful when considering Hypothesis A.

We have

LEMMA 2.10. *Suppose that D is a set of root involutions of G and* $O_2(G) = 1$. *Then*

$$D = \bigcup_{i=1}^{n} D_i, \quad D_i \cap D_j = \emptyset \quad \textit{for} \quad i \neq j$$

such that

$$G = \prod_{i=1}^{n} G_i, \quad \textit{where} \quad G_i = \langle D_i \rangle, \quad [G_i, G_j] = 1$$

for $i \neq j$ *and each* D_i *is a conjugacy class of root involutions of* G_i.

Proof. See [218: (4.1.5), (4.1.6)].

LEMMA 2.11. *Let V be a nondegenerate orthogonal space over* $GF(2)$, $G \subseteq O(V)$ *and D the set of involutions of type* a_2 *in G. Then D is a set of* $\{3,4\}^+$-*transpositions of* $\langle D \rangle$.

Proof. This follows from the fact that both the involutions of type a_2 and the $\{3,4\}^+$-transpositions in $\Omega(V)$ are the unique class of involutions central in a Sylow 2-subgroup of $\Omega(V)$.

Suppose G satisfies Hypothesis A and Q is extraspecial. Let $M = C(z)$, $\bar{M} = M/Q$ and $\tilde{Q} = Q/\langle z \rangle$. If $\bar{t} \in \bar{M}$ acts as an involution of type a_2 on $\tilde{Q}$, then $\bar{D} = \bar{t}^{\bar{M}}$ is a set of $\{3,4\}^+$-transpositions of $\bar{M}_0 = \langle \bar{D} \rangle$. Since $O_2(\bar{M}) = 1$ the Hypothesis of (2.10) is satisfied. Hence $\bar{M}_0$ is a central product of groups generated by a conjugacy class of $\{3,4\}^+$-transpositions, which are by the classification in [67] and [217] essentially groups of Lie type over $GF(2)$. This shows that, if involutions of type a_2 exist, $\{3,4\}^+$-transpositions are an important tool in the classification of groups satisfying Hypothesis A.

We end this section by stating the following trivial and well known lemma, which we will use on many occasions.

LEMMA 2.12. *Suppose the p-groups A and B are TI-subgroups of G and satisfy*

$$(1)\ |A| \leqslant |B|, \quad (2)\ A \cap B = 1, \quad \textit{and} \quad (3)\ B \subseteq N(A).$$

Then $[A,B] = 1$.

Proof. If $AB = P = N_P(B)$ then $[A,B] \subseteq A \cap B = 1$. So assume $P \supset N_P(B)$ and pick $x \in N_P(N_P(B)) - N_P(B)$. Then $B \cap B^x \neq 1$ since $|P| \leqslant |B|^2$, and so $|N_P(B)| < |B|^2$. But then $B = B^x$, contradicting the choice of x.

3. Statement of Results

First we state the results on Hypothesis A, which give a complete characterization of finite simple groups satisfying Hypothesis A.

THEOREM 3.1. *Suppose that G satisfies Hypothesis A. Then one of the following holds.*

(1) $z \in Z^*(G)$.
(2) *G is isomorphic to* $L_2(2^n \pm 1)$, $PGL_2(2^n \pm 1)$ *or G is the extension of* $L_2(9)$ *by* Z_2 *with quasidihedral Sylow 2-subgroups.*
(3) *G is isomorphic to* M_{11}, $L_3(3)$, $U_3(3)$ *or G is* $L_3(3)$ *extended by a graph automorphism.*
(4) *G is isomorphic to* $U_4(3)$ *extended by an automorphism group of order* 2 *or* 4 *or to the Higman–Sims group HS.*
(5) *Q is extraspecial.*

This is Theorem 1 of [13]. The proof will be discussed in Section 6.

THEOREM 3.2. *Suppose that G satisfies Hypothesis A with Q extraspecial and* $|Q| \leqslant 2^5$. *Suppose further that* $L = F^*(G)$ *is simple. Then L is isomorphic to* $U_3(3)$, $U_4(2)$, $L_4(3)$, $U_4(3)$, $G_2(3)$, A_8, A_9, M_{12}, J_2 *or* J_3.

This is a consequence of the sectional 2-rank 4 paper of Gorenstein and Harada [99]. (See [14: (4.6)].)

THEOREM 3.3. *Suppose that G satisfies Hypothesis A with Q extraspecial and* $|Q| \geqslant 2^5$. *Then one of the following holds.*

(1) $z \in Z(G)$.
(2) $\langle z^G \rangle = F^*(G)$ *is isomorphic to* $U_m(2)$, $m \geqslant 4$, *or to* Co_2, *the second Conway group, or to* $L_4(3)$.
(3) *z is not weakly closed in Q.*

Aschbacher proved the above theorem in [14] under the additional hypothesis that $F^*(C(t))$ is a 2-group for each involution $t \in Q$. F. Smith showed in [180: (1.3)] that this hypothesis is superfluous. A proof of (3.3) is discussed in Section 7.

THEOREM 3.4. *Suppose that G is a simple group satisfying Hypothesis A, such that Q is extraspecial with $|Q| = 2^{2n+1}$, $n \geqslant 3$. Then one of the following holds.*
(1) *$G \cong L_{n+2}(2)$, or $n = 3$ and $G \cong M_{24}$ or He, the Held group.*
(2) *$C(z)$ acts irreducibly on $Q/\langle z\rangle$.*

This result is due to Dempwolff and Wong and is the corollary in [48, II]. F. Smith has shown in [180, (1.2)] that $|Q| = 2^5$ if Q is the central product of two $C(z)$-invariant proper subgroups. So one may assume that $A \trianglelefteq C(z)$, $A \neq \langle z\rangle$ and $\Phi(A) = 1$. In [48, I] it is shown that $|A| = 2^{n+1}$ under this hypothesis. If z is not weakly closed in Q, it is easy to see that there are involutions of $C(z)$–Q which induce transvections on A. With this information Demwolff and Wong can first determine the structure of $C(z)$ and then G.

THEOREM 3.5. *Suppose that the finite simple group G satisfies Hypothesis A with Q extraspecial. Let $|Q| = 2^{2n+1}$, $M = C(z)$ and $\bar{M} = M/Q$. Then one of the following holds.*
(1) *$n \leqslant 4$.*
(2) *$G \cong L_{n+2}(2)$ or $U_{n+2}(2)$.*
(3) *$n = 2m$ and $\bar{M} \cong S_3 \times \Omega_{2m}^{\pm}(2)$ or $\bar{M} \cong S_3 \times O_{2m}^{\pm}(2)$.*
(4) *$n = 6$, $E(\bar{M})$ is quasisimple and $|O(\bar{M})| \leqslant 3$. Furthermore, $\bar{M}$ contains an involution $\bar{t}$ such that $\bar{R} \trianglelefteq C_{\bar{M}}(\bar{t})$ and*

$$\bar{R}/O_2(\bar{R}) \cong L_3(2), \quad \Omega_4^{+}(2) \quad or \quad \Omega_6^{\pm}(2)$$

where in the last case $O_2(\bar{R}) = \langle \bar{t}\rangle$ and in any case $\Phi(O_2(\bar{R})) \subseteq \langle \bar{t}\rangle$.
(5) *$n = 10$ and $F^*(\bar{M}) \cong L_6(2)$ or $U_6(2)$.*
(6) *$n = 11$ and $\bar{M} \cong Co_2$, the second Conway group.*
(7) *$n = 12$ and $\bar{M} \cong Co_1$, the first Conway group.*
(8) *$\bar{M}$ is simple and contains an involution $\bar{t}$ such that $\bar{N} = F^*(C_{\bar{M}}(\bar{t}))$ is extraspecial and one of the following holds:*
 (i) *$n = 16$, $|\bar{N}| = 2^{17}$ and $C_{\bar{M}}(\bar{t})/\bar{N} \cong S_3 \times \Omega_8^{+}(2)$, or*
 (ii) *$n = 28$, $|\bar{N}| = 2^{33}$ and $C_{\bar{M}}(\bar{t})/\bar{N} \cong \Omega_{12}^{+}(2)$.*

There is actually more information given in [221], which is difficult to state in a single theorem, but which is helpful in the final treatment of the separate cases of (3.5). The proof of (3.5) is discussed in Section 8, after we have proved some important lemmas in Section 5. The simplicity of G is only assumed for technical reasons and could be avoided by being more careful.

We now assume that the hypothesis of (3.5) is satisfied.

THEOREM 3.6. *Suppose that case* (3) *of* (3.5) *holds. Then $G \cong \Omega_{2m+4}^{\pm}(2)$.*

This is the main theorem of [182].

THEOREM 3.7. *Suppose that case* (5) *or* (8) *of* (3.5) *holds. Then G is isomorphic to* $E_6(2)$, ${}^2E_6(2)$, $E_7(2)$ *or* $E_8(2)$.

This is Theorem A of [183]. Case (5) of (3.5) was also treated by Reifart [167].

THEOREM 3.8. *Suppose that case* (6) *or* (7) *of* (3.5) *holds. Then* $G \cong F_2$ *or G is of type* F_1, *the baby monster or the monster.*

This is Theorem C of [184]. The proof is due to Reifart and Stroth. Existence and uniqueness problems are not touched.

THEOREM 3.9. *Suppose that case* (1) *of* (3.5) *holds and* $n \geqslant 3$. *Then*

(i) $n = 3$ *and* $G \cong L_5(2)$, $U_5(2)$, M_{24}, *He or Sz.*

(ii) $n = 4$ *and* $G \cong L_6(2)$, $U_6(2)$, $\Omega_8^{\pm}(2)$, ${}^3D_4(2)$, $\Omega_8^{+}(3)$, Co_1, Co_2, F_3 *or* F_5.

This is Theorem A of [184]. (Note: The original theorems refer to groups of *type* F_2 and F_5, before uniqueness was established.)

THEOREM 3.10. *Suppose that case* (4) *of* (3.5) *holds. Then* $G \cong L_8(2)$, $U_8(2)$, $\Omega_{10}^{\pm}(2)$ *or* $M(24)'$, *or G is of type* J_4.

This is Theorem B of [184].

The proof of (3.6)–(3.10) will be discussed in Section 9. Notice that (3.1)–(3.10) give a complete classification of simple groups satisfying Hypothesis A.

Next we state some results concerning Hypothesis B.

THEOREM 3.11. *Suppose that the elementary abelian 2-subgroup B of G is a TI-subgroup in G and B is weakly closed in $N(B)$ with respect to G. Let* $G^* = \langle B^G \rangle$. *Then G^* contains an elementary abelian normal subgroup N, such that one of the following holds:*

(1) $G^*/N = Z^*(G^*/N)$, *or*

(2) G^*/N *is a covering group of one of the following groups:* $L_n(2^m)$, $Sz(2^m)$, $U_3(2^m)$, A_6, A_7, A_8, A_9, M_{22}, M_{23} *or* M_{24}.

This is Corollary B of [219]. It was the first important theorem on TI-sets. The proof will be discussed in Section 10.

THEOREM 3.12. *Suppose that G satisfies Hypothesis B. Then one of the following holds.*

(1) *B is cyclic.*

(2) $B \cong Z_4 \times Z_4$ *and* $F^*(G) \cong L_3(4)$.

(3) *B is elementary abelian.*

This is Theorem 1 of [222]. In [192: (1.2)] it is shown that in case (1) either $|B| \leqslant 4$ or a Sylow 2-subgroup of $\langle \Omega_1(B)^G \rangle$ is dihedral or quasidihedral.

THEOREM 3.13. *Suppose that G satisfies Hypothesis B for some maximal abelian normal subgroup B of M. Let $Q = O_2(M)$. Then one of the following holds.*

(1) *Q is of symplectic type and B is cyclic.*
(2) *$B \cong Z_4 \times Z_4$ and $F^*(G) = \langle B^G \rangle \cong L_3(4)$.*
(3) *B is elementary abelian, $|B| > 2$, and one of the following holds:*
   (i) *B is weakly closed in Q, or*
   (ii) *$B = Z(Q) = \Phi(Q) = Q'$ and $Q = \langle B^g \mid B^g \subseteq Q \rangle$.*

This is Theorem 2 of [222]. The proof of (3.12), (3.13) will be discussed in Section 10. By (3.1)–(3.10) the classification of simple groups satisfying the hypothesis of (3.13) amounts to the treatment of case (3) for $|B| > 2$. This obviously splits in the two cases (i) and (ii). We first state the results relevant to the treatment of (i).

THEOREM 3.14. *Suppose that the hypothesis of* (3.13) *is satisfied. Let $V = \Omega_1(B)$. Then one of the following holds:*

(1) *V is weakly closed and therefore $\langle V^G \rangle$ is known by* (3.11), *or*
(2) *$V \subseteq O_2(H)$ for all subgroups H of G satisfying $C_Q(V) \subseteq H \nsubseteq M$ and $O_2(H) \neq 1$.*

This is the main theorem of [223]. The proof uses mainly root involution arguments. (See (4.7).)

THEOREM 3.15. *Suppose that G is a finite group of characteristic* 2-*type, which satisfies* (3)(i) *of* (3.13). *Then one of the following holds:*

(1) *B is weakly closed and $\langle B^G \rangle$ is known by* (3.11), *or*
(2) $\langle B^G \rangle = F^*(G) \cong U_n(q), \quad q = |B|, \quad n > 3.$

This is Corollary 2 of [224]. The hypothesis that G is of characteristic 2 type is only used to guarantee the existence of a subgroup H satisfying $O_2(H) \neq 1$ and $C_Q(B) \subseteq H \nsubseteq M$. The proof of (3.15) will be discussed in Section 10.

THEOREM 3.16. *Suppose that G satisfies* (3)(ii) *of* (3.13). *Then Q is isomorphic to the central product of Sylow* 2-*subgroups of $L_3(q)$, $q = |B|$.*

This is the main theorem of [225]. These 2-groups are strongly connected with extraspecial 2-groups of + type. By Theorem 4 of [25] the automorphism group A of Q is, modulo the normal subgroup

$$C = \{\alpha \in A \mid [Q,\alpha] \subseteq Q'\},$$

isomorphic to $O^+(2l,q)$, where l is the number of factors isomorphic to a Sylow 2-subgroup of $L_3(q)$.

THEOREM 3.17. *Suppose that G satisfies* (3)(ii) *of* (3.13) *and Q is a central product of Sylow* 2-*subgroups of $U_3(q)$ or $L_3(q)$, $q = |B|$. Let $A = B^g \subseteq Q$,*

$A \neq B$ and set $L_A = Q(Q^g \cap M)$, $\bar{M} = M/Q$ and $\bar{M}_0 = \langle \bar{L}_A^{\bar{M}} \rangle$. Then either (I) or (II) holds.

(I) *$\bar{L}_A$ is a TI-set in $\bar{M}$ and $\bar{M}_0 \cong L_2(q)$, $|Q| = q^5$, or $\bar{M}_0 \cong L_2(q^3)$, $|Q| = q^9$.*

(II) *$\bar{L}_A$ is not a TI-set in $\bar{M}$. For $\bar{t} \in \bar{L}_A^{\#}$ set*

$$\bar{R}_{\bar{t}} = \langle \bar{L}_A^{\bar{m}} \mid \bar{t} \in \bar{L}_A^{\bar{m}}, m \in M \rangle \quad \text{and} \quad \bar{N}_{\bar{t}} = O_2(\bar{R}_{\bar{t}}).$$

Then there exists $\bar{t} \in \bar{L}_A^{\#}$ such that one of the following holds:

(1) *$|\bar{N}_{\bar{t}}| = q$, $\bar{R}_{\bar{t}}/\bar{N}_{\bar{t}} \cong \Omega_4^{\pm}(q)$ and $|Q| = q^9$,*
(2) *$|\bar{N}_{\bar{t}}| = q$, $\bar{R}_{\bar{t}}/\bar{N}_{\bar{t}} \cong \Omega_{2n}^{\pm}(q)$, $n \geqslant 3$, $|Q| = q^{2n+1}$ and $\bar{M}_0 \cong L_2(q) \times \Omega_{2n}^{\pm}(q)$,*
(3) *$|Q| = q^{21}$ and $\bar{M}_0 \cong SL_6(q)$ or $U_6(q)$,*
(4) *$|Q| = q^{33}$, $\bar{R}_{\bar{t}}/\bar{N}_{\bar{t}} \cong \Omega_8^{+}(q)$ and $|\bar{N}_{\bar{t}}| = q^{17}$,*
(5) *$|Q| = q^{57}$, $\bar{R}_{\bar{t}}/\bar{N}_{\bar{t}} \cong \Omega_{12}^{+}(q)$ and $|\bar{N}_{\bar{t}}| = q^{33}$, or*
(6) *$\bar{M}_0 \cong SL_n(q)$, $Q/B \cong V_1 \oplus V_2$ where V_1 is the natural module for $SL_n(q)$ and V_2 its dual module. Furthermore, there is $x \in M$ such that $V_1^x = V_2$.*

It should be remarked that case (6) does not occur in a simple group. (3.17) is the main theorem of [203]. Stroth has also shown that in cases (4) and (5) $\bar{M}$ satisfies the hypothesis of (3.17), so that one can get $\bar{M}_0$ inductively. The methods of the proof are essentially the same as in the proof of (3.5). S. Smith is at the moment working on the characterization of G in the separate cases of (3.17), as he has done in the extraspecial case. ((3.6)–(3.10).) If he has finished it, then groups of characteristic 2 type satisfying Hypothesis B for some maximal abelian normal subgroup B of M are, by (3.11)–(3.17), completely classified. This gives a uniform characterization of most of the characteristic 2 type groups.

There are many other theorems concerning the extraspecial and similar problems. In particular, the work of Stroth in [202] should be mentioned.

4. Some Lemmas on *TI*-Subgroups

LEMMA 4.1. *Suppose that the 2-element t acts on the abelian 2-group A and $C_A(t) = C_A(t^2)$. Then $[A,t] = 1$.*

Proof. ([13: (2.1)] and [222: (2.1)]). Let $A_i = \Omega_i(A)$, $i = 0,1,\ldots$ We prove (4.1) by induction on i. Let $C_i = C_{A_i}(t)$. By inductive assumption $A_{i-1} \subseteq C_i \subseteq A_i$, $i \geqslant 1$. Assume $C_i \subset A_i$. Let $y \in A_i - C_i$ such that $z = [y,t] \in C_i$. Since $y^2 \in A_{i-1} \subseteq C_i$ we have

$$y^2 = (y^2)^t = (y^t)^2 = (yz)^2 = y^2 z^2.$$

So $z^2 = 1$. Hence $y^{t^2} = (yz)^t = yz^2 = y$ and so $y \in C_A(t^2) = C_A(t)$, contradicting the choice of y.

COROLLARY 4.2. *Suppose that the abelian 2-subgroup A is a TI-subgroup of G. Let $g \in G-N(A)$. Then one of the following holds:*

(1) *$N_A(A^g)$ is elementary abelian, or*

(2) *$[A,A^g] = 1$.*

Proof. Let $E = N_A(A^g)$ and $F = A^g \cap N(A)$. Then $[E,F] = 1$. Suppose $e \in E$ such that $e^2 \neq 1$. Then

$$A^g \cap C(e) = F = A^g \cap C(e^2)$$

whence, by (4.1), $[e,A^g] = 1$. But then $[A,A^g] = 1$ by (2.12).

The next lemma, [11: (2.5)], was motivated by a remark of Thompson.

LEMMA 4.3. *Let S be a 2-group and Δ an S-invariant collection of subgroups, satisfying*

(1) *$R \in \Delta$ implies $\Phi(R) \neq 1$,*

(2) *if $R,T \in \Delta$ then $|R| = |T|$, and*

(3) *if $R,T \in \Delta$ then either $R \cap T = 1$ or $R = T$.*

Then $[R,T] = 1$ for each pair $R,T \in \Delta$ such that $R \neq T$.

Proof. Let $\Gamma \subseteq \Delta$ be maximal subject to $[R,T] = 1$ for $R,T \in \Gamma$, $R \neq T$. We claim that $\Gamma = \Delta$. Suppose false. Pick $R \in \Gamma$ and $T \in \Delta$ such that $T \subseteq N(\Gamma)$.

Assume $N_T(R) \neq 1$ and pick an involution $x \in N_T(R)$ and $z \in Z(R) \cap C(x)^{\#}$. Then $z \in N(T)$ and $[z,T] \subseteq T \cap C(\Gamma)$, since $z \in C(\Gamma)$. If $[z,T] = 1$ then $T \subseteq N(R)$ and so, by (2.12), $[T,R] = 1$. But if $[z,T] \neq 1$, then $R \subseteq N(T)$ and so in any case $[R,T] = 1$.

Suppose now that $T \notin \Gamma$. Then there is an $R \in \Gamma$ such that $[T,R] \neq 1$ by maximality of Γ. Hence, by the above, $N_T(R) = 1$ and so $P \cap T = 1$ for $P = \langle \Gamma \rangle$. Let $Z = Z(P)$ and $x \in T$ such that $x^2 \neq 1$. Then

$$[C_Z(x^2),x] \subseteq T \cap Z = 1,$$

since $T \subseteq N(Z)$. Hence, by (4.1), $[Z,x] = 1$, contradicting $Z \cap R \neq 1$.

This shows there is no $T \in (\Delta \cap N(\Gamma))-\Gamma$. But then $N(N(P)) = N(P)$ and so $S = N(P)$ and $\Gamma = \Delta$.

COROLLARY 4.4. *Let X be a group in which $F^*(X) = O_2(X) = Q$ and let A be an abelian 2-subgroup of X which is a TI-subgroup. Then $B = \mho^1(A) \subseteq Q$ and, if B is not elementary abelian, then $\langle B^X \rangle$ is abelian.*

Proof. Assume false. Let $W = \langle A^X \cap S \rangle$, $S \in Syl_2(X)$. Then $Q \subseteq N(W)$ and, by (4.3), $W' = 1$. Hence

$$[Q,a,a] \subseteq [W,a] = 1$$

for each $a \in A$. This implies that $[x,a^2] = [x,a]^2$ for $x \in Q$ and so a^2 centralizes $Q/\Phi(Q)$. But then $a^2 \in Q$, since $Q = F^*(X)$. This shows that $B \subseteq Q$. The second part is a consequence of (4.3).

These lemmas show that one has much additional information if an abelian, but not elementary, 2-group is a TI-subgroup of G. The elementary abelian case is more difficult. The most important lemma applying to this case is the following ([220: (2.4)]).

PROPOSITION 4.5. *Suppose that* A *is a 2-subgroup of* G *and* $g \in G - N(A)$ *such that*

(a) $1 \neq L = A \cap N(A^g)$ *and* $\Phi(L) = 1$,

(b) $A \cap A^g = 1$ *and both are* TI*-subgroups of* $X = \langle A, A^g \rangle$, *and*

(c) $|A : L| \leqslant |L|$.

Then either $[A, A^g] = 1$ *or, if* $T = A^g \cap N(A)$ *and* $N = LT$, *the following hold.*

(1) $N \lhd X$, $\Phi(N) = 1$ *and* $\bar{X} = X/N \cong L_2(2^n)$, $Sz(2^n)$ *or* D_{2m}, *m odd.*

(2) $|A : L| = 2$ *if* $\bar{X} \cong D_{2m}$. $|A : L| = 2^n$ *otherwise.*

(3) N *is the direct sum of irreducible* $\mathbf{F}_2\bar{X}$*-modules which are all natural modules if* $\bar{X} \cong L_2(2^n)$ *or* $Sz(2^n)$.

(4) A *is abelian.*

Proof. We have $[L, T] \subseteq A^g \cap A = 1$. Suppose $[A, A^g] \neq 1$. Then by (2.12) $L \neq A$ and $T \neq A^g$. Let $a \in N_A(N) - L$. Then $C_T(a) = 1$ and $[T, a] \subseteq L$. This implies $|T| \leqslant |L|$. By symmetry we get $|T| = |L|$ and N elementary abelian, since $N = L \times L^x$ for $x \in (A^g \cap N(N)) - T$, $x^2 \in T$.

Next we show that $N \trianglelefteq X$. Let $C = N_A(N)$. Since $[N, c] \subseteq L$ and $C_N(c) \cap T = 1$ for each $c \in C - L$, it follows that $[N, c] = L$ and so $C_N(c) = L$ for each such c. Hence, by (4.1), $c^2 \in L$ for each $c \in C - L$ and so $\Phi(C) \subseteq L$. Assume $A \supset C$ and pick $x \in N_A(C) - C$ such that $x^2 \in C$. Since $T = A^g \cap CN$ is a TI-subgroup in CN, we have $T \subseteq N(T^x)$ if $T^x \cap N \neq 1$. But then $[T, T^x] = 1$ and thus $[T^x, N] = 1$. Hence $T^x N \cap C \subseteq L$ and so $T^x \subseteq N$. This implies that $x \in N(N)$, contradicting the choice of x.

So $|T^x N| = |L|^3$ and $|C \cap T^x N| \geqslant |L|^2$ since $T^x N \subseteq CN$. But this contradicts condition (c). This shows that $C = A$ and by symmetry $N \lhd X$. Let $\bar{X} = X/N$. Then, as shown, $\bar{A}$ is elementary abelian. We show that $\bar{A}$ is strongly closed in $C_{\bar{X}}(\bar{a})$ for each $\bar{a} \in \bar{A}^\#$.

Let $h \in X$ be such that $\overline{A^h} \subseteq C_{\bar{X}}(\bar{a})$. Then, by the above, $C_N(\overline{A^h}) = L^h$. On the other hand, $\overline{A^h}$ normalizes $[N, \bar{a}] = L$ and thus $C_L(\overline{A^h}) \neq 1$. Hence $L \cap L^h \neq 1$ and so $A = A^h$. Now (1) and (2) follow readily by Shult's fusion theorem (see [91]) and an inspection of the groups listed there. ($\bar{X} \cong D_{2m}$ if $\bar{X}$ is 2-nilpotent, since in this case $\bar{X}$ is generated by 2 involutions. Further $\bar{X} \not\cong U_3(2^n)$, since this group is not generated by 2 elementary abelian 2-groups.)

The proof of (3) and (4) is more technical and therefore omitted. The reader is referred to [219: (3.2), (3.9)] and [192: (1.7)].

REMARK 4.6. There are many natural cases, where the hypothesis of (4.6) is satisfied. If, for example, the abelian 2-subgroup A is a TI-subgroup of G and

$g \in G - N(A)$ is such that $N_A(A^g) \neq 1$, then (a) and (b) are satisfied. If in addition A is elementary, then (c) is a consequence of (2.3)(2) and therefore always satisfied in this situation.

If A is an abelian 2-group of exponent 4 and a TI-subgroup of G and $g \in G - N(A)$ is such that $N_A(A^g) \neq 1$, it can be shown with somewhat more argument that the hypothesis of (4.6) is satisfied. (See [222: (2.10)].) In addition it turns out in this case that $L = \Omega_1(A)$ and that the extension of $\bar{X}$ by N does *not* split if $\bar{X} \cong L_2(2^n)$, $n \geqslant 3$. This is an important tool in the proof of (3.12).

COROLLARY 4.7. *Suppose that the elementary abelian 2-group A is a TI-subgroup of X and $A \cap Z \neq 1$, where $Z = Z(O_2(X))$. Then*

$$D = \{t | t \sim A^{\#} \text{ in } X\}$$

is a set of root involutions of $\langle D \rangle$.

Proof. ([220: (2.5)]). Since $A \cap Z \neq 1$, the hypothesis of (4.6) is satisfied for each pair A^h, A^g, $g,h \in X$, such that $A^g \neq A^h$. (See remark (4.6).) An inspection of the groups $X = \langle A^h, A^g \rangle$ in (4.5) implies that for each pair $d \in (A^h)^{\#}$, $e \in (A^g)^{\#}$ properties (2.9)(2) and (3) hold. But then by definition D is a set of root involutions of $\langle D \rangle$.

If $A \nsubseteq O_2(X)$ the structure of $\langle D \rangle$ is even more restricted. This is due to the fact that $\bar{A}$ is a TI-set in $\bar{X} = X/O_2(X)$ and $[\bar{A},Z] \subseteq A \cap Z$. The second property restricts the nature of the representation and most of the root involution groups do not have such a representation.

5. Some General Lemmas under Hypothesis A

In this section we assume Hypothesis A and that $Q = F^*(C(z))$ is extraspecial. Introduce the following notation. Let $M = C(z)$, $\tilde{M} = M/\langle z \rangle$ and $\bar{M} = M/Q$. By (2.5) $\tilde{Q}$ is a nondegenerate orthogonal space over $\boldsymbol{F}_2$. Hence $|Q| = 2^{2n+1}$, $n \in N$. This integer n is called the *width* of Q. For $a = z^g$ we set $Q_a = Q^g$. If there exists some $z \neq z^g = a \in Q$ we set $L = (Q_a \cap M)Q$. This 2-group L plays an important role in the treatment of the extraspecial problem. All lemmas in this section are from [221: Section 3].

LEMMA 5.1. *Let $\alpha \in M - Q$ be an involution. Then the following hold.*

(1) *If* $\widetilde{C_Q(\alpha)} \neq C_{\tilde{Q}}(\alpha)$, *then* $\alpha \sim \alpha z$ *in* Q.

(2) *If* $\widetilde{C_Q(\alpha)} = C_{\tilde{Q}}(\alpha)$, *then* $[Q,\alpha]$ *is elementary abelian and* $\alpha^M \cap Q\alpha \subseteq \alpha^Q \cup (\alpha z)^Q$. *Furthermore,*

$$|C_{\bar{M}}(\bar{\alpha}) : \overline{C_M(\alpha)}| \leqslant 2.$$

(3) $[\alpha,Q] = Z(Q_0)$, *where* $\tilde{Q}_0 = C_{\tilde{Q}}(\alpha)$.

Proof. (1) is trivial. Next we show (3). Let $|[\alpha,\tilde{Q}]| = 2^l$. Then by (2.3) $|\tilde{Q} : C_{\tilde{Q}}(\alpha)| = 2^l$, whence $|C_Q(Q_0)| = 2^{l+1}$ by (2.5). Pick $x \in Q$ and $y \in Q_0$. Then

$$[\alpha,x]^y = [\alpha^y,x^y] = [\alpha,x],$$

since $\alpha^y = \alpha$ or αz and $x^y = x$ or xz. This shows that $[\alpha,Q] \subseteq C_Q(Q_0)$ and so by the above $[\alpha,Q] = C_Q(Q_0)$. But

$$\widetilde{[\alpha,Q]} = [\alpha,\tilde{Q}] \subseteq C_{\tilde{Q}}(\alpha) = \tilde{Q}_0$$

by (2.3). Hence $C_Q(Q_0) = [\alpha,Q] = C_Q(Q_0) \cap Q_0 = Z(Q_0)$. This proves (3).

To prove (2) assume $\widetilde{C_Q(\alpha)} = C_{\tilde{Q}}(\alpha)$. By (3) we have $[Q,\alpha] = Z(C_Q(\alpha))$. Suppose $[Q,\alpha]$ is not elementary abelian and pick an element v of order 4 in $[Q,\alpha]$. Since $\alpha \sim \alpha v$ or $\alpha \sim \alpha v^3$ by (2.3)(4), α inverts v. Hence $\tilde{v} \in C_{\tilde{Q}}(\alpha)$ but $v \notin C_Q(\alpha)$, a contradiction. This shows $[Q,\alpha]$ is elementary abelian.

Now let $\alpha h \in \alpha^M \cap Q\alpha$, $h \in Q$. Then $\widetilde{C_Q(\alpha h)} = C_{\tilde{Q}}(\alpha h) = C_{\tilde{Q}}(\alpha) = \widetilde{C_Q(\alpha)}$, whence $h \in Z(C_Q(\alpha)) = [Q,\alpha]$. Since by (2.3) $\alpha[Q,\alpha] \subseteq \alpha^Q \cup (\alpha z)^Q$, this implies the first part of (2).

Let now $g \in M$ such that $\bar{g} \in C_{\bar{M}}(\bar{\alpha})$. Then, by what we have shown, $\alpha^g \in \alpha^Q$ or $\alpha^g \in (\alpha z)^Q$. In the first case $\bar{g} \in \overline{C_M(\alpha)}$. In the second case $\alpha^{gx} = \alpha z$ for some $x \in Q$. This shows each product of 2 elements of $C_{\bar{M}}(\bar{\alpha}) - \overline{C_M(\alpha)}$ lies in $\overline{C_M(\alpha)}$, which proves the last part of (2).

For the next lemma we assume that $n \geqslant 3$, where n is the width of Q, and that there exists some $a = z^g \in Q$, $a \neq z$. The first assumption is clearly justified by Theorem 3.2. If the second case does not hold, this is a special case which will be treated in Section 7.

LEMMA 5.2. *One of the following holds.*

(1) *$\tilde{a}$ is the only singular vector in $C_{\tilde{Q}}(\bar{L})$ different from* 0.

(2) *There exists an $\alpha \in L - Q$, which acts as an involution of type a_2 on $\tilde{Q}$.*

Proof. Suppose (1) is false. Since by (2.4) $z \in Q_a$, it follows that

$$L_0 = M \cap Q_a = \langle z \rangle \times Q_0,$$

Q_0 extraspecial. Hence $\tilde{L}_0$ is extraspecial. This shows $C_{\widetilde{Q \cap Q_a}}(\bar{L}) = \langle \tilde{a} \rangle$.

Now pick $x \in Q$ such that $\tilde{x} \in C_{\tilde{Q}}(\bar{L})$, $\tilde{x} \neq \tilde{a}$ and $\tilde{x}$ a singular vector different from 0. By (2.5) x is an involution. Since $\tilde{x}$ centralizes $\tilde{L}_0$, x normalizes L_0 and thus centralizes a, since $\langle a \rangle = L_0'$. So, by the above, $x \in C(a) - Q_a$. Further $[L_0,x] \subseteq \langle z \rangle$, whence x centralizes a subgroup of index at most 2 in L_0 and thus centralizes a subgroup of index at most 4 in Q_a. By (2.8) this implies that $|Q_a : C_{Q_a}(x)| = 4$, since $x \notin Q_a$.

If $C_{\tilde{Q}_a}(x) \supset \widetilde{C_{Q_a}(x)}$ then x is of type b_1 on $\tilde{Q}_a$ and, by (2.7),

$$\langle v \rangle = [x,Q_a] \cong Z_4.$$

Hence $[L_0,x] \subseteq \langle v\rangle \cap \langle z\rangle = 1$, contradicting $|Q_a : L_0| = 2$. Hence $C_{\tilde{Q}_a}(x) = \widetilde{C_{Q_a}(x)}$ and, by (5.1)(2) and (2.7), x is of type a_2 on $\tilde{Q}_a$. Conjugating by $\bar{g}^1$, this proves (5.2). ($a = z^g$.)

LEMMA 5.3. *Suppose* $\bar{L}^h \cap N_{\bar{M}}(\bar{L}) \neq 1$ *for some* $h \in M$. *Then* $o(aa^h)\,|\,2$.

Proof. Let $b = a^h$ and assume $o(ab) = 4$. By definition of $\bar{L}$, we may pick $x \in Q_b \cap M$ such that

$$1 \neq \bar{x} \in \bar{L}^h \cap N_{\bar{M}}(\bar{L}).$$

By (5.2) $\tilde{a}^x = \tilde{a}$, whence $a^x = a$ or $a^x = az$. Since $b^a = bz$, $x^a = x$ or $x^a = xz$ and $z \in Q_{bz}$; this implies $x \in Q_{bz}$. But then $x \in Q$, since $bz \sim b(bz) = z$ in $Q_b \cap C(x)$, contradicting $\bar{x} \neq 1$.

LEMMA 5.4. *One of the following holds.*

(1) $|\bar{L}| = 2^{n-1}$ *and* $|Q \cap Q_a| = 2^{n+1}$.

(2) *There exists an* $\alpha \in L - Q$, *which acts as an involution of type* a_2 *on* $\tilde{Q}$.

Proof. Suppose false. Since $Q \cap Q_a$ is elementary abelian,

$$|Q \cap Q_a| \leqslant 2^{n+1}$$

by (2.5). Thus $|\bar{L}| \geqslant 2^{n-1}$ since $|Q_a \cap M| = 2^{2n}$. So we have $|\bar{L}| > 2^{n-1}$ and $|Q \cap Q_a| < 2^{n+1}$. Since $\widetilde{M \cap Q_a}$ is extraspecial of width $n-1$, it follows that

$$\bar{A} = C_{\bar{L}}(Q \cap Q_a) \neq 1.$$

Now we have

$$[L, C_Q(a)] = [Q(Q_a \cap M), C_Q(a)] \subseteq \langle z\rangle[Q_a \cap M, C_Q(a)] \subseteq Q \cap Q_a.$$

Hence $[\widetilde{C_Q(a)}, \bar{L}, \bar{A}] \subseteq [\widetilde{Q \cap Q_a}, \bar{A}] = 1$. Since $[\bar{L}, \bar{A}] = 1$ the 3-subgroup lemma implies $[\widetilde{C_Q(a)}, \bar{A}, \bar{L}] = 1$, whence

$$[\widetilde{C_Q(a)}, \bar{A}] \subseteq C_{\widetilde{Q \cap Q_a}}(\bar{L}) = \langle \tilde{a}\rangle.$$

By (2.6) and (2.7), the axis of a transvection on $\tilde{Q}$ is $\widetilde{C_Q(y)}$, $y \in Q$ of order 4. Hence no element in $\bar{A}^\#$ centralizes $\widetilde{C_Q(a)}$. Thus by (2.7) all elements of $\bar{A}^\#$ are of type a_2 or c_2 on $\tilde{Q}$ and, since we assume (5.4) is false, they are all type c_2.

Now $\widetilde{C_Q(a)}/\langle \tilde{a}\rangle$ is again a nondegenerate orthogonal space over $GF(2)$, with quadratic form and scalar product inherited from $\tilde{Q}$. Since all elements of $\bar{A}^\#$ induce transvections on $\tilde{Q}/\langle \tilde{a}\rangle$, it follows from (2.7) that $[\tilde{Q}, \bar{A}]/\langle \tilde{a}\rangle$ is a totally nonsingular subspace of $\widetilde{C_Q(a)}/\langle \tilde{a}\rangle$ of order $|\bar{A}|$. This implies $|\bar{A}| \leqslant 4$, since 4 is the order of a maximal totally nonsingular subspace of an orthogonal space over $\boldsymbol{F}_2$. On the other hand, since $\widetilde{M \cap Q_a}$ is extraspecial and $|\widetilde{Q \cap Q_a}| \leqslant 2^{n-1}$, we have $|\bar{A}| \geqslant 4$. This shows $|\bar{A}| = 4$ and $|\widetilde{Q \cap Q_a}| = 2^{n-1}$.

Assume now $b \sim a$ in M and $b \in C_Q(a) - Q_a$. Since all elements of $\bar{A}^\#$ act as transvections corresponding to $\tilde{a}$ on $\widetilde{C_Q(a)}$, it follows that

$$\widetilde{C_Q(a)} \subseteq \bigcup_{t \in \bar{A}^\#} C_{\tilde{Q}}(t).$$

Hence there is some $t \in \bar{A}^\#$ centralizing b. Now by (2.7) $\tilde{a}$ is the only singular vector different from 0 in $[\tilde{Q},t]$. Hence

$$[\widetilde{Q \cap Q_b}, t] \subseteq \langle \tilde{a} \rangle \cap (\widetilde{Q \cap Q_b}) = 1$$

by (2.4). It follows that $\widetilde{Q \cap Q_b} \subseteq C_{\tilde{Q}}(t) \subseteq \widetilde{C_Q(a)}$. Hence $\langle \tilde{a} \rangle (\widetilde{Q \cap Q_b})$ is a totally singular subspace of dimension n of $\tilde{Q}$ and thus $\tilde{Q}$ is of + type as an orthogonal space over $\boldsymbol{F}_2$.

Now, since $[A, Q \cap Q_a, Q] \subseteq [z,Q] = 1$, the three subgroup lemma implies $[A,Q] \subseteq C_Q(Q \cap Q_a)$. Since $\tilde{Q}$ is of + type,

$$\widetilde{C_Q(Q \cap Q_a)} = (\widetilde{Q \cap Q_a}) \perp \langle \tilde{\alpha}, \tilde{\beta} \rangle,$$

where $\langle \alpha, \beta \rangle \cong D_8$ and $o(\alpha) = o(\beta) = 2$. Set $\bar{A} = \langle t, \tau \rangle$ and pick $x \in Q - C_Q(a)$. Then, as above, $[\tilde{x},t]$ and $[\tilde{x},\tau]$ are nonsingular. Hence

$$[\tilde{x},t] \in \tilde{\alpha}\tilde{\beta}(\widetilde{Q \cap Q_a}) \quad \text{and} \quad [\tilde{x},\tau] \in \tilde{\alpha}\tilde{\beta}(\widetilde{Q \cap Q_a}).$$

But then $[\tilde{x}, t\tau] \in \widetilde{Q \cap Q_a}$, since $[\widetilde{Q \cap Q_a}, \bar{A}] = 1$. This shows that $[\tilde{Q}, t\tau]$ is totally singular, contradicting $t\tau$ of type c_2 on $\tilde{Q}$.

This shows each $b \in C_Q(a)$ satisfying $b \sim a$ in M lies already in $Q \cap Q_a$. Suppose such an element b exists. Then (2.4) implies $b \sim ba$ in M, since $C_{Q_a}(b) \neq Q_a \cap M$ and $C_{Q_b}(a) \neq Q_b \cap M$. But each $c \in a^M$ centralizes one of b, a or ab, whence, by the above, $c \in Q_b$, $c \in Q_a$ or $c \in Q_{ab}$. Thus c centralizes $\langle a,b \rangle$. This shows $a^M \subseteq Q_a$. Now obviously $\bar{M}$ does not act irreducibly on $\tilde{Q}$. Hence an inspection of the groups listed in (3.4) implies $|Q \cap Q_a| = 2^{n+1}$, contradicting our assumption.

This shows $a^M \cap C_Q(a) \subseteq \{a, az\}$. Suppose there is an $h \in M$ such that $\bar{L} \cap \bar{L}^h \neq 1$. Then by (5.3) $o(aa^h) \mid 2$. Hence $a^h \in \{a, az\}$ and thus $\bar{L}^h = \bar{L}$. This shows $\bar{L}$ is a TI-set in $\bar{M}$. By (5.2) obviously $\bar{L}$ is weakly closed in $N_{\bar{M}}(\bar{L})$ with respect to $\bar{M}$. Hence $\bar{L}$ is a weakly closed TI-subgroup in $\bar{M}$ in the sense of (3.11). But then [219: Section 5] implies $N_{\bar{M}}(\bar{L})$ transitive on $\bar{L}^\#$, since $O_2(\bar{M}) = 1$. Since $\bar{A} \trianglelefteq N_{\bar{M}}(\bar{L})$ this implies $2^n \leqslant |\bar{L}| = 4$, contradicting $n \geqslant 3$. This proves (5.4).

Proposition 5.5. *Suppose that no element in $\bar{L}$ acts as an involution of type a_2 on $\tilde{Q}$. Then the following hold.*

(1) *$\bar{L}$ is weakly closed in $N_{\bar{M}}(\bar{L})$ and $C_{\tilde{Q}}(\bar{L}) = \langle \tilde{a} \rangle$.*

(2) *$|\bar{L}| = 2^{n-1}$, $|Q \cap Q_a| = 2^{n+1}$, $\tilde{Q}$ is of + type and $C_{\bar{L}}(Q \cap Q_a) = 1$.*

Proof. All statements, except $C_{\tilde{Q}}(\bar{L}) = \langle \tilde{a} \rangle$, are consequences of (5.2), (5.4) and their proofs. Suppose therefore $C_{\tilde{Q}}(\bar{L}) \neq \langle \tilde{a} \rangle$. Pick $\tilde{y} \in C_{\tilde{Q}}(\bar{L})$ such that

$o(y) = 4$. By the proof of (5.2) and our hypothesis, y induces an involution of type c_2 on $Q_a/\langle a\rangle$. Hence there exists an $\bar{x}\in\bar{L}^{\#}$ acting as an involution of type c_2 on $\tilde{Q}$. Since, for each $\bar{u}\in\bar{L}^{\#}$ of type c_2, $[\widetilde{C_Q(a)},\bar{u}] = \langle\tilde{a}\rangle$, the proof of (5.4) shows $\bar{x}$ is the only involution of type c_2 in $\bar{L}$. In particular, $\bar{x}$ is strongly closed in $\bar{L}$ and so 2-central in $\bar{M}$, since $\bar{L}$ is weakly closed in $N_{\bar{M}}(\bar{L})$.

Let $b\in a^M\cap C(a)$ and suppose $\bar{x}$ centralizes $\tilde{b}$. If $b\notin Q_a$, then as in (5.4)

$$[\bar{x},\widetilde{Q\cap Q_b}]\subseteq(\widetilde{Q\cap Q_b})\cap\langle\tilde{a}\rangle = 1,$$

whence $\widetilde{Q\cap Q_b}\subseteq C_{\tilde{Q}}(\bar{x})\subseteq\widetilde{Q\cap C(a)}$, a contradiction since $a\notin Q_b$ and, by (5.2), $Q\cap Q_b$ is a maximal abelian subgroup of Q. Hence either $b\in Q_a$ or $[\tilde{b},\bar{x}] = \langle\tilde{a}\rangle$. By (2.4) this shows, as in (5.4), that

$$(*)\qquad \tilde{b}\sim\tilde{b}\tilde{a}\text{ in }\tilde{M}\text{ for each }b\in a^M\cap C(a).$$

Let $B = a^M\cap Q_a$. Claim $|\tilde{B}|\leqslant 3$. By (*) $\tilde{B} = \langle\tilde{B}\rangle^{\#}$. So if $|\tilde{B}| > 3$ then each $e = a^g\notin C(a)$, $g\in M$, centralizes a subgroup U of order at least 8 in $\langle B\rangle$. Hence $C_{\tilde{U}}(\bar{x}^{\bar{g}})\neq 1$ and so by the above $\widetilde{U\cap Q_e}\neq 1$. Let $1\neq\tilde{u}\in\widetilde{U\cap Q_a\cap Q_e}$. Then by (2.4) $\{a,e\}\subseteq Q\cap Q_u$, contradicting $o(ae) = 4$. So $a^M\subseteq C(a)$ and $\bar{M}$ is not irreducible on $\tilde{Q}$. But then an inspection of the groups listed in (3.4) shows that there is an involution of type a_2 in $\bar{L}$, contradicting the assumption. This proves the claim.

Suppose $\bar{x}^{\bar{g}}\in C_{\bar{M}}(\bar{x})$, $g\in M$. Then by (5.3) $a^g\in C(a)$, since $\bar{x}\in N_{\bar{M}}(\bar{L}^{\bar{g}})$. Hence by the above $a^g\in a^M\cap Q_a = B$. Since $|\tilde{B}|\leqslant 3$ this implies $|\bar{x}^{\bar{M}}\cap C_{\bar{M}}(\bar{x})|\leqslant 3$. Immediately $\langle\bar{x}^{\bar{M}}\cap C_{\bar{M}}(\bar{x})\rangle$ is elementary abelian. Hence by the above $a^g\in a^M\cap Q_a = B$. Since $|\tilde{B}|\leqslant 3$ this implies $|\bar{x}^{\bar{M}}\cap C_{\bar{M}}(\bar{x})|\leqslant 3$. Immediately $\langle\bar{x}^{\bar{M}}\cap C_{\bar{M}}(\bar{x})\rangle$ is elementary abelian. contradiction to $\bar{L}\trianglelefteq C_{\bar{M}}(\bar{x})$, $\bar{L}$ weakly closed and $\bar{x}$ strongly closed in $\bar{L}$. This proves (5.5).

6. Q of Symplectic Type but not Extraspecial

Before we use Hypothesis A we prove a more general lemma. (All lemmas in this section are from [13]. The proofs are slightly changed.)

LEMMA 6.1. *Let Z be a cyclic subgroup of order 4 in G and z the involution in Z. Assume Z is weakly closed in $N_G(Z)$ and set $X = \langle z^G\rangle$. Then either*

(1) *$z\in Z^*(G)$, or*

(2) *X has dihedral or quasihedral Sylow 2-subgroups and contains a 4-group U with $U^{\#}\subset z^G$.*

Proof. Assume $z\notin Z^*(G)$. Let $Z\subseteq S\in Syl_2(G)$ and $t\in z^G\cap S$, $t\neq z$. Then $\langle t,z\rangle$ is dihedral of order 4. Let $\langle t,z\rangle\subseteq D\subseteq S$ maximal subject to $D = \langle z^G\cap D\rangle$ and D dihedral. Suppose $s\in(z^G\cap N_S(D))-D$ and set $U = C_D(s)$. Then $U\subseteq T\in Syl_2(C(s))$ and s is contained in a conjugate W of Z normal in

T. Since $\langle z\rangle \subseteq Z(D)$ and each involution of $(z^G \cap S)-z$ inverts Z, all involutions of D are contained in z^G. Hence each involution of U inverts W and so $C_U(W) = 1$. This implies $U = \langle z\rangle$ and, by a well known lemma of Suzuki [204], $D\langle s\rangle$ is dihedral since D is dihedral, contradicting the maximality of D.

This shows $\langle z^G \cap S\rangle$ is dihedral. Now (6.1) is a consequence of Corollary B4 in [92].

From now on we assume in this section that Hypothesis A holds and that Q is not extraspecial. We use the following notation; $M = C(z)$, $\bar{M} = M/Q$ and $\tilde{M} = M/Z$ where $Z = Z(Q)$. First we prove the following lemma.

LEMMA 6.2. *One of the following holds.*
(1) $Q = Z \circ P$, $Z \cong Z_4$, *P extraspecial.*
(2) $z \in Z^*(G)$.
(3) *A Sylow 2-subgroup of* $\langle z^G\rangle$ *is dihedral or quasidihedral.*

Proof. Suppose (1) and (2) do not hold. Since Q is not extraspecial, (2.1) implies there is a cyclic characteristic subgroup A of Q with $|A| \geqslant 8$. (Actually $A = Z(C_Q(\Phi(Q)))$.) Let $Z = \Omega_2(A)$. By (6.1) either (3) holds or there exists some $Z \neq Z^g \subseteq C(Z)$. In the second case, by (4.2), $[A,A^g] = 1$ and, by (4.4), $W = \langle \mho^1(A^g)^M\rangle$ is abelian. But then $\Phi(AW)$ is not cyclic, contradicting (2.1).

So, for the classification of groups satisfying Hypothesis A with Q not extraspecial, one may assume that (6.2)(1) holds. By (2.5) $\tilde{Q}$ is a non-degenerate symplectic space. By (6.1) there is some $A = Z^g \subseteq M$, $A \neq Z$. Let $A = \langle a\rangle$. By (4.5) $a^2 \in Q$. But $A \nsubseteq Q$, since otherwise $\langle A^M\rangle$ would be abelian by (4.4), contradicting (2.1).

The following lemma is easy.

LEMMA 6.3. *Either* $\bar{a}$ *induces a transvection on* $\tilde{Q}$ *or* $\bar{a}$ *is of type* c_2 *on* $\tilde{Q}$.

Proof. This follows essentially from the fact that $|Q : C_Q(a^2)| = 2$ and

$$[A,C_Q(a^2)] \subseteq A \cap Q = \langle a^2\rangle.$$

LEMMA 6.4. *Let D be the set of involutions of type* c_2 *in* $Sp_{2n}(2)$, $n \geqslant 2$. *Then* $o(de) \in \{1,2,\ldots,6\}$ *for all* $d,e \in D$.

Proof. This follows from the fact that each involution of type c_2 in $Sp_{2n}(2)$ is the product of 2 commuting transvections. The computation is easy.

A set of $\{3,5\}$-*transpositions* of a group G is a set of odd transpositions of G, where 1,3 and 5 are the only odd numbers occurring as the order of a product de for $e,d \in D$.

LEMMA 6.5. *Let* $\bar{D} = \bar{a}^{\bar{M}}$. *Then one of the following holds.*
(1) $\bar{a}$ *is a transvection of* $\tilde{Q}$ *and* $\bar{D}$ *is a set of* 3*-transpositions of* $\langle \bar{D}\rangle$.
(2) $\bar{a}$ *is of type* c_2 *on* $\tilde{Q}$ *and* $\bar{D}$ *is a set of* $\{3,5\}$*- or* 3*-transpositions of* $\langle \bar{D}\rangle$.

Proof. If $\bar{a}$ is a transvection, obviously (1) holds. So assume $\bar{a}$ of type c_2 on $\tilde{Q}$. By (6.4) we need to show that there is no element $\bar{b} = \bar{a}^g$, $g \in M$, such that $o(\bar{a}\bar{b}) = 4$ or 6.

Assume first $o(\bar{a}\bar{b}) = 4$. Let $B = A^g$. Then $\langle A,B \rangle$ is a 2-group, whence by (4.3) $[A,B] = 1$, contradicting $o(\bar{a}\bar{b}) = 4$.

So assume $o(\bar{a}\bar{b}) = 6$. Let $C \sim B$ in $\langle A,B \rangle$ such that $1 \neq \bar{a}\bar{c} \in Z(\langle \bar{a},\bar{b} \rangle)$, where $\langle c \rangle = C$. As above $[A,C] = 1$. Hence $t = (ac)^2 = a^2c^2 \in z^G$ since, by (4.4), $a^2 \in O_2(C(c^2))$, and so $a^2 \sim a^2c^2$ in $O_2(C(c^2))$. If $[a^2,b^2] = 1$, then $\langle A,B \rangle$ normalizes $\langle a^2,b^2 \rangle$ and so $\langle \bar{a},\bar{b} \rangle \cong S_3$, a contradiction to $o(\bar{a}\bar{b}) = 6$. So $(a^2b^2)^2 = z$.

Let $\bar{h}$ be the element of order 3 in $\bar{X} = \langle \bar{a},\bar{b} \rangle$ and $\tilde{Q}_0 = [\tilde{Q},\bar{h}]$. Then $|\tilde{Q}_0| = 16$ or $|\tilde{Q}_0| = 4$. In the first case $\bar{X}$ centralizes $C_{\tilde{Q}}(\bar{h})$ and $\tilde{Q}_0$ is a nondegenerate symplectic space. Hence $\bar{X} \subseteq Sp_4(2)$ and $\bar{a},\bar{b}$ are of type c_2 on $\tilde{Q}_0$. But then $o(\bar{a}\bar{b}) \neq 6$ by the structure of $S_6 \cong Sp_4(2)$.

So $|\tilde{Q}_0| = 4$. Hence $[\tilde{Q}_0,\bar{a}\bar{c}] = 1$ and so $[\tilde{Q}_0,\bar{a}] = [\tilde{Q}_0,\bar{c}] = \langle \tilde{u} \rangle$. Now $\widetilde{a^2} \in [\tilde{Q},\bar{a}]$, $\widetilde{c^2} \in [\tilde{Q},\bar{c}]$ and both are not centralized by $\bar{h}$. Hence $\widetilde{a^2} \in \tilde{u}C_{\tilde{Q}}(\bar{h})$ and $\widetilde{c^2} \in \tilde{u}C_{\tilde{Q}}(\bar{h})$, whence $\widetilde{(a^2c^2)} \in C_{\tilde{Q}}(\bar{h})$. This implies that h centralizes $t = a^2c^2$, where $o(h) = 3$. But we may choose c so that $c \sim b$ in $\langle h \rangle$. Hence $B \subseteq C(t)$ and so $b^2 \in O_2(C(t))$ by (4.4). Since, for the same reason, $a^2 \in O_2(C(t))$, this implies $z = (a^2b^2)^2 = t$. But then $c^2 = a^2z$ and so $C \sim A$ in Q, contradicting $\bar{a} \neq \bar{c}$. This proves (6.5).

Using the classification and properties of groups generated by 3- or (3,5}-transpositions, together with the structure of Q as a symplectic space over $GF(2)$, one now determines the structure of M. Using known characterization theorems, this determines the groups satisfying Hypothesis A with Q not extraspecial. (See [13].)

7. The Case in which z Is Weakly Closed in Q

We assume in this section that G satisfies Hypothesis A with Q extraspecial. Use the notation introduced in Section 5. The crucial lemma of Aschbacher is the following.

LEMMA 7.1. *Let H be a subgroup of G satisfying*

(a) *$Q \subseteq H \nsubseteq M$ and $O_2(H) \neq 1$, and*
(b) *H minimal subject to* (a).

Let $V = \langle z^H \rangle$, $K = C_H(V)$. Then either

(1) *$H/K \cong S_3$, $|V| = 4$, $V \subseteq Q$ and z not weakly closed in Q, or*
(2) *$H/K \cong A_5 \cong \Omega_4^-(2)$, $|V| = 16$ and V is the natural module for $H/K \cong \Omega_4^-(2)$. Further, if $z \neq z^h$, $h \in H$, then $z^h \notin Q$ and z^h acts as an involution of type a_2 on $\tilde{Q}$.*

Proof. ([14: (2.1), (2.2)]) Let $S = O_2(H)Q$ and $Z = Z(S) \cap O_2(H)$. Then $Z \neq 1$ and $Z \subseteq C(Q)$. Hence $Z = \langle z \rangle$. This implies that V is elementary abelian.

Let $\bar{H} = H/K$. Since $C_V(Q) = \langle z \rangle$, the minimality of H implies $\bar{H} = \langle \bar{Q}, \bar{Q}^{\bar{h}} \rangle$ for each $h \in H - M$. Since $\bar{Q}$ is elementary, $\bar{Q} \cap \bar{Q}^{\bar{h}} \subseteq Z(\bar{H})$ and thus centralizes $V = \langle z^{\bar{H}} \rangle$. Hence $\bar{Q} \cap \bar{Q}^{\bar{h}} = 1$ for each $h \in H - M$ and so $\bar{Q}$ is a TI-subgroup of $\bar{H}$.

Certainly $O_2(\bar{H}) = 1$, since $O_2(\bar{H})$ centralizes z and so centralizes V. Hence (4.5) implies $\bar{Q}^{\bar{h}} \cap N_{\bar{H}}(\bar{Q}) = 1$ for each $h \in H - M$. This shows $\bar{Q}$ is strongly closed in the sense of [91] and is a TI-subgroup of $\bar{H}$. It now follows, as in (4.5), that $\bar{H} \cong L_2(2^n)$, $Sz(2^n)$ or D_{2m} for m odd, using [91]. Further $\bar{Q} = \Omega_1(\bar{S})$, $\bar{S} \in Syl_2(\bar{H})$.

Now $[\bar{Q}, V] \subseteq Q \cap V \subseteq Z(Q \cap K)$. Since $|Q : Q \cap K| = 2^n$ this shows $|[\bar{Q}, V]| \leqslant 2^{n+1}$, using Q extraspecial. This together with $C_V(\bar{Q}) = \langle z \rangle$ now implies $\bar{H} \cong S_3$ or $\Omega_4^-(2)$ and V is the "natural module" for $\bar{H}$, using knowledge of $GF(2)$-representations of $L_2(2^n)$ or $Sz(2^n)$, which is easily deduced from the list of absolutely irreducibly modules over a splitting field in characteristic 2 of these groups. (The proof is easy but technical, and therefore omitted. See [14: (2.1)]).

If $\bar{H} \cong S_3$ and $|V| = 4$, it follows from (2.8) that $V \subseteq Q$, since V centralizes $Q \cap K$. In the second case $N_{\bar{H}}(\bar{Q}) = C_{\bar{H}}(z)$, so $|z^H| = 5$. This shows that the elements of z^H play the role of singular vectors in the representation of $\bar{H}$ on V. Hence, by the structure of the natural $\Omega_4^-(2)$-module, $[\bar{Q}, z^h] = [\bar{Q}, V]$ is of order 8 for each $z^h \neq z$. In particular, $[\tilde{Q}, z^h] = [\widetilde{Q, z^h}]$ is totally singular and so z^h is of type a_2 on $\tilde{Q}$ by (2.7). So (7.1) holds.

Suppose now $F^*(C(t))$ is a 2-group for each involution $t \in Q$. Then it is not too difficult to show (see [14: (6.3)]) that such an H satisfying the hypothesis of (7.1) exists. Hence (7.1)(2) holds. By (2.11) this gives us $\{3,4\}^+$-transpositions in $\bar{M} = M/Q$. Furthermore, if $P = Q^h \cap M$ with $h \in H$ such that $z^h \neq z$, (5.1) shows that in "most cases" $\bar{P} \trianglelefteq C_{\bar{M}}(z^h)$. This greatly restricts, together with the $\{3,4\}^+$-transposition property, the possibilities for $\bar{M}$. The $F^*(G) \cong Co_2$ case occurs when $(\overline{z^h})^{\bar{M}}$ is not a class of 3-transpositions of the group it generates. Aschbacher finally characterizes the unitary groups by showing that z^G is a class of 3-transpositions of $\langle z^G \rangle$. This gives some outline of the proof of (3.3).

F. Smith has shown in [180] that G actually satisfies the additional hypothesis that $F^*(C(t))$ is a 2-group for each involution $t \in Q$ in the above situation. This is done by determining the centralizer $C(t)$ of an involution $t \in Q$ for which $F^*(C(t))$ is not a 2-group, by showing that the fusion of z in $C(t)$ satisfies the hypothesis of Corollary B of [218]. This completes the case z weakly closed in Q.

8. The Proof of Theorem 3.5

In this section we prove a crucial intermediate result in the proof of Theorem 3.5, which introduces some structure into a seemingly unstructured general problem. Moreover we roughly sketch the proof of (3.5).

Assume in this section that G satisfies Hypothesis A, $F^*(G)$ is simple and Q is extraspecial. Furthermore we may assume by (3.2) that the *width* n of Q is greater than 2 and by Section 7 that there is some $z^g = a \in Q$, $a \neq z$. Use the notation introduced in Section 5 and let $\bar{M}_0 = \langle \bar{L}^{\bar{M}} \rangle$. Everything in this section is from [221].

LEMMA 8.1. *Suppose there is no involution of type a_2 in $\bar{L}$ and $\bar{L}$ is a TI-set in $\bar{M}$. Then one of the following holds.*

(1) *$\bar{M}$ is not irreducible on $\tilde{Q}$ and $F^*(G) \cong L_{n+2}(2)$, M_{24} or He, the Held group.*

(2) *$n = 4$ and $\bar{M}_0 \cong L_2(8)$ or A_9.*

Proof. Suppose false. Then (3.4) implies $\bar{M}$ is irreducible on $\tilde{Q}$. The hypothesis implies $\bar{L}$ is a weakly closed TI-subgroup of $\bar{M}$ by (5.5). Let $\bar{Z} = Z(\bar{M}_0)$. Then $|\bar{Z}| \equiv 1(2)$ and $\bar{Z}$ centralizes $\tilde{a}$ since it normalizes $\bar{L}$. So, if $\bar{Z} \neq 1$, then $\tilde{Q} = C_{\tilde{Q}}(\bar{Z}) \times [\tilde{Q},\bar{Z}]$ and both factors are nontrivial, contradicting (5.5)(1). So $\bar{Z} = 1$. Now (3.11) implies $\bar{M}_0 \cong L_m(2^l)$, $Sz(2^l)$, $U_3(2^l)$, A_6, A_7, A_8, A_9 or M_{22}, M_{23}, M_{24}.

If $\bar{M}_0 \cong L_m(2^l)$, $m \geqslant 3$, then $\bar{L}$ corresponds to the set of transvections corresponding to a fixed point in the natural representation of $\bar{M}_0$. Hence (5.3) implies $\langle a^M \rangle$ is abelian. But then (1) holds by (3.4). If $\bar{M}_0$ is a Mathieu group, then $|\bar{L}| = 16$ and so $n = 5$ by (5.5). But comparing group orders, these groups are not contained in $O_{10}^+(2)$, contradicting (5.5). If $\bar{M}_0 \cong A_6$ then $|\tilde{Q}| = 2^6$ and it is easy to see that $\bar{M}_0$ cannot be irreducible on $\tilde{Q}$, contradicting (3.4). If $\bar{M}_0 \cong A_7$ then there is an element of order 3 centralizing $\bar{L}$. This is impossible for the same reason that showed $\bar{Z} = 1$.

So we are left with $\bar{M}_0 \cong A_9$, in which case $n = 4$ by (5), or $\bar{M}_0$ is of Bender type. In the second case $\bar{M}_0$ is doubly transitive on $\tilde{a}^{\bar{M}} = \{\tilde{a}_1, \ldots, \tilde{a}_{k+1}\}$. We claim $\tilde{a}_1, \ldots, \tilde{a}_k$ are linear independent.

Assume false. Since we assume $\bar{M}$ irreducible on $\tilde{Q}$, we have $(\tilde{a}_i, \tilde{a}_j) = 1$ for $i,j \leqslant k+1$, where (,) is the symplectic scalar product on $\tilde{Q}$. By renumbering, we have

$$\tilde{a}_{j+1} = \sum_{i=1}^{j} \tilde{a}_i$$

for $j < k$. If j is even, then

$$(\tilde{a}_{k+1}, \tilde{a}_{j+1}) = \sum_{i=1}^{j} (\tilde{a}_{k+1}, \tilde{a}_i) = 0,$$

a contradiction. So j is odd. But then similarly $(\tilde{a}_1,\tilde{a}_{j+1}) = 0$, again a contradiction. This proves our claim.

Now we have $k = 2^l$, 2^{2l} or 2^{3l} depending on whether $\bar{M}_0 \cong L_2(2^l)$, $Sz(2^l)$ or $U_3(2^l)$. Since $2^{n-1} = |\bar{L}| = 2^l$ and $2n = \dim \tilde{Q} \geqslant k$, we get the inequality

$$2n \geqslant 2^{a(n-1)}, \quad a = 1,2 \text{ or } 3.$$

Since we assume $n > 2$, this implies $n = 3$, $a = 1$ or $n = 4$, $a = 1$. In the first case $\bar{M}_0 \cong L_2(4)$, whence $k+1 = 5$ and $|\tilde{Q}| \leqslant 2^5$, since $\bar{M}$ is irreducible on $\tilde{Q}$. This contradicts $|\tilde{Q}| = 2n$. In the second case $\bar{M}_0 \cong L_2(8)$ and so (2) holds.

The following trivial lemma is [221: (4.3)]. The proof is omitted.

LEMMA 8.2. *Let $\bar{X} = \langle \bar{L},\bar{L}^{\bar{h}} \rangle$, $h \in M$, and suppose $\bar{X}/O_2(\bar{X}) \cong S_3$. Then one of the following holds.*

(1) $a^h \in Q_a$.

(2) *There is an involution of type a_2 in $\bar{L}$.*

(8.3) *Notation.* For $\bar{t} \in \bar{L}^{\#}$, let

$$R_{\bar{t}} = \langle L^h \mid \bar{t} \in \bar{L}^{\bar{h}}, h \in M \rangle,$$
$$\tilde{V}_{\bar{t}} = \langle \tilde{a}^h \mid h \in M \text{ and } \bar{t} \in \bar{L}^{\bar{h}} \rangle,$$

and

$$N_{\bar{t}} = C_{R_{\bar{t}}}(\tilde{V}_{\bar{t}}).$$

If $\bar{t}$ is fixed, we omit the index $\bar{t}$ and simply use the letters R, $\tilde{V}$ and N for the above defined groups.

We now come to the main tool for treating the extraspecial problem.

THEOREM 8.4. *Suppose that no element in $\bar{L}$ acts as an involution of type a_2 on $\tilde{Q}$. Then, for each $\bar{t} \in \bar{L}^{\#}$, one of the following holds.*

(1) $R = R_{\bar{t}} = L$.

(2) $R/N \cong L_m(2)$, $Sp_{2m}(2)$ *(note that $Sp_4(2)' \cong A_6$) or $\Omega^{\pm}_{2m}(2)$ and acts in the natural way on $\tilde{V}$.*

Proof. Suppose (1) does not hold for some $\bar{t} \in \bar{L}^{\#}$. By (5.3) V is elementary abelian. If $V \subseteq Q_a$, then $V \subseteq Q_b$ for all $\tilde{b} \in \tilde{a}^R$. Hence (2.4) implies $\tilde{V}^{\#} \subseteq \tilde{a}^R$ since, for $\tilde{b},\tilde{c} \in \tilde{a}^R$, we have $C_{Q_b}(c) \neq Q_b \cap M$. Since $[\tilde{V},\bar{L}] = \langle \tilde{a} \rangle$ and $|\bar{L}/\bar{L} \cap \bar{N}| = 2^{-1}|\tilde{V}|$ and since $Q_a \cap M$ is extraspecial, $\bar{L}\bar{N}/\bar{N}$ is the set of all transvections corresponding to $\tilde{a}$ on $\tilde{V}$. Now it easily follows that $R/N \cong L_m(2)$, where $m = \dim \tilde{V}$, and acts in the natural way on $\tilde{V}$.

So we may assume $V \nsubseteq Q_a$. Let $W = C_Q(V)$ and

$$W_1 = W \cap \Big(\bigcap_{\tilde{b} \in \tilde{a}^R} Q_b\Big).$$

Assume $\tilde{W}_1 \neq 1$. Since R normalizes $\tilde{W}_1$, (5.5)(1) implies $\tilde{V} \subseteq \tilde{W}_1$. But this contradicts $V \nsubseteq Q_a$. Hence $\tilde{W}_1 = 1$ and thus $[\tilde{W},\bar{t}] \subseteq \tilde{W}_1 = 1$, since

$[\widetilde{C_Q(a)},\bar{L}] \subseteq \widetilde{Q \cap Q_a}$ and $\bar{t} \in \bar{L}^r$ for all $r \in R$. By (2.3) this implies that

$$|[\tilde{Q},\bar{t}]| \leqslant |\tilde{Q} : \tilde{W}| = |\tilde{V}|.$$

But $[\tilde{Q},\bar{t}]$ is $\bar{R}$-invariant and so by (5.5)(1) $\tilde{V} \subseteq [\tilde{Q},\bar{t}]$, whence $\tilde{V} = [\tilde{Q},\bar{t}]$.

By (2.3) this shows that the map

$$\chi : \tilde{x} \to [\tilde{x},\bar{t}], \quad \tilde{x} \in \tilde{Q},$$

is an $\boldsymbol{F}_2\bar{R}$-homomorphism from $\tilde{Q}$ on $\tilde{V}$ with kernel $\tilde{W}$. In particular, $[\tilde{Q},\bar{N}] \subseteq \tilde{W}$ and $\tilde{Q}/\tilde{W}$ and $\tilde{V}$ are equivalent $\boldsymbol{F}_2\bar{R}$-modules.

We now define on $\tilde{V}$ a scalar product in the following way. For $\tilde{x} \in \tilde{V}$ choose $\underline{x} \in Q$ such that $\tilde{\underline{x}}\tilde{W} = \chi^{-1}(\tilde{x})$. Set $(\tilde{x},\tilde{y}) = 0$ if and only if $[x,\underline{y}] = 1$, and $(\tilde{x},\tilde{y}) = 1$ if and only if $[x,\underline{y}] = z$, for $\tilde{x},\tilde{y} \in \tilde{V}$. Since $W = C_Q(V)$ these commutators do not depend on the particular element in the coset $\underline{y}W$, whence (,) is well defined. Suppose $[x,\underline{y}] = 1$ and let t be a preimage of $\bar{t}$. Then the three subgroup lemma implies $[\underline{y},t,x] = 1$, whence $[y,\underline{x}] = 1$. This shows (,) is symmetric. Because $[x,\underline{yu}] = [x,\underline{yu}] = [x,\underline{y}]\,[x,\underline{u}]$, the scalar product is bilinear. We claim that $(\tilde{x},\tilde{x}) = 0$. We have $x^t = x\underline{x}$ or $zx\underline{x}$. Thus $\underline{x}^2 = (x\underline{x})^2 = [x,\underline{x}]\underline{x}^2$, since $x^2 = 1$. This shows $[x,\underline{x}] = 1$ for all $x \in V$, which proves our claim.

So (,) is a symplectic scalar product over $\boldsymbol{F}_2$ on $\tilde{V}$ and, since $C_Q(V) = \langle z \rangle$, $\tilde{V}$ is a nondegenerate symplectic space under (,). In particular, $|\tilde{V}| = 2^{2m}$, $m \in \boldsymbol{N}$. By definition of (,) we get that

$$\tilde{a}^{\perp} = \chi(\widetilde{Q \cap C(a)}) \subseteq \widetilde{V \cap Q_a}.$$

Hence $\tilde{a}^{\perp} = \widetilde{V \cap Q_a}$ and $|\widetilde{V \cap Q_a}| = 2^{2m-1}$. As above, this implies $|\bar{L}\bar{N}/\bar{N}| \geqslant 2^{2m-2}$. Since

$$[\tilde{a}^{\perp},\bar{L}] = [\widetilde{V \cap Q_a},\bar{L}] = \langle \tilde{a} \rangle,$$

the action of the full symplectic group on $\tilde{V}$ implies that one may identify $\bar{L}\bar{N}/\bar{N}$ with a subgroup of the biggest normal 2-subgroup E of the stabilizer of $\tilde{a}$ in the full symplectic group on $\tilde{V}$. (If X is the full symplectic group on $\tilde{V}$, then

$$C_X(\tilde{a}) = E\,.\,R, \quad R \cong Sp_{2m-2}(2),$$

acts in the natural way on $\tilde{a}^{\perp}/\langle \tilde{a} \rangle$ and $E \lhd C_X(\tilde{a})$ is elementary abelian of order 2^{2m-1} and $[\tilde{a}^{\perp},E] = \langle \tilde{a} \rangle$.) Comparing orders, $|E : \bar{L}\bar{N}/\bar{N}| \leqslant 2$.

We have two cases to consider. First $\bar{R}$ is transitive on $\tilde{V}^{\#}$. If $m > 2$ it is easily shown that $\bar{L}\bar{N}/\bar{N}$ contains an involution of type a_2 on $\tilde{V}$. Now $O_2(\bar{R}/\bar{N}) = 1$. Hence (2.10), (2.11) and the transitivity of $\bar{R}$ on $\tilde{V}^{\#}$ imply that all involutions of type a_2 in $\bar{R}/\bar{N}$ are conjugate. Now [217: (4.1)] and the main theorem of [217] imply $R/N \cong Sp_m(2)$ and its action is natural on $\tilde{V}$. If $m = 2$, then $X \cong S_6$ and so easily $\bar{R}/\bar{N} \cong A_6$ or S_6, depending upon whether $|\bar{L}\bar{N}/\bar{N}| = 4$ or $|\bar{L}\bar{N}/\bar{N}| = 8$. This proves (8.4) if $\bar{R}$ is transitive on $\tilde{V}^{\#}$.

So assume R is not transitive on $\tilde{V}^{\#}$. Define a quadratic form q on $\tilde{V}$ by

$q(\tilde{v}) = 0$ if and only if $\tilde{v} = 1$ or $\tilde{v} \in \tilde{a}^R$, and $q(\tilde{v}) = 1$ elsewhere. It is shown in the proof of [221, (4.5)] that q is a quadratic form over $GF(2)$. Hence $\bar{R}/\bar{N}$ may be identified with a subgroup of the full orthogonal group $Y = O^{\pm}_{2m}(2)$ on $\tilde{V}$. But since $|\bar{L}\bar{N}/\bar{N}| \geqslant 2^{2m-2}$, it follows as above that $\bar{L}\bar{N}/\bar{N} = O_2(C_Y(\tilde{a}))$. Now it easily follows that $\bar{R}/\bar{N} = Y' \cong \Omega^{\pm}_{2m}(2)$, since $\Omega^{\pm}_{2m}(2)$ is generated by 2 conjugates of $\bar{L}\bar{N}/\bar{N}$ which correspond to nonperpendicular elements of $\tilde{a}^R$. (See [221: (2.1)(1)].). This proves (8.4).

For the proof of Theorem 3.4, the case where involutions of type a_2 exist in $\bar{L}$ has to be considered separately. This is done in Section 12 of [221] and is not too difficult. The proof uses mainly $\{3,4\}^+$-transposition arguments.

So assume no involutions of type a_2 exist in $\bar{L}$. Then the hypotheses of (5.4), (5.5) and (8.4) are satisfied. By (8.1) we may assume $\bar{L}$ is not a TI-set in $\bar{M}$; that is, at least for one $\bar{t} \in \bar{L}^\#$, (8.4)(2) holds. For simplicity assume $R/N \cong \Omega^{\pm}_{2m}(2)$, but R/N not isomorphic to $\Omega^+_4(2)$. In the proof of (8.4) it was shown that $\tilde{Q}$ has 2 composition factors isomorphic to the natural R/N-module, namely $\tilde{V}$ and $\tilde{Q}/\tilde{W}$ where $\tilde{W} = \overline{C_Q(V)}$. Further $[\tilde{Q},\bar{t}] = \tilde{V}$. In Section 6 of [221] it is shown that $[\tilde{W},N] \subseteq \tilde{V}$ and that $\tilde{W}/\tilde{V}$ is "big" considered as (R/N)-module; that is, $|\tilde{W}/\tilde{V}| = 2^{l \cdot 2^{m-1}}$, $l \in N$, and the commutator space of $\tilde{W}/\tilde{V}$ with an involution $\bar{\alpha} \in \bar{L}-\bar{N}$ is similarly big in terms of m. Now it is shown that $\bar{t} \sim \bar{L}-\bar{N}$ in $\bar{M}$. Since $|[\tilde{Q},\bar{t}]| = 2^{2m}$, which is relatively small, this gives an upper bound on m. Further, for the smaller m's, one gets all possibilities for l and thus for the width n of $\tilde{Q}$.

Of course the case $\tilde{W}/\tilde{V} = 1$ has to be treated separately. But then $\bar{N} = \langle \bar{t} \rangle$ and it is easy to see that $\bar{M}_0 \cong S_3 \times \Omega^{\pm}_{2m}(2)$. (See Section 7 of [221].) The above explained argument works best if $R/N \cong \Omega^{\pm}_{2m}(2)$ or $Sp_{2m}(2)$ and m is big. In case of smaller m, additional fusion argument is required, making the proof longish. The case $R/N \cong \Omega^+_4(2)$ is different. It is shown in this case that either $n = 4$ or 6 or that $O(\bar{M}) \cong Z_3$ and then that $\bar{R}_{\bar{\tau}} \cong Z_2 \times \Omega^{\pm}_{2m}(2)$ for some other involution $\bar{\tau} \in \bar{L}^\#$. Now we have again $\bar{M}_0 \cong S_3 \times \Omega^{\pm}_{2m}(2)$.

Finally one may assume that $R_{\bar{t}} = L$ or $R_{\bar{t}}/N_{\bar{t}} \cong L_m(2)$, $m \in N$, for all $\bar{t} \in \bar{L}^\#$, but $\bar{L}$ is not a TI-subgroup in $\bar{M}$. It is shown in Section 11 of [221] that this case does not occur. The proof consists essentially of fusion arguments which first show $t \sim z$ for $t \in L^\#$ such that $R_{\bar{t}}/N_{\bar{t}} \cong L_m(2)$, m maximal. Next one shows $V = Q \cap Q_t$. The final contradiction is obtained by showing first $t \not\sim tz$ in G for all such t and then producing another τ with the same properties, which fuses to τz.

9. The Identification of Groups with $F^*(C(z))$ Extraspecial

In this section we will sketch the proofs of (3.6)–(3.10). We carry on with the notation of Section 8. In particular, we again use the letters M, Q, L, z and M_0 and the two homomorphisms denoted by $^-$ and $^\sim$.

Most of the theorems discussed in this section as well as their explanation here are due to S. Smith.

The completion of the extraspecial problem divides into two cases, the general one corresponding to groups of Lie type over $\boldsymbol{F}_2$ (i.e. the proofs of (3.6) and (3.7)) and the exceptional case corresponding to the groups of small width, (3.9) and (3.10), or to the monster groups, (3.8). The groups of Lie type are identified by showing that z^G is a class of $\{3,4\}^+$-transpositions of G. In the other case one determines the structure of M and, if necessary, that of some other centralizer, to quote some characterization theorem.

Outline of the proofs of (3.6) and (3.7)

(a) Determine fusion of z in Q. This is essentially already accomplished by [221], since there the action of $\bar{M}$ on $Q/\langle z\rangle$ is described.

(b) Determine fusion of z in M. If $x \in z^G \cap (M-Q)$, we study the group $Q \cap Q_x$, where by Q_x we mean $F^*(C(x))$. This can be shown to be an elementary group of suitable size. The main technical result (based on an argument of F. Smith) is the following.

LEMMA. $(Q \cap Q_x)^\# \subseteq z^G \cup (xz)^G$.

With this restriction to just two classes, it can be shown that $\bar{x} \in \bar{L}$ (for suitable a), and then that conjugates of z like x must lie exactly where they are expected in the groups of the conclusion. In particular, $z^G \cap M$ is a set of $\{3,4\}^+$-transpositions.

(c) Show that z^G is a class of $\{3,4\}^+$-transpositions. Here for $x,y \in z^G$, we study $|xy|$. When $4 \mid |xy|$, it is rather easy to apply the earlier work to obtain $|xy| = 4$ exactly, and $(xy)^2 \in z^G$ (the $(^+,\ )$-condition). To deal with the other case, we introduce the usual graph on z^G by connecting involutions which commute. Here the critical subconfiguration is a chain $x—z—w—y$ where $|xw| = |zy| = 3$. Using the work in (b), we find an involution $v \in z^G$ to replace z, reducing to a configuration which can be handled. Thus every chain of length 3 can be reduced to one of length 2, so that connected components of the graph have diameter 2. But we may apply the result of Holt [129] (and, independently, F. Smith [181]) to conclude that the graph is connected. This means we can put the pair x,y inside the centralizer of a third conjugate like z, and obtain $|xy|$ using part (b) above. Now the group G can be identified by the work in [217].

Proofs of (3.8)–(3.10) (See [184])

There remain a number of cases of smaller width. We remark that the cases of width 1 and 2 are covered by the work of Gorenstein and Harada (3.2). The cases of width 3 and 5 are essentially handled by earlier work, leaving only the cases of width 4, 6, 11, and 12 to be dealt with. Here the aim is to start with the information provided by [221], and lead into one of the established classification theorems. Several methods are used:

(a) *Structure of* $\bar{M}$: By (8.4), for $\bar{t} \in \bar{L}^{\#}$, a section of $C_{\bar{M}}(\bar{t})$ called $\bar{R}/\bar{N}$ is determined. Furthermore $\bar{L}$ is weakly closed in a Sylow 2-group of $\bar{M}$, and $\bar{M}'$ can be shown to have sectional 2-rank at most 4. With this information, we can use [99] to determine M_0, and finish.

Width n	*Section* $\bar{R}/\bar{N}$	$\bar{M}_0$	*Group G*	*Identified by work of*
4	trivial (as $\bar{L}$ is a *TI*-set)	A_9	F_3	Thompson [216], Parrott [160]
4		$L_2(8)$	${}^3D_4(2)$	Timmesfeld [217]
4	$\Omega_4^+(2)$	solvable $3^4.2^3$	$\Omega_8^+(3)$	Aschbacher [15]
6	$\Omega_6^-(2)$	$3.U_4(3).2$	$M(24)'$	Parrott [161]
6	$L_3(2)$	$3.M_{22}.2$	J_4	Janko [135]

The case of ${}^3D_4(2)$ was first handled by Reifart [166]. In some cases, additional fusion information must be determined before the final classification can be applied.

(b) *Structure of* $C(x)$ *for* x *a non-2-central involution*: This argument was made by Stroth for the case $n = 11$ of (iii); it appears in [167]. We let x be an involution of Q *not* in z^G. Then $C_Q(x)/\langle x\rangle$ is extraspecial of smaller width, and we can hope to identify $C(x)$ using the theory already established.

Width n	$C_M(x)/C_Q(x)$	$C(x)$	*Identified by work*	G	*Identified by work*
4	S_5	$2.HS.2$	Aschbacher [13]	F_5	Harada [116]
11	$U_6(2).2$	$2.{}^2E_6(2).2$	Reifart [167] or Smith [183]	F_2	Stroth [202]
12	Co_2	$2.F_2$	as above	F_1	Griess [109]

(c) *Use of Thompson order formula*: In this method, study of all involution fusion allows us to compute $|G|$ by Thompson's formula. Then the representation of G on a class of involutions or 3-elements may be described, as a means of identifying G.

Results of	*Width n*	$\bar{M}_0$	*Group G*	*Identified by work of*
Patterson and S. K. Wong [163]	3	$\Omega_6^-(2)$	Sz	Stellmacher [199]
Patterson [162]	4	$\Omega_8^+(2)$	Co_1	Stellmacher [199]
Reifart [167]	10	$U_6(2)$	${}^2E_6(2)$	Timmesfeld [217]
	10	$L_6(2)$	$E_6(2)$	Timmesfeld [217]

It should be possible to apply the methods of [167] for the cases of $E_7(2)$ and $E_8(2)$, but the calculation would be rather tedious.

Remark. The work described above does not address the questions of existence of J_4 and F_1.

10. The Theorems on TI-Subgroups

In this section we will sketch the proof of some of the theorems on TI-subgroups listed in Section 3. The sketches have to be short. It can only be hoped that they give some impression of how the proofs look. For more detail the reader is referred to the original papers.

(10.1) The proof of (3.11)

Assume the hypothesis of (3.11) holds and use the notation introduced there. By (2.12) the hypothesis implies that B is weakly closed in $N(B)$, whence B is weakly closed in each 2-subgroup of G containing B. This implies $B^G = B^{G^*}$. Hence one may assume $G = G^* = \langle B^G \rangle$. Further, if $M = O_2(G)$, then $M \subseteq N(B)$ and so $N = \langle (B \cap M)^G \rangle$ is elementary abelian and $M/N \subseteq Z(G/N)$. Case (3.11)(1) occurs of course if $|BN/N| = 2$. So assume $|\bar{B}| > 2$, where $\bar{G} = G/N$. Then $\bar{B}$ satisfies the hypothesis of (3.11) in $\bar{G}$ and so, to prove (3.11), one may assume $G = \bar{G}$. Since $O_2(G) \subseteq Z(G)$ and $O(G) \subseteq Z(G)$ as $|B| > 2$, it follows easily that G is a perfect central extension of a simple group. Now trivially $G/Z(G)$ satisfies the hypothesis of (3.11). So we may assume G is simple.

If B is strongly closed, then $G \cong L_2(q)$, $Sz(q)$ or $U_3(q)$, where $q = |B|$, by [91]. So we may assume there is some B^g such that $1 \neq N_B(B^g) \neq B$, since B is weakly closed. Now by (4.6) the pair B, B^g satisfies the hypothesis of (4.5). So for $X = \langle B, B^g \rangle$ and N as in (4.5), (4.5)(1)–(4) hold. This is actually the main tool for proving (3.11). Indeed we can compute all the information desired in the group X.

The next important step is to show that $X/N \not\cong Sz(q)$ and if $X/N \cong D_{2m}$ that $m = 3$, so that we have in any case $X/N \cong L_2(2^m)$, $m \geqslant 1$. Now let $H = N(B)$ and $\bar{H} = H/B$. Pick $B^g \neq B$ such that $|A|$ is maximal, where $A = B^g \cap H$. Then it can be shown that $\bar{A}$ satisfies the hypothesis of (3.11) in $\bar{H}$. (See [219: Section 4].) This means that we can apply induction to the group $\bar{K} = \langle \bar{A}^{\bar{H}} \rangle$.

It is visible that the case $|\bar{A}| = 2$ is special and has to be treated separately. But in this case $|B| = 4$ and the treatment is fairly standard. Namely one shows that a Sylow 2-subgroup of K and then of G is dihedral of order 8. So assume $|\bar{A}| \geqslant 4$.

To apply induction it is important to have $O_2(\bar{K}) \subseteq Z(\bar{K})$. Actually in Section 5 of [219] it is shown that H is transitive on $B^{\#}$, which is of course much stronger. The proof consists of fusion arguments. After some more work one comes up with the cases $\bar{K} \cong SL_{n-1}(2^m)$ and $|B| = 2^{m(n-1)}$, or $\bar{K} \cong A_6$, A_7, A_8 and $|B| = 16$. Finally G is identified either by showing $K = H$, which is done in the cases $\bar{K} \cong A_6$, A_7, A_8 or $L_3(2)$, or by showing $D = \{t \mid t \sim B^{\#} \text{ in } G\}$ is a class of root involutions of G, which is done in the cases of "bigger" K. This determines G in any case by known classification theorems. The first case corresponds to the exceptional cases $G \cong M_{22}$, M_{23}, M_{24} or A_9.

(10.2) The proof of (3.12)

Use the hypothesis and notation of (3.12) and in addition let $B_0 = \Omega_2(B)$, $B_1 = \Omega_1(B)$ and $Q = O_2(M)$. Suppose first that $[B_0, B_0^g] = 1$ if $N_{B_0}(B_0^g) \neq 1$. Then, using (4.4), it can be shown that

$$W = \langle B_1^g \mid B_0^g \subseteq M \rangle \subseteq Q$$

and W is elementary abelian. (Here one uses the fact that $[B_1, B_1^g] = 1$ if $N_{B_0}(B_0^g) \neq 1$ by [222: (2.10)]. (See (4.6).) But the assumption implies that $W \subseteq M^g$ for each $B_1^g \subseteq W$. Now the maximality of M implies there is no B_0^g such that $N_{B_0}(B_0^g) \neq 1$, whence B_0 is strongly closed and so G is known. It turns out that this is impossible.

So we find B_0^g such that $B_0 \neq N_{B_0}(B_0^g) \neq 1$. Now by (4.6) we have for B_0, B_0^g the hypothesis of (4.5). In the notation of (4.5), it can be shown again that $X/N \not\cong Sz(2^n)$ or D_{2m}. So we have $X/N \cong L_2(2^n)$, $2^n = |B_0 : B_1|$, by the remark (4.6). Moreover, if $n > 2$, the extension does not split. This turns out to be impossible.

This leaves us with $X/N \cong L_2(4)$ and, after some more argument, $B_0 \cong Z_4 \times Z_4$. Now, with some detailed argument, it is shown that

$$D = \{t \mid t \sim B_1^{\#} \text{ in } G\}$$

satisfies the hypothesis of Corollary B of [36]. Inspecting the groups listed there yields $\langle B_1^G \rangle = F^*(G) \cong L_3(4)$.

Before saying something about the proof of (3.13), we will state the crucial lemma which is a generalization of (2.4).

LEMMA 10.3. *Let G be a group, V an elementary abelian 2-subgroup of G and* $M = N(V)$. *Suppose the following hold:*

(1) $F^*(M) = O_2(M) = Q$,

(2) *V is a maximal abelian normal subgroup of M, and*

(3) *V is a TI-subgroup of G and there is some g such that* $V \neq V^g \subseteq Q$.

Let $P = C_Q(V)$, $W = VV^g$ *and* $H = \langle P,P^g \rangle$. *Then the following hold:*

(a) $V \subseteq Q^g$,

(b) $V^G \cap W$ *is a partition of W and* $H/C_H(W) \cong L_2(q)$, $q = |V|$, *and acts in the natural way on W, and*

(c) $P = Q$ *and* $V = \Phi(Q) = Z(Q) = Q'$.

This is (2.18) of [222]. The hypothesis that M is a maximal 2-local subgroup given there is only used to get $N(V) \subseteq M$. By [222: (2.9)] we have $V = C(P)$ and $\Phi(P) \subseteq V$, which is actually trivial to prove. But this shows that part (c) is a consequence of (b), since we have a cyclic group of order $q-1$ acting regularly on $V^{\#}$.

(10.4) Proof of (3.13)

Use the notation of (3.13). It is easy to show that Q is of symplectic type if B is cyclic. So we may by (3.12) assume B is elementary abelian. If B is not weakly closed in Q, then the hypothesis of (10.3) is satisfied. So by (10.3) the only thing remaining to show is that $Q = \langle B^g \mid B^g \subseteq Q \rangle$. Here the proof consists of longish fusion arguments. Several of them are generalizations of the arguments of [180], where a similar statement was proved for extra-special Q.

(10.5) Proof of (3.15)

We use the notation of (3.15) and assume B is not weakly closed. Let

$$\mathscr{H} = \{H \mid O_2(H) \neq 1 \text{ and } P = C_Q(B) \subseteq H \nsubseteq M\}.$$

Then by (3.14) $B \subseteq O_2(H)$ for each $H \in \mathscr{H}$. The first important result one derives from (3.14) is that B is of *root type* in G, where by [220] an elementary abelian TI-subgroup A of G is of root type in G if, whenever $N_A(A^g) \neq 1$, then $[A,A^g] = 1$, $g \in G$. Namely, suppose this is not the case. Pick B^g such that $B \neq N_B(B^g) \neq 1$ and set $H = \langle B^g,P \rangle$. Then it is shown that $O_2(H) \neq 1$, whence $H \in \mathscr{H}$. But then by the above $B \subseteq O_2(H)$, contradicting $B \nsubseteq O_2(\langle B,B^g \rangle)$ by (4.5).

This shows B is of root type in G. Now pick $H \in \mathscr{H}$ and set $W = \langle B^H \rangle$, $K = C_H(W)$ and $\bar{H} = H/K$. Since $B \subseteq O_2(H)$ and $B \cap Z(O_2(H)) \neq 1$, and since $C(P) = B$ as in (10.3), (4.5) implies W is elementary abelian. Further, $\bar{P}$

is elementary abelian since $\Phi(P) \subseteq B \subseteq W \subseteq K$. It is easy to see that $\bar{P}$ is weakly closed in $\bar{H}$, since $C_W(\bar{P}) = B$ and B is a TI-subgroup of G. Now, if B is weakly closed in P, one can show that $\bar{P}$ is a TI-subgroup of $\bar{H}$ since B is of root type. By (3.11) this determines the structure of $\langle \bar{P}^{\bar{H}} \rangle$ for each $H \in \mathscr{H}$. This is the most important tool for proving (3.15).

Now let

$$\mathscr{H}^* = \{H \mid H \in \mathscr{H} \text{ and } H = \langle P, P^g \rangle \text{ for some } g \in G\}.$$

Obviously $\mathscr{H}^* \neq \emptyset$ if $\mathscr{H} \neq \emptyset$. The next step is the precise description of a group $H \in \mathscr{H}^*$. Namely, if $H \in \mathscr{H}^*$ it is shown that $H/K \cong L_2(q^2)$, $q = |B|$, W is the orthogonal H/K-module (coming from $H/K \cong \Omega_4^-(q)$) and either $K = W$ or K/W is the direct sum of natural H/K-modules. Further, in the latter case, $W = \Phi(K) = Z(K)$. To prove this, I found it necessary to compute many details on $\boldsymbol{F}_2$-representations of $L_2(2^n)$ and $Sz(2^n)$.

Now let $\Sigma = \{B^g \mid \langle P, P^g \rangle \in \mathscr{H}^*\}$. Then it is shown that for $A, C \in \Sigma$ either $[A, C] = 1$ or $\langle A, C \rangle \cong L_2(q)$. The way of showing this is different from Aschbacher's approach in [14], since we neither know the exact structure of Q (i.e., do not have Q/B as an orthogonal space over $\boldsymbol{F}_q$) nor can we obtain root involutions as involutions of type a_2. (It is shown by other arguments that the case $\langle A, C \rangle$ special of order q^3 does not arise.)

In the case corresponding to $U_n(q)$, $n \geqslant 6$, one uses induction and Aschbacher's extension process of [14] to show that the above property holds for all $A, C \in B^G$. This of course identifies $\langle B^G \rangle$ by the characterization of groups generated by odd transpositions.

In the cases corresponding to $U_4(q)$ and $U_5(q)$ one shows that $\Sigma = (B^G \cap M) - B$. With this information one is able to show again that the above property holds for all $A, C \in B^G$.

7

Quasithin Groups

GEOFFREY MASON

1. Introduction

A *quasithin* group is a group G satisfying $e(G) \leqslant 2$. For the definition of the invariant e and the relevance of the problem of finding all finite simple quasithin groups we refer the reader to Chapter 1.

In this chapter we shall report on the progress made on the classification of the finite quasithin groups. Although this work is by no means complete, it is probably fair to say that one now knows that a minimal counterexample to the classification is "small", in a sense to be made more precise below.

The techniques we employ are largely based on, and motivated by, the monumental work of Thompson [214] and Aschbacher's classification of the finite simple thin groups [16]. (A group is called *thin* if $e(G) = 1$.) Evidently a thin group is also quasithin, so the classification of the latter will include Aschbacher's result; in fact we never need to quote the classification of thin groups. (At least, such a necessity has not yet appeared.) We would like to make it clear, however, that our work has been greatly influenced by Aschbacher's, and the material presented below should be viewed as an attempt to understand [16] and reconcile its point of view with a "more general" type of simple group.

2. The Groups

There are a large number of known finite simple quasithin groups, which we list below.

Chev(2):

$L_2(q)$, $U_3(q)$, $Sz(q)$;

$L_3(q)$, $Sp_4(q)'$, $U_4(q)$, ${}^3D_4(q)$, ${}^2F_4(q)'$, $G_2(q)'$;

$L_4(2)$, $L_5(2)$, $Sp_6(2)$, $U_5(4)$.

Chev(odd):

$L_2(p^n)$, $U_3(p)$, $U_3(p^2)$;

$L_3(p), L_3(p^2), G_2(p), PSp_4(p)$;
$L_4(p)$, p Fermat prime; $U_4(p)$, p Mersenne prime.
Alternating:
A_7, A_9.
Sporadic:
$M_{11}, M_{12}, M_{22}, M_{23}, M_{24}$,
J_1, J_2, J_3, J_4,
HS, He, Ru, Mc.

Here, q is a suitable power of 2, and p is a suitable odd prime. Notation for the groups is standard.

Of these, the most important as far as the local analysis is concerned are those which can occur as sections of 2-locals. This is equivalent to requiring that all odd Sylow subgroups have rank at most 2, which eliminates the following: $U_4(q)$, $Sp_6(2)$, $U_5(4)$; all groups in $Chev$(odd) except $L_2(p)$, $L_2(p^2)$, $U_3(p)$, $L_3(p)$; A_9, J_3, Mc.

In a classification as general as we are considering, one first must take into account the possibility that a minimal counterexample G is of component type. If t is an involution in G such that $C = C(t)$ is not 2-constrained, then one easily sees that either $E(C)$ covers $E(C/O(C))$ or else $C/O(C)$ has a component isomorphic to $SL_2(p)$, p odd. The latter case has been handled by Aschbacher [15], so one is reduced to the problem of standard components. We refer the reader to Chapter 2 for the current state of this theory; suffice it to say here that most of the possible standard components that we must consider have already been handled in full generality, whilst the outstanding cases are surely not difficult if one further assumes that $e(G) \leqslant 2$.

From now on, then, we shall assume that G is a minimal counterexample to the classification of finite simple quasithin groups, and that G is of characteristic 2 type.

Finally, we introduce some notation. $\mathscr{M}$ is the set of maximal 2-local subgroups of G; for $X \subseteq G$, we put

$$\mathscr{M}(X) = \{M \in \mathscr{M} \mid X \subseteq M\}.$$

3. The Main Result

Let us first list some assumptions that we may make about the group G.

(3.1) (a) *G has a nonsolvable 2-local subgroup.*
(b) *If T is a Sylow 2-subgroup of G, then $|\mathscr{M}(T)| \geqslant 2$.*

Part (a) follows from the work of Janko [134] and F. Smith [179]; part (b) follows from Aschbacher [17], Solomon [191] and some work of F. Smith.

The known groups satisfying (3.1) are the rank 2 groups of $Chev(2)$ defined over fields with at least four elements, and the groups $U_4(2)$, ${}^3D_4(2)$, $L_4(2)$, $L_5(2)$, $Sp_6(2)$, M_{22}, M_{23}, M_{24}, and J_4. Thus we must characterize the rank 2 groups of Lie type defined over fields of characteristic 2 and order at least 4, together with a small number of exceptions. In this chapter, we shall discuss the ideas which are involved in the proof of the following; because the proof is not yet completed, we state it as a conjecture. Also, to explain the essential ideas, we shall ignore certain technical restrictions.

(3.2) CONJECTURE‡. *Let G be a simple quasithin group of characteristic 2 type, minimal subject to not appearing in the list in Section 2. Assume that* (3.1) *holds and that every proper subgroup of G is a $\mathcal{K}$-group. Then the nonabelian simple sections of 2-local subgroups are amongst the following:* $L_2(4)$, $L_3(2)$, $Sp_4(2)'$, $G_2(2)'$, $L_4(2)$, $L_5(2)$, *and* A_7.

From now on, assume that G is also a counterexample to this conjecture. Let T be a Sylow 2-subgroup of G.

4. Subgroups of Parabolic Type

From the condition that $e(G) \leqslant 2$, it is easily verified that if M is a nonsolvable 2-local subgroup of G, then there is (at least) one normal subgroup L of M satisfying the following:

(4.1) (a) $O_2(L) = F^*(L) \subset L = O^{\infty}(L)$.
(b) *If $\tilde{L} = L/O_2(L)$, then one of the following holds;*
(i) *$\tilde{L}$ is quasisimple,*
(ii) *$\tilde{L}$ is the direct product of two isomorphic simple (thin) groups interchanged in M, or*
(iii) *$F^*(\tilde{L}) = O(\tilde{L})$, $\tilde{L}/F(\tilde{L}) \cong SL_2(p)$ for some odd prime $p \geqslant 5$, and $O(\tilde{L})$ is abelian if $p \geqslant 7$.*

With this in mind we define, for each 2-group $S \subseteq G$, the following set:

(4.2) $\mathscr{L}(S) = \{L \subseteq G \mid$ (a) L satisfies (4.1)
(b) $L \trianglelefteq M$ for some $M \in \mathscr{M}(S)$
(c) S is a Sylow 2-subgroup of $M\}$.

Note that the group M of (b) is uniquely determined by L, namely

$$M = N(L) = N(O_2(L)).$$

Now in general we may have $\mathscr{L}(S) = \emptyset$, but there are certainly 2-groups S for which $\mathscr{L}(S) \neq \emptyset$. Namely, let M be a nonsolvable element of $\mathscr{M}$ with Sylow 2-subgroup S (M exists by (3.1)(a)), and choose $L \trianglelefteq M$ as in (4.1). Then certainly $L \in \mathscr{L}(S)$.

‡ (3.2) has now been proved. In addition, all the remaining possible simple sections of 2-local subgroups have been eliminated. Thus a minimal counterexample to the classification of quasithin simple groups is of component type.

Define a partial order on $\mathscr{L}(S)$ as follows: if $L_1, L_2 \in \mathscr{L}(S)$, write $L_1 \ll L_2$ if $L_1 S \subseteq L_2 S$. Let

(4.3) $\mathscr{L}^*(S) =$ maximal elements of $\mathscr{L}(S)$ with respect to the partial order $\ll$.

The first thing one needs to show is that

(4.4) $\mathscr{L}^*(T) \neq \emptyset$.

This assertion is a quite simple consequence of the fact that G is a counterexample to (3.2), and is discussed more fully in the appendix to this chapter.

We next define the parabolic type subgroups of G, following some of Aschbacher's ideas in [16]. Let

(4.5) $\mathscr{P}(S) = \{L \in \mathscr{L}^*(S) \mid \{N(L)\} = \mathscr{M}(J)$ whenever $JO_2(LS) = LO_2(LS)$ and $O_2(J) \neq 1\}$.

Elements of $\mathscr{P}(S)$ are called the *weak parabolic type* subgroups, and those of $\mathscr{P}(T)$ the *parabolic type* subgroups of G. It should be noted that if P is a suitable maximal parabolic subgroup of a group of Lie type of characteristic 2, then $O^{2'}(P)$ has the properties required in (4.5). The goal of the whole analysis outlined here is in some sense to establish the converse, namely to prove that

(4.6) $\mathscr{P}(T) \neq \emptyset$,

and then to try to show that $L \in \mathscr{P}(T)$ is "similar" to $O^{2'}(P)$ for some maximal parabolic subgroup P of an appropriate group of Lie type. At this point the identification of G will be straightforward. Of course we shall sometimes have to account for exceptional situations which have no Lie type analogue; for example, the sporadic group J_4 arises when $L \in \mathscr{P}(T)$ satisfies $L/O_2(L) \cong M_{24}$. However the analysis outlined below takes all such anomalies nicely into account, and the whole procedure works out rather well! Again we have Aschbacher to thank for the insight he has given us.

Let us now sketch the proof of (4.6). At least after (4.4) we may choose $L \in \mathscr{L}^*(T)$. It is reasonable to ask that every such L lies in $\mathscr{P}(T)$ (which holds in the known groups), but current pushing-up techniques do not permit such a conclusion, so we go for the weaker (4.6). We do at least have the following, which is essentially a consequence of the definitions:

(4.7) $\{N(L)\} = \mathscr{M}(LT)$.

We must show that (4.7) still holds when LT is replaced by a subgroup J satisfying (4.5). Proceeding by contradiction, we choose such a J with $R = O_2(J)$ maximal subject to $|\mathscr{M}(J)| \geqslant 2$. From (4.7) and maximality of R, one immediately gets the following.

(4.8) (a) *$N(R_0) \subseteq N(L)$ for all $1 \neq R_0$ char R.*
(b) *$R \in \mathscr{U}^*_X(J;2)$ for $X \in \mathscr{M}(J) - \{N(L)\}$.*

These observations prove to be very powerful, and allow us to reduce to the following situation:

$$L/O_2(L) \cong L_2(q), \quad q = 2^n \geqslant 4,$$

and, if $X \in \mathscr{M}(J) - \{N(L)\}$ and $K = \langle J^X \rangle$, then

$$K/O_2(K) \cong SL_3(q), \quad Sp_4(q) \quad \text{or} \quad G_2(q)$$

and $J/O_2(K) = O^{2'}(P)$ for some maximal parabolic subgroup P of $K/O_2(K)$; furthermore, R is normal in some Sylow 2-subgroup S of X, so that $S \subseteq N(L)$ by (4.8)(a). We may assume that $S \subseteq T$, whence $S \subset T$ by (4.7).

Thus we find ourselves confronted with a pushing-up situation, where something along the lines of problem 3 of our appendix would be useful. To extricate ourselves from the present predicament, however, we use a device which is important at a number of other places in the analysis also. First, it is easy to see that in the above situation we do at least have

(4.9) $O^2(K) \in \mathscr{P}(S)$.

One can then ask the following.

(4.10) *Suppose $H \in \mathscr{P}(S)$. Under what conditions is it true that $\{N(H)\} = \mathscr{M}(H_0)$ for a subgroup $H_0 \subseteq H$ satisfying*

$$O_2(H_0) \neq 1 \quad \textit{and} \quad \{O_2(HS)\} = \mathscr{U}^*_{N(H)}(H_0{:}2)?$$

To say that $H \in \mathscr{P}(S)$ is precisely the same as the assertion that (4.10) holds in case H_0 covers $HO_2(HS)/O_2(HS)$ and $O_2(H_0) \neq 1$. Roughly speaking, (4.10) asks: "When can one replace H by a suitable proper subgroup H_0 and still retain the uniqueness result $\{N(L)\} = \mathscr{M}(H_0)$?" The answer should be: "almost always". As usual the result we actually have is somewhat weaker, the difference between reality and conjecture again being a shortage of good pushing-up theorems. But here at least is one crucial case when (4.10) has a positive answer, namely

(4.11) *Let $H \in \mathscr{P}(S)$ be such that $H/O_2(H) \in Chev(2)$ is either of twisted type, or defined over a field of at least four elements (excluding $L_2(4)$).*

If $H_0 \subseteq H$ covers a Cartan subgroup of $H/O_2(H)$ and satisfies the other conditions of (4.10), *then $\{N(H)\} = \mathscr{M}(H_0)$.*

Let us explain why this result is so powerful. In all cases except

$$H/O_2(H) \cong (S)L_3(q) \quad \text{or} \quad (S)U_3(q),$$

there is an involution t in the Weyl group of $H/O_2(H)$ which inverts a Cartan subgroup B, and if we choose $H_0 \subseteq H$ to satisfy $O_2(H_0) \neq 1$ and

$H_0 = O_2(H_0)B\langle t\rangle$, then H_0 satisfies the conditions, and hence the conclusions, of (4.10) and (4.11). Similar remarks apply to $L_3(q)$ and $U_3(q)$, but here we must be more conservative in our choice of H_0. Apart from these two cases, however, application of (4.11) will yield the following corollary.

(4.12) *Let $1 \neq B_0 \subseteq B$ where $B \subseteq H$ has odd order and is incident with a Cartan subgroup of $H/O_2(H)$. Then one of the following holds:*

(a) $N(B_0) \subseteq N(H)$, *or*

(b) $O_2(N(B_0)) = 1$.

If we set $D = C(B_0) \cap O_2(HS)$, with B_0 as in (4.12), then either $D = 1$ or else $N(D) \subseteq N(L)$ by (4.11). Of course the first possibility imposes strong conditions on the action of H on $O_2(HS)$ (usually strong enough to yield a contradiction), whereas if the second possibility holds then either (a) of (4.12) holds or else (b) holds and we can virtually write down the complete structure of $N(B_0)$. The upshot is that either (a) of (4.12) holds or $N(B_0)$ is "known".

Finally, let us return to the proof of (4.6). We may apply the above considerations when B is the Cartan subgroup of $O^2(K)/O_2(O^2(K))$ since $O^2(K) \in \mathscr{P}(S)$ by (4.9). We may even take $B \subseteq SJ = SK \cap M$, so B sits in both M and X. As we have such strong control over $N(B_0)$ for $1 \neq B_0 \subseteq B$, in particular when $K/O_2(K) \cong Sp_4(q)$ or $G_2(q)$, it is not difficult to argue to a contradiction. If $K/O_2(K) \cong SL_3(q)$ a little more effort is required, but eventually all is resolved and (4.6) is established.

We shall see more applications of the philosophy of (4.10) and (4.11) below.

5. First Main Reduction

From now on we fix $L \in \mathscr{P}(T)$, as we may by (4.6). Let $M = N(L)$ and $Q = O_2(LT)$.

We further subdivide $\mathscr{P}(T)$ as follows‡.

(5.1) $\mathscr{P}_f(T) = \{L \in \mathscr{P}(T) \mid [L, Z(O_2(L))] \neq 1\}$.

$\mathscr{P}_t(T) = \mathscr{P}(T) - \mathscr{P}_f(T)$.

The same kind of subdivision occurs in Thompson's work [214] and Aschbacher's thin groups paper [16]. The reason appears to be something along the following lines: we handle both cases by performing weak closure arguments on certain nonidentity normal elementary abelian subgroups E of M, and if possible we prefer that $m(E)$ be large (which means $m(E) \geqslant 3$), that E be 2-reducible in M, and that $[E,L] \neq 1$. Now if both of the last two

‡ The subscripts f, t stand for *faithful* and *trivial* respectively.

conditions fail the analysis gets quite intricate, and in general one needs knowledge about all elements of $\mathscr{L}(T)$ in order to get by. Thus we must first get the requisite information about the groups in $\mathscr{P}_f(T)$ before doing any of the "level two" analysis, which is the term Aschbacher uses in [16] to describe the weak closure analysis of non-2-reduced normal subgroups E of M.

We show that G is known if $\mathscr{P}_f(T)$ contains any "large" groups, so that when analysing $\mathscr{P}_t(T)$ one knows, roughly speaking, that the elements of $\mathscr{P}_f(T)$ are as in (3.2).

Throughout the remainder of this section, then, we assume

(5.2) $L \in \mathscr{P}_f(T)$.

It is not hard to see that (5.2) is equivalent to the existence of a 2-subgroup V satisfying

(5.3) (a) $1 \neq V = [V,L] \trianglelefteq M$, *and*
(b) V *is 2-reducible in* M; *i.e.*, $O_2(M/C(V)) = 1$.

For the following discussion we set $C = C(V)$ and $\bar{M} = M/C$. There are the following important invariants associated with this situation, namely

(5.4) $m = m(\bar{M},V)$ and $l = m_2(\bar{M}/C_{\bar{M}}(\bar{L}))$.

We refer the reader to Sections 5 and 6 of Chapter 5 for a discussion of the relevance of these invariants. Recall that m is the minimal $\boldsymbol{F}_2$-codimension of $C_V(t)$ in V as t ranges over the involutions of $\bar{M}$.

The first thing one must do is establish

(5.5) (*First main reduction*). *We have* $m \leqslant l$, *or else either* $\bar{L} \cong L_3(2)$, $m = 3$ *and* $l = 2$, *or* $\bar{L} \cong L_2(p)$ *and* $F^*(L/O_2(L)) = O(L/O_2(L))$ *as in* (4.1).

(5.5) says that V is in general a "small" module relative to the action of $\bar{L}$. In fact after the results of [45] one knows all chief $\bar{L}$-factors of V, a most useful fact.

There is an important preliminary to the proof of (5.5), involving the following set.

(5.6) $\Gamma = \Gamma(\bar{M};V) = \{U \subseteq V \mid |C_{\bar{M}}(U)| \text{is odd}\}$.

It was an important observation of Aschbacher in [16] that using (4.6) one can establish the following.

(5.7) *If* $U \in \Gamma$ *then* $C(U) \subseteq M$.

For let $D = C(U)$. As $U \in \Gamma$ we may take a Sylow 2-subgroup R of $C_M(U)$ to be L-invariant, whence (4.6) yields $N(R_0) \subseteq M$ for $1 \neq R_0$ char R. This establishes that R is 2-Sylow in D, and also that if $D \nsubseteq M$ then

$$\langle N_D(R_0) \mid 1 \neq R \text{ char } R\rangle \subseteq M \cap D \subset D.$$

Then the theory of blocks [17b] comes into play and a contradiction can be reached.

The point is that in doing weak closure arguments on normal elementary groups E of M one needs results of the form $C(E_0) \subseteq M$ for "many" subgroups E_0 of E (optimality being reached when this holds for *all* nonidentity subgroups—in other words $C(e) \subseteq M$ for all involutions $e \in E^{\#}$); (5.7) provides a reservoir of such "good" subgroups of V, namely Γ, free of charge.

Turning to the proof of (5.5), the practical meaning of the hypothesis $m \geqslant l+1$, which we assume by way of contradiction, is that in this case Γ provides enough good subgroups of V to implement the weak closure machinery. In fact it is an immediate consequence of (5.7) that‡

(5.8) $V(ccl_G(V);T) \subseteq Q$,

and a little more effort establishes the following.

(5.9) *If there is* $g \in G$ *with* $|V^g : V^g \cap M| = 2$, *then* $\bar{L} \cong L_3(2)$, $m = 3$ *and* $l = 2$.

So in proving (5.5) we can assume that $V^g \subseteq Q$ whenever $|V^g : V^g \cap T| \leqslant 2$. By (4.6) and some general results this gives

(5.10) *If* $T \subseteq S$ *with* S *solvable, then* $S \subseteq M$.

Now one chooses a subgroup $K \subseteq G$ with $T \subseteq K \nsubseteq M$, with $O_2(K) \neq 1$, and with $|K|$ minimal subject to this property; K exists by virtue of (3.1)(b). By (5.10), K is nonsolvable, and the only difficulty in obtaining a contradiction arises when $E(K/O_2(K))$ is a Bender group. In this case one again has by (5.10) that $K \cap M$ covers a Cartan subgroup B of $E(K/O_2(K))$, and here the idea is to locate B within M and show that the action of B on V is not compatible with the structure of K. The details are somewhat tedious but not difficult.

We conclude our discussion of (5.5) with the remark that the overall goal in each of the successive reductions one makes in analysing Hypothesis (5.2) is always to attempt to establish the analogues of (5.8) and (5.9), and then to study the group K above. Those who know better may forgive us if we say that this is the essence of the method of weak closure.

Finally, let us record a corollary of (5.5).

(5.11) *Assume that* (5.2) *holds. Then* $\bar{L}$ *is one of the following:* $SL_2(q)$, $SL_3(q)$, $Sp_4(q)'$, $G_2(q)'$, $(q = 2^n)$; $SL_4(2)$, $SL_5(2)$; A_7, M_{11}, M_{12}, M_{22}, M_{23}, M_{24}; $\hat{A}_6$, $\hat{A}_7$, $\hat{M}_{22}$ *(3-fold covers); or* $L_2(p)$ *as in* (5.5).

(5.11) follows immediately from (5.5) and results of [18] and [45]. Hence the (relatively) simple proof of (5.5) has already eliminated many possibilities for $L \in \mathscr{P}_f(T)$.

‡ $V(ccl_G(V);T)$ denotes the weak closure of V in T with respect to G.

6. A Characterization of J_4

Next we consider the following case.

(6.1) *Assume that* $L \in \mathscr{P}_f(T)$ *is sporadic. Then* $L/O_2(L) \cong M_{24}$ *and* $G \cong J_4$.

Surprisingly perhaps, this result is not at all difficult to establish if one follows the weak closure philosophy. We will assume that $L/O_2(L) \cong M_{24}$ as it is quite representative of the arguments. What we shall do is show that G has a central involution z such that $C(z)$ is isomorphic to the corresponding subgroup of J_4, and then quote Janko's definition [135].

Now by (5.5) we already know that V is one of two (dual) $\boldsymbol{F}_2\bar{L}$-modules of dimension 11. (Actually, one can only assert this about $V/C_V(\bar{L})$, but it is readily established that $C_V(\bar{L}) = 1$.). Using (5.7) as a springboard one establishes the following properties.

(6.2) (a) $C(U) \subseteq M$ *whenever* $U \subseteq V$ *and* $|V:U| \leqslant 2^7$.
(b) V *is uniquely determined.*
(c) *There is* $v \in V^\#$ *with* $C(v) \nsubseteq M$.

It turns out that V must be what is sometimes called the Todd module: for our purposes this means that $\bar{M} \cong M_{24}$ has two orbits in its action on $V^\#$, with representatives z, t, and moreover

$$C_{\bar{M}}(z) \cong E_{64} . \hat{S}_6, \quad \text{and} \quad C_{\bar{M}}(L) \cong Aut\,(M_{22}).$$

Referring back to (4.10), with $L \in \mathscr{P}(T)$ playing the role of H, we may take H_0 to be $C_M(t)$, and (4.10) then yields that $C(t) \subseteq M$. With this result available it is not surprising that (a) of (6.2) holds. What is surprising, perhaps, is that J_4 exists (conjecturally, at least) in spite of it!

Not surprisingly, (6.2)(a) forces the analogues of (5.8)–(5.10) to hold, whereas by (6.2)(c) we must have $H = C(z) \nsubseteq M$, where z is as above and $z \in Z(T)^\#$ without loss. H will be an element of $\mathscr{M}(T) - \{M\}$ guaranteed by (3.1)(b).

As $E(H/O_2(H))$ must have $\hat{A}_6$ involved in a 2-local,

$$E(H/O_2(H)) \cong M_{24} \quad \text{or} \quad \hat{M}_{22},$$

and as (5.10) holds the first possibility is out. In fact (5.10) forces H to be "correct" modulo $O_2(H)$, and further analysis of the weak closure of V in H yields the desired isomorphism type for H.

7. Rank Two Elements of $\mathscr{P}_f(T)$

We further restrict $\mathscr{P}_f(T)$ by eliminating some cases.

(7.1) *Suppose* $L \in \mathscr{P}_f(T)$ *and* $L/O_2(L) \in Chev\,(2)$ *has Lie rank* $\geqslant 2$. *Then* $L \cong L_3(2), L_4(2), L_5(2), Sp_4(2)'$ *or* $G_2(2)'$.

Proceeding by way of contradiction, we obtain from (5.11) that

$$L/O_2(L) \cong SL_3(q), \quad Sp_4(q) \quad \text{or} \quad G_2(q), \quad q \geqslant 4.$$

The value of the latter inequality is that the Cartan subgroup of $L/O_2(L)$ becomes nontrivial, so that (4.12) comes into play.

First we identify V.

(7.2) *V is a "standard" $\boldsymbol{F}_2\bar{L}$-module.*

This means that V is irreducible as an $\bar{L}$-module and $|V| = q^3, q^4, q^6$ according as $\bar{L} \cong SL_3(q), Sp_4(q)$ or $G_2(q)$ respectively.

We let $\bar{P}_1$, $\bar{P}_2$ be the two maximal parabolics of $\bar{L}$ permutable with $\bar{T}$. After (7.1) we may choose notation so that

$$V_i = C_V(O_2(\bar{P}_i)) \quad \text{and} \quad |V_1| = q, \quad |V_2| = q^2.$$

Then $[V_1, P_1] = 1$ whilst $P_2/O_2(P_2) \cong Z_{q-1} \times L_2(q)$ is faithful on V_2. Now we supplement the "good" subgroups Γ of (5.7) by proving (7.3).

(7.3) $N(V_2) \subseteq M$.

With (7.3) we then get that

(7.4) $N(V_1) \nsubseteq M$.

Notice that $V_1 \trianglelefteq T$, so that $N(V_1)$ lies in some element of $\mathscr{M}(T) - \{M\}$ as required by (3.2)(b). But what does $N(V_1)$ look like? As $N_M(V_1)$ contains T and involves $L_2(q)$, either the $L_2(q)$ "blows up" to a larger quasisimple group, which is almost always a rank 2 Lie type group defined over $\boldsymbol{F}_q$, or else there is some solvable group in $N(V_1)$ (but not in M) centralizing the $L_2(q)$-section. The upshot is that if $B \cong Z_{q-1} \times Z_{q-1}$ lies in L and is incident with a Cartan subgroup of $L/O_2(L)$ then, for various nonidentity subgroups B_0 of B, we will in general have $C(B_0) \cap N(V_1) \nsubseteq M$. As discussed in Section 4, this strongly restricts the structure of $C(B_0)$, and playing off the groups M, $N(V_1)$ and $C(B_0)$ for appropriate B_0 finds its ultimate reward in a contradiction. As usual the case that $\bar{L} \cong SL_3(q)$ is rather more difficult in view of the more restricted applicability of (4.11), and at the moment our proof involves a rather messy analysis of the chief $A_G(V_1)$-factors of $O_2(N(V_1))$ which we do not propose to go into here.

8. Construction of Some Groups of Lie Type of Rank Two

Here we assume the following hypothesis.

(8.1) $L \in \mathscr{P}_f(T)$ and $L/O_2(L) \cong L_2(q)$, $q = 2^n \geqslant 8$.

Again, after (5.5) we know the precise structure of V as an $\boldsymbol{F}_2\bar{L}$-module;

namely, $V/C_V(\bar{L})$ is either the standard module or the orthogonal module, the latter arising from the isomorphism $L_2(q^2) \cong O_4^-(q)'$.

Unlike the situation of Section 6, the possibility that $C_V(\bar{L}) \neq 1$ is not so easily dismissed. In fact we must prove

(8.2) *Suppose that* $C_V(\bar{L}) \neq 1$. *Then* $G \cong Sp_4(q)$.

Recall that in $Sp_4(q)$ the maximal parabolics P are such that $O_2(P) = [O_2(P),P]$ is elementary abelian of order q^3 with $O_2(P) \cap C(O^{2'}(P))$ of order q.

If we set $Z = C_V(\bar{L})$, then $C(z) \subseteq M$ for all $z \in Z^{\#}$ by (4.6); in particular, we get in conjunction with (5.7) that

(8.3) $C(U) \subseteq M$ *whenever* $U \subseteq V$ *and* $|V:U| < q|Z|$.

(8.3) having provided us with plenty of "good" subgroups of V, we readily deduce

(8.4) (a) $V(ccl_G(V);T) \subseteq Q$, *and*
(b) $V^g \subseteq M$ *whenever* $|V^g : V^g \cap M| < q$.

From (8.4) we get as a corollary the analogue of (5.10), and can then turn to a minimal group $K \nsubseteq M$ with $T \subseteq K$ as before. It turns out that $K/O_2(K)$ must have 2-rank less than $l+m_2(Z)$, from which it follows quite easily that $E(K/O_2(K)) \cong L_2(q)$. As $K \nsubseteq M$, we have $[K,z] \neq 1$ for all $z \in Z^{\#}$, and, as $Z \cap Z(T) \neq 1$, also that $E(K/O_2(K))$ is faithful on $Z(O_2(K))$.

Moreover from (8.4) we get that $VO_2(K)/O_2(K)$ is a Sylow 2-subgroup of $E(K/O_2(K))$. From these facts we readily find that K and L are isomorphic to the maximal parabolics of $Sp_4(q)$, and the group can be identified quite easily.

The next step is the proof of

(8.5) *Suppose that* V *is the orthogonal module. Then* $G \cong SU_4(q^{1/2})$.

In discussing (8.5) we change notation so that $\bar{L} \cong L_2(q^2)$, $q \geqslant 4$, and $|V| = q^4$. Now $\bar{L}$ has two orbits on $V^{\#}$, say $\mathscr{I}$ and $\mathscr{J}$, with representative i,j and point stabilizers of the form $L_2(q)$ and $E_{q^2}.Z_{q+1}$ respectively.

As $C_{\bar{L}}(i)$ is a "large" subgroup of $\bar{L}$ it should be no surprise that we first establish

(8.6) *If* $x \in \mathscr{I}$ *then* $C(x) \subseteq M$.

By using a little linear algebra, one knows that if $U \subseteq V$ and $|U| > q$ then $U \cap \mathscr{I} \neq \emptyset$, so (8.6) yields

(8.7) $C(U) \subseteq M$ *if* $U \subseteq V$ *and* $|U| > q$.

This result gives us the good subgroups of V, and (8.4) follows easily. This time we show that if $y \in \mathscr{J}$ then $C(y) \nsubseteq M$, choose such y in $Z(T)$, and consider the structure of $H = C(y)$ in some detail. One shows

(8.8) *$E(H/O_2(H)) \cong L_2(q)$, $Sz(q)$ or $(S)U_3(q)$.*

At this point we know that $H \cap L$ covers a Cartan subgroup of both $E(H/O_2(H))$ and $\bar{L}$ and, by utilizing these two cyclic groups in conjunction with (8.4), we are eventually able to force the precise structure of H and L; namely, either $E(H/O_2(H)) \cong L_2(q)$ and $Q = V = O_2(M)$, or‡ $E(H/O_2(H)) \cong SU_3(q)$ and $O_2(H)$ is special of order q^7. At this point a recent fusion theorem of McBride [145] shows that L contains T, and then the results of Timmesfeld [219] on TI-sets allow us to identify G, proving (8.5).

At this point we may assume that V is the standard $F_2\bar{L}$-module. We know that $U \subseteq V$ lies in Γ if $|U| > q$, and we supplement this by showing

(8.9) *If $U \subseteq V$, $|U| = q$ and $C(U) \nsubseteq M$, then $C_{\bar{L}}(U)$ has even order.*

Now all subgroups U satisfying (8.9) are $\bar{L}$-conjugate since such a U is centralized by a 2-Sylow of $\bar{L}$. We let Δ be the set of such subgroups of V. They correspond, of course, to root subgroups.

(8.10) *If $\Delta = \emptyset$, then $G \cong SL_3(q)$.*

The hypothesis of (8.10) gives us enough good subgroups of V to argue as before, and we say no more about this relatively simple case.

Thus we now fix $U \in \Delta$ with $U \subseteq T$. After (8.10) we know that $N(U) \subseteq X \in \mathscr{M}(T) - \{M\}$ as required by (3.2)(b).

At this point we must tackle the odd order subgroups of $C = C(V)$. Up until now we have rather avoided this topic, for the following reason: if P is a nonidentity subgroup of C of odd order, then one will almost always know that $C_M(P)$ covers $\bar{L}$. (This certainly holds under Hypothesis (8.1).) Since $[P,V] = 1$, either $N(P) \subseteq M$ or else the structure of $N(P)$ is closely determined as $L \in \mathscr{P}(T)$. Up until now we have been able to get by with such relative vagueness, but the ensuing analysis now demands that this point be taken up in more detail. Namely, we prove

(8.11) *If P is a nonidentity subgroup of C of odd order and $N(P) \nsubseteq M$, then either $G \cong {}^3D_4(q)$ or q is a square and $G \cong G_2(q)$.*‡‡

As remarked above, if $N = N(P) \nsubseteq M$ then the structure of N is very restricted. We choose P of maximal order, and first show without too much work that

(8.12) *$E(N) \cong (S)L_3(q)$; moreover P is cyclic.*

‡ This case should correspond to $U_5(q)$, but this group is not quasithin for $q \geqslant 8$.
‡‡ $|P| = q^2+q+1$ if $G \cong {}^3D_4(q)$ and $|P| = 3$ if $G \cong G_2(q)$, q a square.

Now with $U \in \Delta$ as above, we set $H = N(U)$ and $\tilde{H} = H/O_2(H)$, noting that $P \subseteq N$. We next show

(8.13) $\langle \tilde{P}^{\tilde{H}} \rangle = L_2(q_0)$, $Sz(q_0)$ *or* $(S)U_3(q_0)$ *and* $|P|$ *divides* $q_0 - 1$.

The next observation is crucial and is the kind of thing one always looks for in these situations.

(8.14) *There is* $g \in \langle P^H \rangle$ *such that* $W = V^g \trianglelefteq O_2(H)$ *and* $[W,V] = U$.

Namely, by (8.13) there is a 2-element $g \in \langle P^H \rangle$ which inverts P: we claim that such a g works! Indeed $\langle P^H \rangle \subseteq \langle T,P,g \rangle$ by (8.13), so that $g \notin M$ and hence $W = V^g \neq V$. As $V \trianglelefteq O_2(H)$, also $W \trianglelefteq O_2(H)$. Moreover, g inverts P; hence g acts on $E(N)$ and also $E(N) \cap H$. As $L \in \mathscr{P}(T)$, $V \subseteq E(N)$ and $E(N) \cap M$ covers $\bar{L}$, we see that g must induce an outer automorphism on $E(N)$. As $E(N) \cong SL_3(q)$ by (8.12), it follows that VW is a Sylow 2-subgroup of $E(N)$ and (8.14) follows immediately.

The effect of (8.14) is very strong. For example, as $C_T(U) = QW$ and

$$O_2(H) = (O_2(H) \cap C(W))\, V \trianglelefteq T,$$

we get $[Q,W,W] \subseteq U$, which says that $\bar{W}$ (a Sylow 2-subgroup of $\bar{L}$) is quadratic on each noncentral chief $\bar{L}$-factor of Q; hence each such factor is a standard $\boldsymbol{F}_2\bar{L}$-module. Moreover the size of $|Q : C_Q(\bar{W})|$ is effectively bounded above by (8.13). Putting these facts together eventually yields that $\langle \tilde{P}^{\tilde{H}} \rangle \cong L_2(q_0)$ where $q_0 = q^3$ or q, and that $Q = O_2(M)$ has order q^{11} or q^5 respectively. Then G is readily recognized, again by the results of McBride and Timmesfeld mentioned above.

We are now prepared to assault the more difficult configurations. First fix $U \subseteq T$ with $U \in \Delta$ as before, with $H = N(U) \nsubseteq M$. Note that there is $B \subseteq L \cap H$ with $B \cong Z_{q-1}$ acting faithfully on U. Moreover, by the philosophy of (4.10) and (4.11), we have strong control over $N(B_0)$ for $1 \neq B_0 \subseteq B$ in case this latter group is not in M. This is crucial in proving the following.

(8.15) *N is nonsolvable.*

(8.16) $E(N/O_2(N)) \cong L_2(q)$ *or* $Sz(q)$.

We must construct the groups $G_2(q)$ and ${}^2F_4(q)$ from these two possibilities. What one has to do is differentiate between the cases $V(ccl_G(V);T) \nsubseteq Q$ and $V(ccl_G(V);T) \subseteq Q$. In the first case it is easy to find a conjugate $W = V^g \subseteq T$ with properties as in (8.14), and a similar analysis goes through. In the latter situation we are given free of charge a good factorization of LT, and one can use this in the usual way to investigate H. The details are not unlike the foregoing, and we shall finish our analysis of Hypothesis (8.1) at this point.

9. Level Two Analysis of *Chev*(2)

After Sections 5–8 we know that if $L \in \mathscr{P}_f(T)$, then either $L/O_2(L)$ is as in (3.2) or is a 3-fold cover $\hat{A}_6$ or $\hat{A}_7$, or else $L/S(L) \cong L_2(p)$ as in (5.5) with $[S(L),V] = 1$. This latter case presents few problems, for if R is a Sylow r-subgroup of $S(L)$, r odd, we can derive from $L \in \mathscr{P}(T)$ that $N(R) \subseteq M$. Moreover there are certain nonidentity characteristic subgroups T_0 of T with $T_0 \trianglelefteq LT$; for example, if $p \geqslant 7$ we may take T_0 as two of those subgroups occurring in the Thompson triple factorization [214]. Using this, we can control the subgroups of $\mathscr{M}(RT)$, and thereby prove that $C(z) \subseteq M$ for $z \in Z(T) \cap V^{\#}$. Now we easily get that $\{M\} = \mathscr{M}(T)$, in contradiction to (3.7)(b).

At this point we have established that the conclusions of (3.2) hold for the elements of $\mathscr{P}_f(T)$, so one turns to the set $\mathscr{P}_t(T)$ and assumes the following hypothesis.

(9.1) $L \in \mathscr{P}_t(T)$.

It is natural to first take up the case

(9.2) $L/O_2(L) \in Chev(2)$, either twisted or defined over a field of at least four elements (excluding $L_2(4)$).

Here the ideas discussed in (4.10)–(4.12) really come into their own. For if $B \subseteq L$ has odd order and is incident with a Cartan subgroup of L permutable with T, then of course $[B,Z] = 1$, where we have set

$$Z = \Omega_1(Z(O_2(L))),$$

and $C(z) \subseteq M$ for $z \in Z^{\#}$ since $L \in \mathscr{P}_t(T)$. Coupled with (4.10)–(4.12), this proves to be very powerful.

To date there are certain problems associated with the case where

$$L/O_2(L) \cong (S)L_3(q)$$

which necessitate a delay in the solution of this part of (9.2). (These arise, as usual, from the restricted applicability of (4.10)–(4.12) in this case.) Thus, to illustrate how (9.2) is attacked, we will assume that $L/O_2(L) \not\cong (S)L_3(q)$. As a further simplification, we will assume $T \trianglelefteq TB$. In the contrary case one has to utilize various factorization theorems at certain points and the arguments are a little more complicated, though the overall idea is the same. We start by obtaining the following, which is quite easy in view of the discussion of the last paragraph.

(9.3) *One of the following holds.*

(a) $N(B) \subseteq M$.

(b) $E(N(B))$ *is a Bender group.*

(c) $E(N(B)) \cong L_2(p^n)$, *p odd.*

(d) $E(N(B)) = 1$ *and* $C_Q(B)$ *is isomorphic to a 2-subgroup of* $GL_2(p)$, *some odd prime p.*

Now choose $X \in \mathcal{M}(T) - \{M\}$, as we may by (3.1)(b). As the elements of $\mathcal{P}_f(T)$ are so restricted then all elements of $\mathcal{L}_f(T)$ are similarly restricted, and X will in general be generated by its solvable subgroups containing T, or else involve $L_2(4)$. In any case we find a nonidentity characteristic subgroup T_0 of T such that $N_X(T_0) \nsubseteq M$. As $T \trianglelefteq TB$, we may assume that

(9.4) *There is* $X \in \mathcal{M}(T) - \{M\}$ *with* $TB \subseteq X$.

The goal is to analyse the subgroup X of (9.4) using (9.3) and variations on (9.3) which describe $N(B_0)$ for suitable nonidentity subgroups B_0 of B, and then to construct the groups $U_4(4)$, $U_5(4)$, ${}^3D_4(4)$ and ${}^3D_4(2)$ which arise from (9.2).

(9.5) *If* X *is nonsolvable, then* $G \cong U_4(4)$, $U_5(4)$ *or* ${}^3D_4(4)$.

In this case we must have $E(X/O_2(X)) \cong L_2(4)$, and

$$B_0 = C_B(E(X/O_2(X))) \neq 1.$$

With the description of $N(B_0)$ as a subgroup of M available, we get precise information on $|C_Q(B_0)|$, and the exact structure of L and X is forthcoming.

(9.6) *If* X *is solvable, then* $G \cong {}^3D_4(2)$.

Here we find there is a nonidentity odd order subgroup P of X with $P \nsubseteq M$ and $[P,B] = 1$. (Actually, one gets along the way that B is necessarily cyclic, and then establishes that P exists.) Thus $C(B) \nsubseteq M$. As $Z \subseteq O_2(X)$ and $P \nsubseteq M$, it follows that $[Z,P]$ is a nontrivial 2-group in $C(B)$. The only possibility is that (9.3)(c) holds with

$$E(N(B)) \cong L_2(7) \cong L_3(2).$$

At this point we are very close to the situation of (8.11) and, with the precise knowledge of $|C_Q(B)|$ available here, we have no trouble in again constructing the appropriate group.

10. Remaining Level Two Analysis

Almost the last major task we must carry out in proving (3.2) is the analysis of $\mathcal{P}_t(T)$ in the remaining cases. So we continue to assume (9.1). After the results of Section 9, we have

(10.1) *If* $L \in \mathcal{P}_t(T)$, *then one of the following holds.*
  (a) $L/O_2(L) \in Chev(odd)$.
  (b) $L/O_2(L) \in Chev(2)$, *defined over* $\mathbf{F}_2$.

(c) $L/O_2(L) \cong L_2(4)$, $(S)L_3(q)$ *for q even, or a sporadic group.*
(d) $L/S(L) \cong L_2(p)$ *as in* (4.1).

Vague allusions were made to the case $L/O_2(L) \cong (S)L_3(q)$ in Section 9, and these can be expanded on now. If, in this case, q is an even power of 2 and $L/O_2(L) \cong SL_3(q)$, then we have $N(A) \subseteq M$ where $Z_3 \cong A \subseteq L$ is incident with $Z(L/O_2(L))$, this because $L \in \mathscr{P}_t(T)$. In this situation the analysis of the last section holds up, as it also does in case $N_L(T)$ covers a Cartan subgroup of $L/O_2(L)$. So we may assume in dealing with $L/O_2(L) \cong (S)L_3(q)$ that $Z(L/O_2(L)) = 1$ and some 2-element of T induces an outer involution of $L/O_2(L)$.

The point of these reductions is that the minimal faithful $\boldsymbol{F}_2(LT/Q)$-modules are now forced to be "large": for example, the standard 3-dimensional $\boldsymbol{F}_q SL_3(q)$-modules cannot admit such an action. All of the remaining groups, excluding $L_2(5)$ and $L_2(7)$ of course, have all nontrivial $\boldsymbol{F}_2$-modules "large". (For example, the invariant m of (5.4) is large compared with the 2-rank of $L/O_2(L)$; cf. [18].) All of this means that the level two analysis of these groups is much more comfortable than one might have originally believed.

Of course, the difficult groups with their small 2-modular representations have, by and large, been handled in Section 9 without the necessity of doing either any weak closure arguments or prolonged analysis of TI-groups. We only need resort to these unpleasant tactics when we must, and the groups of (10.1) are reasonable in this regard.

11. Concluding Remarks

It is reasonable at this time to assume that all elements of $\mathscr{P}(T)$ satisfy the conclusions of (3.2). If there is some element of $\mathscr{M}$ which does not, then we readily find some group $H \in \mathscr{P}_f(S)$ which does not, where S is some (proper) subgroup of T, without loss. The existence of such an H implies that a quasisimple section of H cannot be pushed-up. In fact we have

(11.1) $E(H/O_2(H)) \cong SL_3(4)$, $Sp_4(4)$, or $G_2(4)$.

Now the ideas embodied in (4.10)–(4.12), applied to

$$O^2(E(H \bmod O_2(H))) \in \mathscr{P}_f(S),$$

will help to yield a contradiction.

What remains to be done? First, of course, one has to be sure that the foregoing is actually watertight. That process is still going on, slowly but surely. If we return to the elements of $\mathscr{P}(T)$, assuming this set nonempty, by (3.2) any $L \in \mathscr{P}(T)$ is either $L_2(4)$ or generated by those of its solvable subgroups which permute with T. Thus we expect the analysis to revolve strongly around the control of the solvable subgroups of G permutable with T. There is no doubt, however, that these small cases are problematical. Similar remarks apply when $\mathscr{P}(T) = \emptyset$.

12. Appendix

Here we shall briefly discuss the role that pushing-up plays in the proof of (3.2), and some further directions that this theory might take.

1. The proof of (4.4) is essentially that of showing that $\mathscr{M}(T)$ contains nonsolvable groups. Because of (3.1) there are certainly nonsolvable groups in $\mathscr{M}(S)$ for some $1 \neq S \subseteq T$, and one needs to push-up in the usual sense of the term.

The obstruction to this process, that is the groups which in our situation are not readily pushed-up, are those in *Chev*(2) defined over $\boldsymbol{F}_2$. As these are excluded by the hypothesis of (3.2), (4.4) is easily obtained.

The analogous problem in the general quasithin classification is not so easily dismissed. In fact it is not much less extensive than

Problem 1. Let G be a simple group of characteristic 2-type with $T \in Syl_2(G)$. Show that either all 2-locals are solvable or some group in $\mathscr{M}(T)$ is nonsolvable. Assume G is a $\mathscr{K}$-group if necessary.

2. Further pushing-up problems occur in the course of establishing (4.6). These can be finessed to some extent, but it would be a saving in both space and effort if something along the following lines were available.

Problem 2. X is such that $F^*(X) = O_2(X)$ and $X/O_2(X) \cong SL_3(2^n)$ or $Sp_4(2^n)$. Provide a pushing-up theorem analogous to the Baumann–Glauberman–Niles work on $L_2(2^n)$ (see Chapter 9, Section 2).

Somewhat weaker, but still difficult, is

Problem 3. X as above and $X/O_2(X) \cong Sp_4(2^n)$. Provide an analogue to Campbell's pushing-up of a parabolic, achieved in the case

$$X/O_2(X) \cong SL_3(2^n), \quad [38].$$

8

Odd Standard Form Problems

ROBERT L. GRIESS, Jr.

1. Introduction

One might describe an "odd standard form problem" or an "odd component problem" in the following way. Let p be an odd prime. Suppose that G is a finite group with

$$O_{p'}(G) = 1 \quad \text{and} \quad O_p(G) = 1$$

and suppose that x is an element of G of order p such that $L = L_{p'}(C_G(x))$ is quasisimple, $m_p(C_G(L)) = 1$, and $G = \langle x^G \rangle$. Then, given the isomorphism type for L and some additional hypotheses, identify G. (Here, as in the remainder of this chapter, we use the notation of Chapter 4.)

This sort of problem is an analogue of the problem of classifying finite groups, given the centralizer of an involution in standard form. Actually, the importance of odd component problems is that they arise as a way to deal with finite groups of characteristic 2 type. In a series of lectures given at the University of Chicago conference on finite group theory in the summer of 1972, Gorenstein suggested that after groups of component type were handled, attention would shift to centralizers of elements of order 3 as a way to deal with groups of characteristic 2 type. Now, a few years later, it turns out that all odd primes (not just 3) figure in the relevant odd component problems. The recent and extremely important Main Theorem of Gorenstein and Lyons [102] indicates the family of odd component problems to be attacked for $e(G) \geqslant 4$. In Chapter 4 of this volume, they have given a slightly simplified version of their results; their theorem is the taking off point for most of the recent work on odd standard form problems where, of course, the full strength of their results can be assumed.

Before proceeding to the follow-up of the Gorenstein–Lyons Theorem, we mention some earlier work on odd standard form problems for $p = 3$. Miller [151] characterized the groups $SL(4,4^n)$, $n \geqslant 2$, by the centralizer of an element of order 3. Chermak [40] characterized $G_2(4^n)$ by the centralizer of an element of order 3. O'Nan [159] characterized various simple groups by centralizers of elements of order 3 which have a component L such that $L/Z(L) \cong L_2(q)$, for various q. Finally, we mention that Stafford [193] characterized J_4 by the centralizer of an element of order 3, a perfect central extension of M_{22} by Z_6. Other authors have encountered special cases of component problems in their classification activity.

There are 3-local characterizations of finite groups not in the sphere of component problems. In particular, we mention the characterization of finite groups having a self-centralizing subgroup of order 3 by Feit and Thompson [52], and the work of various authors on $C\theta\theta$-groups (see, for example, [126]) and groups having no elements of order 6 (e.g. [68]). Also there has been the recent work of Glauberman which, in particular, characterizes S_4-free groups as a consequence of his factorization theorems [84].

We now turn to the general odd standard form problem. In the theorems that have been proved, it has usually been assumed that G is simple, or at least $O_{p'}(G) = Z(G) = 1$. In each case, the hypotheses imply that the subgroup $\langle x \rangle$ of odd prime order is not weakly closed in $C_G(x)$. This is the most striking difference between standard form problems for odd primes and for the prime 2. When the prime is 2, Glauberman's Z^*-theorem [77] implies that $\langle x \rangle^G \cap C_G(x) \neq \{\langle x \rangle\}$. No such general theorem exists for odd primes, even though the statement appears to be true in groups whose composition factors are known simple groups of p-rank at least 3.

The hypothesis that $\langle x \rangle$ is not weakly closed in $C_G(x)$ can be realized when $e(G) \geqslant 4$ by quoting the Main Theorem of Gorenstein and Lyons. For $e(G) \leqslant 3$, it is not clear when one will be able to obtain it. The classification of finite simple groups G with $e(G) \leqslant 3$, when completed, may be the only way around this problem. We remark that the thin groups have been completely classified by Aschbacher [16], while the status of work in the cases $e(G) = 2$ or 3 is discussed elsewhere in this volume.

2. Statement of Results when $m_{2,3}(G) \leqslant 3$

We are aware of four general results about odd standard form problems for $m_{2,3}(G) \leqslant 3$. First, we present the following result due to Reifart and Stroth. Their hypothesis is that G be a $\mathscr{K}^*$-*group* is weaker than the assumption that all proper simple sections of G be known simple groups. Namely, it consists of the assumptions that whenever X is a nonabelian proper simple section of L with $m_{2,3}(X) \leqslant 3$ and Schur multiplier of order 3 or 9, then X is known, and that if X is a finite simple group with $m_{2,3}(X) \leqslant 2$ and X has abelian Sylow 3-subgroups, then X is known.

THEOREM 1 [168]. *Let G be a finite group of characteristic 2 type satisfying the following conditions:*

(a) $m_{2,3}(G) \leqslant 3$;

(b) $G = O^3(G)$; *and*

(c) *G is a* $\mathscr{K}^*$*-group containing a subgroup L which is a perfect central extension of a known sporadic group such that*

 (i) *3 does not divide* $|C_G(L) \cap C_G(L^g)|$ *for* $g \in G - N_G(L)$,

 (ii) $C_G(L)$ *has a nontrivial cyclic Sylow 3-subgroup, and*

 (iii) *if* $S \in Syl_3(C_G(L))$, *then S is not strongly closed in SL.*

Then $L/Z(L) \cong M_{22}$, $Z(L) \cong Z_6$ *and* $G \cong J_4$.

Next, we state two results of Finkelstein and Frohardt, who considered the following hypotheses.

Hypothesis A. G is a finite group containing an element b of order 3 such that, if $C = C_G(b)$, the following conditions hold:

(A1) C has a normal subgroup L isomorphic to a group of Lie type defined over $\boldsymbol{F}_2$;

(A2) $C_C(L)$ has cyclic Sylow 3-subgroups;

(A3) $m_{2,3}(G) = m_{2,3}(C)$;

(A4) $\langle b \rangle$ is not weakly closed in C with respect to G.

Hypothesis A'. Hypothesis A holds, and also

(A5) If $B^* \in \mathscr{B}_{\max}(C;3)$, then B^* acts nontrivially on a 2-subgroup of C.

Hypothesis A''. Hypothesis A holds, together with the following.

(A6) Either b fuses to no element y of L for which $C_L(y) \cong Z_3 \times A_5$ or the subgroup

$$\alpha(B) = \langle [\theta(C(y)),B] \mid y \in B^\# \rangle \langle \bigcap_{y \in B^\#} \theta(C(y)) \rangle$$

has odd order for all subgroups B of C isomorphic to $Z_3 \times Z_3$, where $\theta(X) = O_{3'}(S(X))$ and $S(X)$ is the largest normal solvable subgroup of the group X.

(A7) The nonabelian simple sections of $C(y)$ are known for all elements y of order 3 in C.

The following theorems are those proved by Finkelstein and Frohardt. We shall say that a quasisimple group X has *type* Y if Y is a nonabelian simple group and $X/Z(X) \cong Y$.

THEOREM 2 [62]. *If G is a finite simple group of characteristic 2 type which satisfies Hypothesis A'' and with L of type* $Sp(6,2)$, *then* $G \cong B_4(2)$ *or* $F_4(2)$.

THEOREM 3 [63]. *If G is a finite simple group of characteristic 2 type which satisfies Hypothesis A' with L of type* $A_3(2)$, *then* $G \cong A_5(2)$ *or* ${}^2D_4(2)$.

This last result can probably be obtained if (A7) is dropped from Hypothesis A″. To complete the necessary standard form problems for $e(G) = 3$, they have proved the following.

THEOREM 4 [60]. *If G is a finite simple group of characteristic 2 type which satisfies Hypothesis A with L of type $A_4(2)$, then $G \cong A_6(2)$.*

3. Statement of Results When $e(G) \geqslant 4$

The simple groups of characteristic 2 type with $e(G) \geqslant 4$ having a standard component for a suitable odd prime (in a sense to be made precise below) have now been determined by Gilman and Griess. Their work completely settles the odd standard form problems indicated by the Main Theorem of Gorenstein and Lyons. (By this we shall understand Theorem A of Chapter 4, together with the more detailed information that is given in [102].)

We shall also discuss three other results which follow up the Gorenstein–Lyons theorem. They all have some overlap with the Gilman–Griess result.

Notation and definitions in this section are either standard or may be found in the discussion in Chapter 4.

The first result is due to Finkelstein and Frohardt. See Section 2 for a statement of Hypothesis A.

THEOREM A [61]. *Let G be a finite simple group of characteristic 2 type which satisfies Hypothesis A and suppose that L has type $A_n(2)$, $n \geqslant 5$. Then $G \cong A_{n+2}(2)$, or $n = 5$ and $G \cong E_6(2)$.*

Secondly, we give a characterization of $B_{n+1}(2)$ by Finkelstein and Solomon.

THEOREM B [65]. *Let G be a finite simple group of characteristic 2 type which satisfies Hypothesis A′ and suppose that L has type $B_n(2)$, $n \geqslant 4$. Then $G \cong B_{n+1}(2)$.*

Thirdly, we present a characterization of all groups $G \in Chev(2)$ with $e(G) \geqslant 4$ by Gilman and Griess [75]. It can be proven under a weaker hypothesis than is stated below, but is, in effect, dependent on the work of Gorenstein and Lyons to limit the number of possible configurations to deal with.

Hypothesis C. The hypotheses and alternative (II) of the conclusion of the Main Theorem of Gorenstein and Lyons hold.

This conclusion involves a triple (B,x,L) as in Chapter 4. It is the subgroup L which is to be regarded as the odd standard component and which appears in the next result.

THEOREM C. *Assume Hypothesis C and that the type of L is a member of $Chev(2)$. Then $G \in Chev(2)$.*

Finally, we mention an early and rather special result of Guterman [112], who studies groups with a standard 3-component of type $B_3(q)$, where $q = 4^n$ and $n \geqslant 1$; that is, if $Y = C_G(L)$, then Y has cyclic Sylow 3-subgroups, $m_{2,3}(C) = m_{2,3}(YL)$ and $L \lhd C_G(b_0)$ whenever b_0 is an element of order 3 in Y. He determines the possible fusion patterns in a maximal elementary abelian 3-group of YL. He then characterizes $F_4(q)$ as the only simple group of characteristic 2 type with such a 3-component, under the additional assumptions that either $n > 1$ or $C = YL$ plus some hypotheses on proper sections relative to their 3-locals.

Hypotheses A, B and C all follow from the Main Theorem of Gorenstein and Lyons, but Hypotheses A and B are weaker. They do not, for example, require proper sections of G to be $\mathscr{K}$-groups.

Theorem C suffices to complete the characterization of simple groups of characteristic 2 type with $e(G) \geqslant 4$ which have standard type with respect to some (B,x,L) satisfying the appropriate conditions. In this sense, simple groups of characteristic 2 type with $e(G) \geqslant 4$ having a standard component for a suitable odd prime are known. The existence of such a triple is the assertion of alternative (II) in the Main Theorem of Gorenstein and Lyons.

We can discuss the proofs of Theorems A, B and C together because their outlines are similar. First, let B be an elementary abelian p-group of G containing the given element of order p with $И_G(B;2) \neq \{1\}$ and $m_p(B) = m_{2,p}(G)$. Embed B in B^*, where B^* is elementary abelian and contains every element of order p in its centralizer. Our hypotheses on L imply that $|B^* : B|$ is 1 or p.

Step 1 is the determination of the possibilities for

$$A_G(B^*) = N_G(B^*)/C_G(B^*).$$

Here, $A_G(B^*)$ is a linear group in which the stabilizer of a 1-dimensional subspace $\langle b \rangle$ acts essentially as a Weyl group on $B^*/\langle b \rangle$. See Proposition 4 in Section 4 for the possibilities. The last possibility mentioned is

$$O_p(A_G(B^*)) \neq 1.$$

Assuming this, $B_1 = [B^*, O_p(A_G(B^*))]$ is a hyperplane of B^*, and we can quickly get a contradiction in the following way. (Think of this case arising by viewing x as inducing a field automorphism of order p on $E(G) \in Chev(2)$ and taking $G = E(G)\langle x \rangle$; note that the "wreathed case" $L \wr Z_p$ does not occur here since $m(B) = m_{2,p}(G)$ and p is odd.) By taking $P \in Syl_p(C_G(B_1))$ we have $P = Q\langle x \rangle$, where Q is the unique maximal abelian subgroup of P (Q is "nearly" homocyclic). Then, it is not hard to show that Q is weakly closed in a Sylow p-subgroup of $N = N_G(Q)$. Since p is odd and Q is abelian, the Hall–

Wielandt transfer theorem implies that G/G' and N/N' have isomorphic Sylow p-subgroups. Since $N/C_G(Q)$ looks roughly like $Z_p \times A_{C_G(x)}(B_1)$ with $x \notin N'$ in any case, we contradict the simplicity of G. Thus, $O_p(A_G(B^*)) = 1$ and $A_G(B^*)$ is "essentially" a Weyl group.

Step 2 is concerned with the construction of a subgroup G_0 of G. In the proofs of Theorems A and B, G_0 is defined to be

$$E(\langle L, N_G(B^*)\rangle).$$

In the proof of Theorem C, we are given a triple (B,x,L). We then choose a standard subcomponent‡ (D,K) and let $K_1,\ldots,K_r$ be the complete set of distinct groups occurring as third components in (B,x,L) and in all neighbors of (B,x,L) with respect to (D,K). A consequence of Hypothesis C is that $r \geqslant 2$. The group G_0 is defined to be $\langle K_1,\ldots,K_r\rangle$. It turns out that $G_0 = E(\langle L, N_G(B^*)\rangle)$ in this case too. Next, G_0 is identified as an appropriate member of *Chev*(2). Generators and relations are required for members of *Chev*(2) in order to identify G_0. For Theorems A, B and C, we use, respectively, Propositions 1, 2, and 3 of Section 4. To get the "root elements" needed in those Propositions, we transform appropriate elements of L or of the K_i by conjugation with elements of $N_G(B^*)$.

Step 3 is the proof that $G_0 = G$. Since, by now, we have $G_0 \in Chev(2)$, this will complete the identification of G. Set $M = N_G(G_0)$ and assume that $M \neq G$. The goal is to use Holt's theorem [129] which characterizes finite groups having a 2-central involution which fixes only one point in some transitive permutation representation. Thus, we want to prove statements like $C_G(t) \subseteq M$ and $M \cap t^G = t^M$ for certain involutions t of M. As an intermediate step, we prove statements like $N_G(P) \subseteq M$ for certain p-subgroups P of M. We then get that $C_G(t)$ is generated by suitable subgroups of the form $C_G(t) \cap N_G(P)$. This step requires proving that the p'-cores of certain p-components are trivial, checking some generation results in the known groups and studying the interaction between some 2-local and some p-local subgroups.

This completes the discussion of odd standard form problems.

4. Appendix

We record here the auxiliary results needed in the discussion of Section 3.

In the first Proposition, we say that $\tilde{K}_i$ is a fundamental subgroup of $\tilde{G} \in Chev(2)$ if $\tilde{K}_i = \langle U_\alpha, U_{-\alpha}\rangle$ where α is a fundamental root in a root system for $\tilde{G}$ and $U_\alpha, U_{-\alpha}$ are the root groups corresponding to α, $-\alpha$, respectively. A family of subgroups $\{K_i \mid i = 1,\ldots,n\}$ of a group G is called an $A_2(2)$ *generating system for* G if $G = \langle K_i \mid i = 1,\ldots,n\rangle$, each K_i is isomorphic to

‡A subcomponent (D,K) is the pair obtained from Definition 2.2(6) of Chapter 4, taking $K = L_0$.

$A_2(2)$, and, for each pair $i,j \in \{1,\ldots,n\}$, $i \neq j$, either $[K_i,K_j] = 1$ or K_i,K_j generate $A_2(2)$ as a pair of "root $A_1(2)$'s", i.e. as the $\tilde{K}_i$ above for $\tilde{G} \cong A_2(2)$. Moreover, such a system is said to have *type* A_n,E_6, respectively, if the indexing set $\{1,2,\ldots,n\}$ corresponds to the nodes of a Dynkin diagram of type A_n,E_6, respectively, in such a way that $[K_i,K_j] = 1$ or $\langle K_i,K_j \rangle \cong A_2(2)$ according to whether nodes i and j are disconnected or connected.

PROPOSITION 1. *Let* $\{\tilde{K}_i\}$ *be an* $A_2(2)$ *generating system of type* A_n *or* E_6 *for* X. *Let* $Y \cong A_n(2)$ *or* $E_6(2)$ *and* $K_i = \langle U_{\alpha_i}, U_{-\alpha_i} \rangle$, $1 \leqslant i \leqslant n$, *the fundamental subgroups of* Y. *Then there exists an epimorphism* $\pi : X \to Y$ *such that* $\tilde{K}_i\pi = K_i$, $1 \leqslant i \leqslant n$, *and* $\ker \pi \subseteq Z(X)$. *Also, if* Y *has type* $A_n(2)$, Y *acts transitively on the set of* $A_2(2)$ *generating systems of type* A_n *for* Y.

Proof. See [61].

PROPOSITION 2. *Let* V *be a vector space of dimension* $2n+2$, $n \geqslant 4$, *over a finite field* K *such that* V *has a nondegenerate symplectic or quadratic form. Suppose*

$$V = V_0 \perp V_1 \perp \ldots \perp V_n, \quad dim\, V_i = 2, \quad i = 0, 1, \ldots, n,$$

and $n \geqslant 5$ *if* V *is quadratic.*

Let $T \cong S_{n+1}$ *act by natural coordinate permutations on* $V_0 \perp V_1 \perp \ldots \perp V_n$ *(respectively,* $T \cong S_n$ *on* $V_1 \perp \ldots \perp V_n$ *if* V *is quadratic) and let* $t \in T$ *correspond to the transposition* $(n-1, n)$. *For each* $S \subset \{0, 1, \ldots, n\}$, *let* G_S *be the subgroup of* $G = Sp(V)$ *or* $\Omega^{\pm}(V)$ *fixing pointwise*

$$\bigoplus_{i \notin S} V_i.$$

Then $G = \langle G_{\{0,1,\ldots,n-1\}}, T \rangle$. *Moreover,* G *has a presentation consisting of generators*

$$G_{\{0,1,\ldots,n-1\}} \cup T$$

and relations the multiplication tables of $G_{\{0,1,\ldots,n-1\}}$ *and* T *together with* $[t,G_{01}] = 1$ *if* $G = Sp(V)$ *and* $[t,G_{012}] = 1$ *if* $G = \Omega^{\pm}(V)$.

Proof [65]. This extends the earlier work of Wong [233] which establishes the result when K has odd characteristic.

PROPOSITION 3. *Let* Σ *be an indecomposable root system of rank at least* 2 *and let* $<$ *be an ordering on* Σ [195]. *Let* $p > 0$ *be a prime. To each* $\alpha \in \Sigma$, *let there be associated a "root group"* X_α *of p-power order. Suppose that for any pair of roots* α,β *with* $\alpha \neq -\beta$ *the following holds: whenever* $x_\alpha \in X_\alpha$ *and* $x_\beta \in X_\beta$ *there are elements* $x_\gamma \in X_\gamma$ *for every* $\gamma \in \Sigma$ *of the form* $\gamma = i\alpha + j\beta$, i, j *nonnegative integers, such that*

$$(*) \qquad [x_\alpha, x_\beta] = \prod_\gamma x_\gamma$$

where the order of the product is given by $<$.

Let G be the group generated by all the X_α, $\alpha \varepsilon \Sigma$, subject to the relations in X_α (relations of type (A)) and all relations (*) (relations of type (B)).

Suppose that $\bar{G}$ is a finite, quasisimple group of Lie type over a field of characteristic $p > 0$ such that $\bar{G}$ is generated by elements $\bar{x}_\alpha$, one for each $x_\alpha \in X_\alpha$, $\alpha \in \Sigma$, such that there is a homomorphism $\phi : G \to \bar{G}$ satisfying $x_\alpha \mapsto \bar{x}_\alpha$ for all $x_\alpha \in X_\alpha$, and $\phi|_{X_\alpha}$ is a monomorphism for all $\alpha \in \Sigma$. Then $\ker \phi \subseteq Z(G)$ and $|\ker \phi|$ is finite of order prime to p.

Proof. This proposition is equivalent to the relevant theorems in [195] and [196] (see also [197]), the only difference being that we do not exhibit the multiplicative structure of the root groups and the explicit form of the relations (*) in terms of root elements. (A precise description of the relations for the twisted groups is rather tricky.)

In the next proposition, a *reflection* shall mean a diagonalizable linear transformation in characteristic not 2 which has eigenvalues $-1, 1, 1, \ldots, 1$.

PROPOSITION 4. *Let **F** be a field of characteristic $p \neq 2$ and B an **F**-vector space of dimension $n+1$, $n \geqslant 3$. Suppose that $b_0 \in B$ and that $H \subseteq H^* \subset K$ are finite subgroups of $\mathrm{Aut}_F(B)$ with the following properties.*

(i) $H^* = N_K(\langle b_0 \rangle)$.

(ii) $C_K(b_0)$ *contains* $H(C_K(b_0) \cap C_K(H))$ *with index* 1, 2 *or* 3, *where H is a Weyl group of type A, $B = C$, D, F_4, E_6, E_7 or E_8; also $B/\langle b_0 \rangle$ is a nontrivial subquotient of the natural **Z**-module for H reduced modulo p.*

Then, if $K_1 = \langle g \in K | g$ induces a reflection on $B \rangle$, one of the following holds:

(a) *K_1 is isomorphic to the Weyl group of an indecomposable root system;*

(b) $p = 3$, $n = 3$, *and* $K_1 \cong S_6 \times Z_2$;

(c) $n = 3$, $Z(K_1) = \{\pm 1_B\}$ *and* $K_1/Z(K_1) \cong W^*/Z(W^*)$, *where W^* has index 1 or 2 in a Weyl group of type F_4; or*

(d) $p > 0$, $K_1 = O_p(K_1)(H^* \cap K_1)$ *and K_1 leaves invariant a hyperplane complementing Fb_0.*

Proof. [75]; special cases were proven in [61]. These sources contain the complete list of possibilities for K given one of the above H.

We remark that Wagner [227] has recently announced some classification of finite primitive groups in characteristic $p > 0$ generated by reflections.

Acknowledgement. For their helpful comments and suggestions in the preparation of this article, we thank Daniel Frohardt, Robert Gilman, Patrick McBride, Stephen D. Smith, Ronald Solomon, and, especially, Larry Finkelstein.

9

The Revision Project and Pushing-Up

GEORGE GLAUBERMAN

1. The Revision Project and the Odd Order Paper

In this work we will discuss two subjects: first, current revisions of papers on simple groups, including the Odd Order Paper of Feit and Thompson [53]; second, some recent "pushing-up" theorems. (Since one purpose of revising a paper is to reduce its length, we might describe these two subjects as "slimming down" and "pushing-up".)

In this section, we will discuss "slimming down". One outstanding example of this is given by Thompson's second paper [211] on normal p-complements. It stands as a model for anyone trying to revise a paper. First of all, it cut the length of the proof from 23 pages in his original paper [210] to 4 pages. At the same time, it reduced the complexity of the proof from a level which was difficult for experts to a level suitable for graduate student seminars. Both the reduction in length and the reduction in complexity were real, in the sense that they were not achieved by quoting other long or difficult references.

Thompson's revision is valuable not only for what was taken out, but for what was put in. In the revised paper, Thompson introduced the concept of J-subgroups and showed that the centre and the J-subgroup could be used together to obtain strong information.

We should also note what Thompson's revision did *not* do. Although it did not use certain innovations from the original paper, it did not make them obsolete. These innovations, such as technique for applying weak closure and the Hall–Higman Theorem B ([114]; see also [G: Ch. 11]) reappeared and were further developed in the Odd Order Paper, and in other papers up to the present day. Indeed, let us suppose that Thompson had discovered his revised proof instead of his original proof of four years earlier. Then group theory would probably not be four years ahead of where it is now; more likely, it would be four years behind. Sometimes it is better to succeed the hard way first.

Now let us consider the Odd Order Theorem.

ODD ORDER THEOREM (Feit–Thompson [53]). *Every finite group of odd order is solvable.*

This result had appeared as a conjecture in Burnside's book [B: Note M] over fifty years earlier. Very little work had been done on it until the 1950's when important special cases were proved by Suzuki [205] and Feit, M. Hall and Thompson [51]. Suzuki handled the case in which the centralizer of each nonidentity element of the group is abelian ((CA)-groups), while Feit, Hall and Thompson required only that centralizers be nilpotent ((CN)-groups). As it turned out, the general outlines of the (CN)-paper and of the Odd Order Paper itself both followed the outline of the (CA)-paper. Therefore, we will discuss the (CA)-paper first.

Suzuki begins by assuming that the theorem is false and then takes a counterexample G of minimal order. A short argument shows that

(1) *G is simple and every proper subgroup of G is solvable.*

Using (1), Suzuki shows easily that

(2) *every maximal subgroup of G is a Frobenius group whose Frobenius kernel is a trivial intersection set in G.*

At this point, Suzuki introduces exceptional characters (see, for example, [G: pp. 146–7, 409–10]) and shows by a series of calculations that

(3) *exceptional characters yield detailed information about G which leads to a contradiction.*

In their paper, Feit, Hall and Thompson obtain (1) by a short argument. However, (2) is more difficult, and requires the fact that solvable (CN)-groups have a rather restricted form. Using a somewhat different argument from that of Suzuki, the authors eventually obtain (3).

Although it is quite short by present standards, the paper of Feit, Hall and Thompson has been significantly shortened. The proof of (2) in the paper uses the methods of Thompson's original normal p-complement paper; a much simpler proof, based on Thompson's second normal p-complement paper, appears in [G: Theorem 14.2.3]. In addition, some recent work of Silbey [176] permits one to use Suzuki's original argument for (3) if desired.

The Odd Order Paper is divided into six chapters as follows:

Chapters I–III. General results about p-groups, solvable groups, and characters (70 pages).

Chapter IV. Local analysis of maximal subgroups in the minimal counterexample

Sections 15–25 Uniqueness Theorem (50 pages)

Section 26 Weak analogue of (2) (45 pages)

Chapter V. Weak analogue of (3) (70 pages)
Chapter VI. Contradiction by generators and relations (15 pages).

It should be emphasized that Chapters I–III do not consist merely of quoted results, but contain many innovations which have been applied and developed in later papers. Chapters IV–VI treat the minimal counterexample G. It is trivial to obtain condition (1) of Suzuki's paper. However, the maximal subgroups of G are not necessarily (CA)-groups or (CN)-groups; *a priori*, they can be any solvable groups of odd order. Therefore, the authors can obtain only a weak analogue of Suzuki's step (2) in Chapter IV, and they need to do more work to achieve it. At the end of Chapter IV, they show that the maximal subgroups are almost, but not quite, Frobenius groups, and that the maximal subgroups contain "tamely imbedded" subsets instead of trivial intersection sets. This makes it necessary to use an extensive character-theoretic analysis in Chapter V that does not quite reach a contradiction, but gives remarkably precise information about the maximal subgroups of G. Finally, a complicated and ingenious argument in Chapter VI yields a contradiction.

An excellent summary of the Odd Order Paper is given in Chapter 16 of [G]. In Chapter 14, Gorenstein gives a good, short proof of the Feit–Hall–Thompson Theorem. Based on personal experience, I would strongly recommend reading this proof at least twice before embarking on the Odd Order Paper itself.

Three years ago, I taught a two-quarter class based on a revision of Chapter IV of the Odd Order Paper. At about the same time, Sibley began working on a revision of Chapter V. Although we have had several delays in the past, we are hoping to have our work printed soon.

Both Sibley's revision and mine have substantially reduced some parts of the original proof. However, these reductions are derived partially from innovations in the original paper. For example, the arguments involving weak closure and factorizations in Chapter IV of the Odd Order Paper have been used, in turn, in Thompson's second theorem on normal p-complements, in further results on fusion, and then in the remarkably short proof of the Uniqueness Theorem obtained by Bender [27] which simplifies Chapter IV of the Odd Order Paper.

Thanks to Bender's proof, my revision of Chapter IV starts off with a bang: Bender's work covers Sections 14–25 in about 10 pages. Thus, in working on my revision, I concentrated on the remainder of Chapter IV, i.e. Section 26. Here I had two major goals.

(1) To find some new approach which would reduce the length substantially.
(2) To reorganize and simplify the proof, so that it would become accessible to non-specialists.

Unfortunately, I could not manage (1), but I had better luck with (2) than I had expected. In order to ensure that my class was well prepared, I had required rather heavy prerequisites. But as I went through the course, I found that only selected topics from the first eight chapters of Gorenstein's book were really necessary. These topics not only prepared the students for the class, but also simplified many arguments in the original proof, and thus reduced the difficulty of the proof substantially.

I would like to say a few words about Sibley's work on Chapter V. He has succeeded in both goals (1) and (2), mainly by generalizing to tamely imbedded subsets some of his own earlier results on coherence for trivial intersection subsets [176]. In addition, he has used the simplification of the isometry of characters in Chapter V given by Dade [47]. Sibley has used the same background in character theory as the original proof; in particular, he does not need block theory (except possibly for a gap in the original paper). As a result of his work, he has cut the length of Chapter V approximately in half.

Although Chapter VI is only fifteen pages long, it seems to be the most mysterious and intriguing part of the proof. Gorenstein describes it as an "absolute tour de force". Unfortunately, to my knowledge there have been no substantial reductions in this chapter and no applications of it outside the Odd Order Paper. Recently I have managed to cut it by two pages, but only by placing one of its arguments a few pages earlier, not by adding any new ideas.

Altogether, I think that the present revisions would cut the proof from 250 pages to roughly 150 pages and would make it accessible to advanced graduate students. However, part of my revision includes repetitive arguments, which leads me to suspect that a shorter proof is possible. Hence, I would like to suggest as an open problem:

Problem 1.1. Find a shorter, unified proof that the maximal subgroups of G have the structure described in Section 26.

Even if Problem 1.1 can be solved, there still remains a serious defect; the present revisions do not change the basic outline of the proof. It seems to me that there should be some key idea in the original proof or some totally new idea which could be exploited to give a relatively short proof of the theorem. Despite some similarities between the Odd Order Paper and some classification papers published shortly afterwards, there seem to be several essential arguments peculiar to the former. In this sense, the Odd Order Paper stands alone. Therefore, I think that the needed idea should depend essentially on the fact that $|G|$ is odd, and might have some connection with the mysterious Chapter VI.

Problem 1.2. Figure out what is happening in Chapter VI.

Problem 1.3. Find a really short proof of the whole theorem.

The Odd Order Paper was the first of many long major papers on simple groups. Gorenstein gives an account of such papers written before 1969, in Chapter 16 of his book. They cover the following results:

1. Odd Order Theorem;
2. Dihedral Sylow 2-subgroups Theorem (Gorenstein–Walter [103]);
3. C-group Theorem (Suzuki [207]);
4. N-group Theorem (Thompson [214]);
5. Abelian Sylow 2-subgroups Theorem (Walter [228]).

Two of the above papers have been completely revised by Bender. Shortly after the original abelian Sylow 2-subgroup paper appeared, he applied his "Bender method" to cut about 100 pages from the proof [28]. (By using similar methods, Goldschmidt [91] later extended Bender's work to groups with a strongly closed abelian 2-subgroup.) More recently, Bender [31] has also applied his methods to substantially shorten the Gorenstein–Walter dihedral Sylow 2-subgroups paper. (He uses some preliminary character-theoretic results of Bender–Glauberman [32] derived from the Gorenstein–Walter paper and the work of Brauer.)

Suzuki's C-group Theorem classifies the groups in which the centralizer of each involution has a normal Sylow 2-subgroup. It is actually proved in a series of papers by Suzuki (culminating in [207]). Part of the early material and some related theorems appear in a set of lecture notes by Ito [131] on Frobenius groups and Zassenhaus groups. It would be quite desirable to have a unified revision of Suzuki's work and such closely related theorems; no doubt, there is a great deal of duplication in the literature. Recently, Suzuki has included some portions of these results and other important results in his new book [S] in Japanese.

Another major theorem, obtained shortly after Gorenstein's book was published, was the classification of groups with quasidihedral and wreathed Sylow 2-subgroups by Alperin, Brauer and Gorenstein [2]. Because these groups can involve sections isomorphic to $SL(2,p)$ for odd primes p, one cannot use the "ZJ-subgroup" for "pushing-up" (as described in the introduction of Section 2). Therefore, they present difficulties that do not occur in the earlier classification of groups with dihedral Sylow 2-subgroups. Hence, the following question arises.

Problem 1.4. Can Bender's methods be adapted to yield a substantial reduction in the Alperin–Brauer–Gorenstein paper?

Thompson's N-group paper has been a model for many later papers which have generalized sections of it and formalized some of its arguments. In addition, the N-group Theorem as a whole has been generalized by

Gorenstein–Lyons [100] in their classification of groups in which all 2-local subgroups are solvable. Most of these papers have followed Thompson's original approach, but have generalized or simplified his arguments. In his talk in the Park City, Utah Conference, Mason [143] discussed some general thoughts on revising the N-group paper. We will discuss "pushing-up" in the next section, but will now describe three particular unpublished results related to "pushing-up" which could be applied in a revision of the N-group paper. In particular, each of them yields information about the Sylow 2-subgroups of N-groups.

THEOREM 1.1 (Hayashi [122]). *Suppose S is a Sylow 2-subgroup of a group G and T is a nonidentity subgroup of S. Assume that*

(1) $T \lhd N_G(S)$, *and*

(2) *for every subgroup U of T, $N_G(U)/C_G(U)$ is solvable.*

Then there exists a nonidentity subgroup $W(T)$ of T such that

(a) $W(T) \lhd N_G(T)$, *and*

(b) *whenever T is a Sylow 2-subgroup of a solvable $3'$-subgroup H of G, then $W(T)O(H) \lhd H$.*

The above statement, which is only a very special case of Hayashi's result, extends a result of mine on S_4-free groups (Proposition II.6.1 of Glauberman [84]). For odd primes, one can obtain an analogous result with $W(T) = Z(J(T))$. One would suspect that $W(T)$ can always be chosen to be a characteristic subgroup of T in Theorem 1.1, but this is not known, which makes Hayashi's proof especially remarkable.

I obtained the next result by following the approach of Lemma 13.2 of the N-group paper and by applying a key idea of Baumann's work (see Theorem 2.3) and some earlier results of mine.

THEOREM 1.2. *Suppose p is a prime and S is a Sylow p-subgroup of a simple group G. Assume that*

(1) *every p-local subgroup of G is solvable,*

(2) *S (or merely $N_G(S)$) is contained in a unique maximal p-local subgroup M of G, and*

(3) *$SCN_3(S)$ is not empty.*

Then G satisfies at least one of the following conditions:

(a) *whenever $Z(S) \subseteq U \subseteq S$, then $N_G(U) \subseteq M$;*

(b) *$p = 3$ and S is isomorphic to a Sylow 3-subgroup of $Sp_4(3)$;*

(c) *$p = 3$ and S is isomorphic to a Sylow 3-subgroup of $G_2(3)$.*

Note that, if $p = 2$, then Alperin's work on fusion (in particular, Theorem 5.2 of [1]) yields that M controls the fusion of S in G. Therefore $S \cap Z(O_{2'2}(M) \bmod O_{2'}(M))$ is a strongly closed abelian subgroup of S, and G is known by Goldschmidt's Theorem [91].

Most of the N-group paper is concerned with the situation in which

Theorem 1.2(2) fails, that is, in which a Sylow 2-subgroup of G is contained in at least two maximal 2-local subgroups of G. A new result of Goldschmidt covers an important special case of this situation (Section 18 of Thompson's N-group paper [214]).

THEOREM 1.3 (Goldschmidt [93]). *Suppose M_1 and M_2 are subgroups of a group G. Assume that*

(a) *$M_1 \cap M_2$ is a Sylow 2-subgroup of M_1 and M_2,*

(b) *$M_1/O_2(M_1) \cong M_2/O_2(M_2) \cong S_3$ (the symmetric group of degree 3), and*

(c) *M_1 and M_2 have no common nonidentity normal subgroup.*

Then $|M_1 \cap M_2| \leqslant 2^7$.

This result will be discussed more fully in Chapter 10.

2. Pushing-Up

In this section we will discuss recent results on "pushing-up". Let us recall that for a prime p, a *p-local subgroup* of a group G is a subgroup of the form $N_G(R)$ for some nonidentity p-subgroup R of G. We may describe the "pushing-up" problem as follows:

Suppose p is a prime and H is a p-local subgroup in a group G. Let T be a Sylow p-subgroup of H. Assume that

(2.1) $C_H(O_p(H)) \subseteq O_p(H)$

and

(2.2) $N_G(T) \nsubseteq H$.

Can H be "pushed-up" to (i.e., be shown to be contained in) some strictly larger p-local subgroup of G?

This problem arises in the study of various aspects of group theory, such as fusion and the structure of simple groups. In Section 4, we will discuss an application to fusion in the special area of transfer. In Section 1, we mentioned an application to simple groups, namely, Theorem 1.2. Note that conditions (2.1) and (2.2) are always satisfied when G is a group of characteristic p type (as defined in Definition I.5.22 of Gorenstein [98]) and T is not a Sylow p-subgroup of G.

The pushing up problem can sometimes be solved and sometimes not. For example, it can be solved if there exists a nonidentity characteristic subgroup T^* of T such that $T^* \lhd H$; then

$$H \subset \langle H, N_G(T)\rangle \subseteq N_G(T^*).$$

This occurs, for example, if p is odd and $SL(2,p)$ is not involved in H (Theorem II.7.6 of Glauberman [84]), in which case we can take $T^* = Z(J(T))$. In order to search for a similar "good" characteristic subgroup T^*

in other situations, we can restrict our attention to H and forget about the larger group G. In order to conform with our previous notation, we shall now write G and S for H and T.

In pages 52–5 of Glauberman [84], we discussed the pushing-up problem and related problems and showed that they usually lead to groups of Lie type involved in G. To simplify matters, let us assume that $G/O_p(G)$ itself is a group of Lie type of characteristic p. Then $G/O_p(G)$ is generated by the parabolic subgroups of rank one which contain the Sylow p-subgroup $S/O_p(G)$. In proving certain theorems, one can eventually reduce by induction to the case in which $G/O_p(G)$ is itself of rank one and is isomorphic to $SL(2,p^n)$ for some n. (This is approximately what happens for Theorem 2.1 of Chapter 5 and Theorem 4.1 of this chapter.) For this reason, we consider the following hypothesis.

(E_0) Let G be a group, p be a prime, and S be a Sylow p-subgroup of G. Assume that

(i) $C_G(O_p(G)) \subseteq O_p(G)$,
(ii) S is contained in a unique maximal subgroup of G, and
(iii) for some $K \lhd G$ and some natural number n, $G/K \cong PSL(2,p^n)$.

As explained above, the group G in (E_0) may arise as a subgroup or a section of a larger, possibly simple, group. In any case, we wish to find a nonidentity characteristic subgroup of S that is normal in G. However, we must first take note of a well known counterexample (for any given prime power p^n), described in pages 48–9 of Glauberman [84] and denoted by $Qd(p^n)$. Here

$$O_p(G) \text{ is elementary abelian of order } p^{2n},\ G/O_p(G) \cong SL(2,p^n),$$
$$\text{and } O_p(G) \text{ is a \textit{standard module} (or \textit{natural module}) for } G/O_p(G),$$

i.e., $G/O_p(G)$ acts on $O_p(G)$ in the same way that $SL(2,p^n)$ acts on a 2-dimensional vector space over $GF(p^n)$.

Curiously enough, there are groups which are similar to $Qd(p^n)$ but slightly more complicated, for which there is no difficulty in finding a characteristic subgroup. In particular, one can construct G such that

$$O_p(G) \text{ is extraspecial of order } p^3,\ Z(O_p(G)) = Z(G), \text{ and } G/Z(G) \cong Qd(p).$$

(Let $G = GL(2,3)$ if $p = 2$, and use Lemma 4.2 of Glauberman [79] otherwise.) In this case, $Z(S) = Z(G) \lhd G$ and, of course, $Z(S)$ is a characteristic subgroup of S.

Although one example does not prove much, it turns out by a theorem of Niles that the above example is typical, i.e., that a characteristic subgroup always exists, provided that S is at least slightly more complicated than a Sylow p-subgroup of $Qd(p^n)$.

THEOREM 2.1 (Niles [155]). *Assume* (E_0) *and suppose that* $p \neq 3$ *and that no nonidentity characteristic subgroup of S is a normal subgroup of G. Then*

(a) *S has nilpotence class* 2,
(b) *S has exponent* 4 *if* $p = 2$ *and exponent p otherwise,*
(c) *G has only one noncentral chief factor within* $O_p(G)$,
(d) $G/O_p(G) \cong SL(2,p^n)$, *and*
(e) *the chief factor in* (c) *is a standard module for* $G/O_p(G)$.

Actually, Niles also handles the case in which $p = 3$, which is more complicated and will be included in the next section. In addition, he obtains [155: Theorem A] much more precise information for all p in the original pushing-up problem. Note that if $p = 2$, then the conclusion of Theorem 2.1 yields that $O^p(G)$ is *short* and that $O^p(G) \in \mathscr{X}$ in the terminology and notation of Chapter 5.

Baumann [24] independently obtained Theorem 2.1 in the special case in which $p = 2$ and $G/O_2(G) \cong SL(2,2^n)$. His methods are quite different from Niles' and actually suggest some ways to construct characteristic subgroups. By using his methods, Niles and I have been able to construct two nonidentity characteristic subgroups S_1 and S_2 of S which depend only on S but not on G, and have the following property.

THEOREM 2.2 (Glauberman and Niles [86].) *Assume* (E_0) *and suppose that* $p \neq 3$, $S_1 \ntriangleleft G$, *and* $S_2 \ntriangleleft G$. *Then S and G satisfy conditions* (a)–(e) *of Theorem* 2.1.

Theorem 2.2 allows one to strengthen (as in Theorem 4.1) some applications of Theorem 2.1. Section 3 will be concerned with the proof of a sharper form (Theorem 3.1) of Theorem 2.2. Now we describe some of the background and key ideas.

Let

$$\begin{aligned} d(S) &= \max\{|A| \mid A \text{ is an abelian subgroup of } S\} \\ \mathscr{A}(S) &= \{A \mid A \text{ is an abelian subgroup of } S \text{ and } |A| = d(S)\} \\ J(S) &= \langle \mathscr{A}(S) \rangle, \quad ZJ(S) = Z(J(S)), \quad \tilde{J}(S) = C_S(ZJ(S)). \end{aligned}$$

Thompson's many results on factorizations (e.g., [213]) suggest that we should try to take S_1 and S_2 close to $Z(S)$ and $J(S)$. In fact, we do take S_1 to be a subgroup of $Z(S)$ and S_2 to be a characteristic subgroup of $\tilde{J}(S)$. Part of the reason for such a choice is that we can apply a beautiful result of Baumann.

THEOREM 2.3 (Baumann [23: (2.11.1.4)]). *Assume* (E_0) *and let*

$$N = \langle \tilde{J}(S)^G \rangle.$$

Suppose that $Z(S) \ntriangleleft G$. *Then* $N \triangleleft G$ *and* $\tilde{J}(S)$ *is a Sylow p-subgroup of N.*

It is this theorem that first demonstrated the importance of $\tilde{J}(S)$ by showing that certain results (e.g., Theorem 2.2) could be reduced to the case in which $S = \tilde{J}(S)$.

Going back to our original pushing-up problem, we recall that under suitable restrictions one always has $ZJ(T) \lhd H$. Similarly, we might suspect that one could improve Theorem 2.2 by taking only one subgroup instead of two, e.g., by requiring that $S_1 = S_2$. Unfortunately, this is not possible, as the following examples show.

Let L be any untwisted simple group of Lie type of rank $r \geqslant 3$ over $GF(p^n)$, e.g., $PSL(r+1, p^n)$. Let S be any Sylow p-subgroup of L and let $P_1, \ldots, P_r$ be the parabolic subgroups of rank one in L that contain S. For $i = 1, \ldots, r$, let $G_i = O^{p'}(P_i)$, so that

$$O_p(G_i) = O_p(P_i) \quad \text{and} \quad G_i/O_p(G_i) \cong SL(2,p^n).$$

It is easy to show that $G_1, \ldots, G_r$ satisfy (E_0) and that

$$L = \langle P_1, \ldots, P_r \rangle = \langle N_G(S), G_1, \ldots, G_r \rangle.$$

Since $O_p(L) = 1$, no nonidentity characteristic subgroup of S can be normal in every group G_i. However, Theorem 2.2 shows that, for suitable characteristic subgroups S_1 and S_2, each G_i normalizes S_1 or S_2.

If $r \geqslant 4$ above, then $Z(J(S)) \neq Z(S)$. The proof of Theorem 2.2 shows that whenever $ZJ(S) \neq Z(S)$, one can take $S_1 = Z(S)$ and $S_2 = \tilde{J}(S)$. Thus, if $r \geqslant 4$, each rank one parabolic subgroup of $SL(r+1, p^n)$ containing S must have $Z(S)$ or $\tilde{J}(S)$ as a normal subgroup.

Using the Chevalley commutator formulas, one can show that in each of the above groups L, there is some G_i for which all of the noncentral chief factors within $O_p(G_i)$ are standard modules for $G_i/O_p(G_i)$. Thompson has raised the question of whether Theorem 2.2 can be improved by excluding such configurations. In particular, he has asked (for odd p):

Problem 2.1. Assume (E_0). Suppose that $G/O_p(G) \cong SL(2,p^n)$ and that some noncentral chief factor of G within $O_p(G)$ is not a standard module for $G/O_p(G)$. Is $ZJ(S) \lhd G$?

Problem 2.1 is related to two open problems (Questions 16.3 and 16.4) in Glauberman [FSG]. Aschbacher has raised another problem which would extend Theorem 2.2:

Problem 2.2. Assume (E_0). Do there exist three nonidentity characteristic subgroups S_1,S_2,S_3 of S, which depend only on S, such that either

(i) at least two of S_1,S_2,S_3 are normal subgroups of G, or

(ii) S and G satisfy conditions somewhat weaker than conditions (a)–(e) of Theorem 2.1?

If a solution to Problem 2.2 could be obtained, it would probably yield

Goldschmidt's new result (Theorem 1.3). Examples involving graph automorphisms (similar to the examples in pages 53–4 of Glauberman [84]) show that S_1,S_2,S_3 do not exist if S is isomorphic to a Sylow p-subgroup of $L_5(p^n)$ or (if $p = 2$) $F_4(2^n)$.

3. A Pair of Characteristic Subgroups for Pushing-Up

In Section 2, we mentioned a joint result (Theorem 2.2) with Niles which uses some techniques of Baumann. In this section, we will discuss its proof.

We shall be concerned mainly with the hypothesis (E_0), which we repeat for convenience.

(E_0) Let G be a group, p be a prime, and S be a Sylow p-subgroup of G. Assume that

(i) $C_G(O_p(G)) \subseteq O_p(G)$,
(ii) S is contained in a unique maximal subgroup of G, and
(iii) for some $K \lhd G$ and some natural number n, $G/K \cong PSL(2,p^n)$.

(It follows easily that $K/O_p(G)$ is a p'-group. In many cases, we shall know that

$$G/O_p(G) \cong SL(2,p^n),$$

so that $|K/O_p(G)| = 1$ or 2.)

Given any p-group S, let $\Phi(S)$ be the Frattini subgroup of S and let

$$1 = Z_0(S) \subseteq Z(S) = Z_1(S) \subseteq Z_2(S) \subseteq \ldots$$

be the upper central series of S; i.e., $Z_i(S)/Z_{i-1}(S) = Z(S/Z_{i-1}(S))$ for $i = 1,2,3,\ldots$

Niles and I obtain the following more precise version of Theorem 2.2.

THEOREM 3.1. *Suppose p is a prime and S is an arbitrary p-group. Define*

$$S_0 = \begin{cases} [\Phi(S),S]\Phi(\Phi(S)) & \text{if } p = 2,\\ [[\Phi(S),S],S]\Phi(\Phi(S)) & \text{if } p = 3,\\ [\Phi(S),S]\mho^1(S) & \text{if } p > 3. \end{cases}$$

Assume that $[ZJ(S),S] \neq 1$ or that $S_0 \neq 1$. Then there exist nonidentity characteristic subgroups S_1,S_2 of S having the following properties:

(3.1) *$S_1 \subseteq Z(S)$,*
(3.2) *S_2 is a characteristic subgroup of $\tilde{J}(S)$, and*
(3.3) *whenever (E_0) is satisfied for some group G, then $S_1 \lhd G$ or $S_2 \lhd G$.*

Note that only S is given here; there may be many groups G satisfying (E_0). The hypothesis on S guarantees that S is not "too small". To see this, suppose S violates the hypothesis. Then $S = C_S(ZJ(S)) = \tilde{J}(S)$, S has exponent p^2, and, if $p \neq 3$, S has nilpotence class at most two. For $p = 3$, the proof shows that we can take $S_0 = [\Phi(S),S]\mho^1(S)$ if we weaken our restrictions on S_2 to require only that S_2 be invariant under $O^2(Aut\, \tilde{J}(S))$ (and thus under every automorphism of S of order a power of 3). This is relevant to the pushing-up problem at the beginning of Section 2 if $p = 3$ and T is not a Sylow 3-subgroup of G.

For part of our proof, and for an application in Section 4, we must consider the following notation and hypothesis.

(E) G, p, S, K, and n satisfy (E_0).
$M = O_p(G)$.
$Z(S) \neq Z(G)$ and $\tilde{J}(S) \ntrianglelefteq G$.

Now take S and p as in Theorem 3.1. We will discuss some aspects of the proof. Our first result is an easy exercise.

PROPOSITION 3.2. *Suppose S_1 and S_2 satisfy* (3.1), (3.2), *and*

(3.3′) *whenever* (E) *is satisfied for some group G, then $S_1 \triangleleft G$ or $S_2 \triangleleft G$.*

Then S_1 and S_2 satisfy the conclusion of Theorem 3.1.

Now we can eliminate some special cases.

PROPOSITION 3.3.

(a) *If $[ZJ(S),S] \neq 1$, then the conclusion of Theorem* 3.1 *is satisfied for*

$$S_1 = [ZJ(S),S] \cap Z(S) \quad \text{and} \quad S_2 = \tilde{J}(S).$$

(b) *If $Z(S)$ is not elementary abelian, then the conclusion of Theorem* 3.1 *is satisfied for*

$$S_1 = \mho^1(Z(S)) \quad \text{and} \quad S_2 = \tilde{J}(S).$$

In the proof of Proposition 3.3, we first note that, by Proposition 3.2, both (a) and (b) reduce to the case of a group G that satisfies (E). Here,

$$G/C_G(Z(M)) \cong SL(2,p^n)$$

and $Z(M)/Z(G)$ is a standard module for $G/C_G(Z(M))$. In particular, $Z(M)/Z(G)$ is elementary abelian. This gives an easy proof of (b), and (a) follows by a variant of the proof of Theorem 2.3. (Note that $[ZJ(S),S] = 1$ if and only if $S = \tilde{J}(S)$.)

Proposition 3.3 allows us to assume henceforth that

(3.4) $Z(S)$ is elementary abelian and $\tilde{J}(S) = S$.

We will say that a pair of groups (G^*,M^*) is an (E)-*embeddable pair* for S if condition (E) is satisfied with G and M replaced by G^* and M^*. In this case, we say that M^* is an (E)-*embeddable subgroup* of S. Thus, every (E)-embeddable subgroup of S is normal in S and has an elementary abelian factor group. Let

$$\mathcal{M} = \{M^* \mid M^* \text{ is an (E)-embeddable subgroup of } S \text{ that contains } \Omega_1(Z_2(S))\}.$$

PROPOSITION 3.4. *Suppose*

(a) *$\mathcal{M}$ is empty, or*

(b) *$p = 3$ and no element of $\mathcal{M}$ contains $\Omega_1(Z_3(S))$.*

Then the conclusion of Theorem 3.1 is satisfied for

$$S_1 = S_0 \cap Z(S) \quad \textit{and} \quad S_2 = S = \tilde{J}(S).$$

The proof of Proposition 3.4 reduces to elementary commutator calculations. Now we can assume that

(3.5) $\mathcal{M}$ is not empty and, if $p = 3$, $\mathcal{M}$ contains an element that contains $\Omega_1(Z_3(S))$.

PROPOSITION 3.5. *Suppose M and M^* are distinct (E)-embeddable subgroups of S. Then*

(a) *$J(M) \nsubseteq M^*$,*

(b) *there exists $A \in \mathcal{A}(M) - \mathcal{A}(M^*)$, and*

(c) *for every A as in (b), $AM^* = S$ and $(A \cap M^*)Z(M^*) \in \mathcal{A}(S)$.*

Proposition 3.5 is the first main step toward Theorem 3.1. The difficult part is (a), and its proof is taken from Baumann's paper [24]. Then (b) follows immediately, while (c) follows from the facts mentioned in our discussion of Proposition 3.3. Of course, the same statements are valid with M and M^* interchanged. The next major step is also quite difficult and is taken from Baumann's paper.

PROPOSITION 3.6. *Suppose (G,M) and (G^*,M^*) are (E)-embeddable pairs for S and $M \neq M^*$. Assume that $M \in \mathcal{M}$ or $M^* \in \mathcal{M}$. Let*

$$Q^* = [Z(M^*),G^*] \quad \textit{and} \quad E = \langle Z(M^*)^G \rangle.$$

Then

$$[E,M] = [Z(M^*),S] = Q^* \cap Z(M) = Q^* \cap Z(S) \subseteq Z(G) \subseteq Z(M).$$

Given (3.4) and Propositions 3.2 and 3.6, the following result is an easy exercise. (One must note that, for every $\alpha \in \mathrm{Aut}\, S$, M_1^α and M_2^α are (E)-embeddable and $M_1^\alpha \in \mathcal{M}$.)

PROPOSITION 3.7. *Suppose M_1 and M_2 are (E)-embeddable subgroups of S and $M_1 \in \mathcal{M}$. Assume that, for every $\alpha \in \mathrm{Aut}\, S$, $M_1^\alpha \neq M_2$. For $i = 1,2$, let*

$$Y_i = \langle [Z(M_i^\alpha),S] \mid \alpha \in \mathrm{Aut}\, S \rangle.$$

Then the conclusion of Theorem 3.1 is satisfied for $S_1 = Y_1$ and $S_2 = Y_2$. Moreover, if $S_1 \cap S_2 \neq 1$, the conclusion of Theorem 3.1 is satisfied for

$$S_1 = S_2 = Y_1 \cap Y_2.$$

By (3.5) and Proposition 3.7, we may assume that *all* the (E)-embeddable subgroups of S are of the form M_1^α, $\alpha \in Aut\, S$, for some *fixed* $M_1 \in \mathcal{M}$. Moreover, if $p = 3$, then $M_1 \supseteq \Omega_1(Z_3(S))$. Now it follows from one of the main results of Niles' Ph.D. Thesis (see Theorem 6.9 of [155]) that the conclusion of Theorem 3.1 is satisfied for

$$S_1 = S_0 \cap Z(S) \quad \text{and} \quad S_2 = \bigcap_{\alpha \in Aut\, S} M_1^\alpha.$$

We have now completed the proof of Theorem 3.1. However, for some applications (e.g., in Section 4) the groups Y_1, Y_2 in Proposition 3.7 are not convenient, and we need to know more information. By using the methods in the second half of Baumann's paper (after he proves Proposition 3.6 above) and by using an unpublished argument for $p = 2$ kindly provided by Campbell (from [38]), Niles and I obtain the following result.

THEOREM 3.8. *Assume the hypothesis and notation of Theorem* 3.1. *Suppose* $Z(S)$ *is elementary abelian and* $\tilde{J}(S) = S$. *Then the conclusion of Theorem* 3.1 *is satisfied for* $S_1 = S_0 \cap Z(S)$ *and for* S_2 *defined as follows:*

If $p \neq 3$, *let* $S_2 = \bigcap O_p(G)$, *where* G *ranges over the family of all groups for which* (E) *is satisfied and* $O_p(G) \supseteq \Omega_1(Z_2(G))$.

If $p = 3$, *let* $S_2 = \bigcap O_p(G)$, *where* G *ranges over the family of all groups for which* (E) *is satisfied and* $O_p(G) \supseteq \Omega_1(Z_3(G))$.

(*In either case, let* $S_2 = S$ *if there are no groups* G *in the family described.*)

Note that in Theorem 3.8, $S_2 \supseteq \Phi(S)$.

4. An Application of Pushing-Up to Transfer

In Section 3, we discussed a joint result with Niles on pushing-up. In the section, we will describe an application of this work to transfer [85].

Suppose p is a prime and S is a nonidentity Sylow p-subgroup of a finite group G. It is easy to see that the largest abelian p-factor of G is isomorphic to $S/(S \cap G')$. Theorem 12.4 of (Glauberman, [FSG]) shows that, if $p \geqslant 5$, then this factor group is determined by the normalizer of a single nonidentity characteristic subgroup K_∞ of S; in fact,

$$S \cap G' = S \cap (N_G(K_\infty))'.$$

For a long time, we have tried to prove an analogous result for $p = 3$. So far, we have not succeeded. However, by using the pushing-up results described in the previous sections one can show that $S \cap G'$ is determined jointly by the normalizers of *two* characteristic subgroups of S.

THEOREM 4.1. *Suppose* S *is a nonidentity Sylow* 3-*subgroup of a group* G. *Then there exists a nonidentity characteristic subgroup* K *of* S *such that*

$$S \cap G' = (S \cap (C_G(Z(S)))')(S \cap (N_G(K))').$$

Moreover, K depends only on S, not on G.

The definition of K is similar to the definition of one of the characteristic subgroups described in Section 3, and will be given later.

Theorem 4.1 and the above-mentioned result for $p \geqslant 5$ have the following consequence:

COROLLARY OF THEOREM 4.1. *Suppose p is an odd prime and S is a nonidentity Sylow p-subgroup of a group G. Then*

$$S \cap G' = \langle S \cap (N_G(K^*))' \mid K^* \text{ is a nonidentity characteristic subgroup of } S \rangle.$$

By taking G to be the symmetric group of degree four, one sees that the corollary is not valid if we allow $p = 2$.

The corollary shows that, for p odd, $S \cap G'$ is determined by p-local subgroups which contain S. This can be used to investigate the internal structure of G. For example, if G is a nonabelian simple group and $p = 3$, then Theorem 4.1 yields that

$$(S \cap (C_G(Z(S)))')(S \cap (N_G(K))') = S \cap G' = S.$$

This has been used by Rickman [169] in his work on groups with a fixed-point-free automorphism of order r^2 for some prime r.

For technical reasons, we prove an intermediate result from which Theorem 4.1 follows quickly. In order to state it, we require some notation.

Suppose p is a prime, k is a natural number, and R is a p-group. We let $\mathscr{J}(R,k)$ be the set consisting of every subgroup Q of R for which there exists a group G that satisfies the following conditions:

(4.1) Condition (E) is valid for some groups K and M, with R and k in place of S and n,

(4.2) $[F,R,R] \neq 1$ for some chief factor F of G within $O_p(G)$,

(4.3) $Q = \bigcap_{\alpha \in Aut\, R} (O_p(G))^\alpha$.

For p odd, define

$$\begin{aligned} \mathscr{J}(R) &= \{R\} \cup \mathscr{J}(R,1) \cup \mathscr{J}(R,2), && \text{if } p = 3; \\ \mathscr{J}(R) &= \{R\} \cup \mathscr{J}(R,1), && \text{if } p = 5; \\ \mathscr{J}(R) &= \{R\}, && \text{if } p > 5. \end{aligned}$$

THEOREM 4.2. *Suppose p is an odd prime and S is a Sylow p-subgroup of a finite group G. Then*

$$S \cap G' = \langle S \cap (C_G(Z(S)))', S \cap (N_G(K))' \mid K \in \mathscr{J}(\tilde{J}(S)) \rangle.$$

The proof of Theorem 4.2 is a blend of old and new methods. The first step is to reduce to the case when G is p-constrained. This is done mainly by using the methods of Alperin and Gorenstein in Glauberman [FSG]. We assume that the result is false and that G is a minimal counterexample. Let

$$R = \langle S \cap (C_G(Z(S)))', S \cap (N_G(K))' \mid K \in \mathscr{J}(\tilde{J}(S)) \rangle,$$

$T = O_p(G)$ and $M = N_G(R \cap T)$. Clearly, we have $R \subset S \cap G'$. By using the concept of a section conjugacy functor, we obtain the following result from Theorem I.5.10 of Glauberman [84].

PROPOSITION 4.3. *The group G satisfies the following conditions:*

(a) $C(T) \subseteq T$;

(b) *M is the only maximal subgroup of G that contains S;*

(c) $[T,M] \subseteq T \cap M' \subseteq T \cap R$;

(d) *for each $x \in S - T$, there exists a chief factor F_x of G within T such that $[F_x,x,x] \neq 1$.*

Moreover,

(e) *for every subgroup A_0 of S that is not contained in T, there exist $A \subseteq S$, $x \in G - M$, a transversal $\mathscr{R}$ of $A \cap S^x$ in A, a chief factor F_1 of G within T, and an element v of F_1 such that*

 (i) *A is conjugate to A_0 in G,*

 (ii) *$A \cap S^x = A \cap M^x \subset A$, and*

 (iii) *$\Pi_{h \in \mathscr{R}} V^{h^{-1}} \not\equiv 1$, modulo $[F_1, A \cap S^x]$.*

The induction hypothesis and the properties of $\tilde{J}(S)$ then yield the next result.

PROPOSITION 4.4. *The following conditions are satisfied:*

(a) $M \supseteq N_G(Z(S))$;

(b) *$M \supseteq N_G(K)$ for all $K \in \mathscr{J}(\tilde{J}(S))$;*

(c) *M is the unique maximal subgroup of G that contains $\tilde{J}(S)$; and*

(d) $J(S) \nsubseteq T$.

Conditions (a) and (b) show that we have a strong form of "failure of factorization", a condition that is closely related to the pushing-up problem. As mentioned in Section 2, this sort of condition usually yields some group of Lie type of characteristic p which is involved in G. Now we try to verify this in our present situation.

Under conditions (a) and (d) of Proposition 4.4, $J(S)$ acts nontrivially on some noncentral chief factor F of G within $\Omega_1(Z(T))$. Thompson's Replacement Theorem and an argument on the elements of $\mathscr{A}(S)$ (Theorem 3(b) of Glauberman [80]) then yield $A_0 \in \mathscr{A}(S)$ such that

(4.4) $A_0 \nsubseteq C_G(F), \quad [Z(T),A_0,A_0] = 1,$

and

(4.5) $[X,A_0,A_0,A_0,A_0,A_0,A_0] \subseteq Y$ for every chief factor X/Y of G within T.

We take F and A_0 in this way (with some further restrictions) and let $C = C_G(F)$ and

$$\mathscr{A}^* = \{A \in \mathscr{A}(S) \mid AT \text{ is conjugate to } A_0T \text{ in } G\}.$$

Clearly,

(4.6) the elements of $\mathscr{A}^*$ satisfy analogues of (4.4) and (4.5).

It is at this point that the new methods come into play. If $p \geqslant 7$, then (4.6) and condition (e) of Proposition 4.3 easily yield a contradiction (by a variation of Lemma A1.8, page 56, of Glauberman [FSG]). This is as far as the earlier methods take us. In particular, we would have liked to obtain improved commutator conditions as suggested in Question 16.4 of Glauberman [FSG], but we have not succeeded. However, with an eye toward "pushing-up", we continue to analyse the case in which $p = 3$ or $p = 5$. Here, (4.6), Proposition 4.3(e), and induction eventually yield that

(4.7) $|AC/C| = 3,\ 3^2$, or 5,

and

(4.8) $|F/C_F(A)| = |[F,A]| = |AC/C|$.

Since F is a chief factor of G, G/C operates faithfully and irreducibly on F as a vector space over $\boldsymbol{Z}_p$. By (4.4), "quadratic pairs" enter the situation. In the case in which some element of G/C acts as a transvection on F, the work of McLaughlin [149] shows that some normal subgroup of G/C is a group of Lie type over $GF(p)$. Otherwise, (4.7), (4.8), and some further arguments yield a quadratic pair for the prime 3 with a root group of order 3^2. Then the work of Ho [128] shows that some normal subgroup of G/C is a group of Lie type over $GF(3^2)$. (Actually, in order to avoid using the full strength of Ho's paper, we reduce to some special cases.) Thus, in all cases, G/C has a normal subgroup of Lie type. By applying Proposition 4.3(b), we eventually show that

(4.9) $G/C \cong SL(2,p^k)$ for $p^k = 3, 3^2$, or 5.

By (4.9), Proposition 4.4(c), and Theorem 2.3, $\tilde{J}(S) = S$. At this point, the result of Niles mentioned in Section 3 (Theorem 6.9 of Niles [155]), shows that $K \lhd G$ for some $K \in \mathscr{J}(S)$. This yields a contradiction which completes the proof of Theorem 4.2.

It is not difficult to obtain Theorem 4.1 from Theorem 4.2 by using Theorem 3.8. We let

$$K = \mho^1(Z(\tilde{J}(S))) = \mho^1(ZJ(S))$$

if $ZJ(S)$ is not elementary abelian; otherwise, we let $K = S_2$ for S_2 as in Theorem 3.8.

10

Pushing-Up in Finite Groups

DAVID M. GOLDSCHMIDT

1. Concepts and Results

Generally speaking, "pushing-up" refers to a collection of techniques, ideas and theorems which are concerned with the study of inclusion relations among a given set of p-local subgroups of a group. Although some of the ideas are perhaps foreshadowed in early work of Frobenius, the subject begins in earnest with Thompson's thesis [209]. There we find the following partial order defined on the set of p-subgroups of a group G.

(1.1) For nonidentity p-subgroups P_1 and P_2 of G, define $P_1 \lesssim P_2$ if one of the following holds:

(a) $|N_G(P_1)|_p < |N_G(P_2)|_p$,

(b) $|N_G(P_1)|_p = |N_G(P_2)|_p$ and $|P_1| < |P_2|$, or

(c) $P_1 = P_2$.

This partial order, or a minor variant of it, is usually referred to as the *Thompson order*. In standard terminology, a p-local subgroup $N(P_1)$ can be *pushed-up* if it can be shown that there is another p-local subgroup $N(P_2)$ containing $N(P_1)$ with P_2 strictly above P_1 in the Thompson ordering. Notice that if $P_2 \subseteq N(P_1)$ and $P_1 = O_p(N(P_1))$, conditions which usually hold in practice, then condition (a) must apply if $N(P_1)$ can be pushed up to $N(P_2)$. Thus, if $\langle P_1, P_2 \rangle \subseteq P$ where $P \in Syl_p(N(P_1))$, then $P \notin Syl_p(G)$, and so p divides $|N(P)/P|$. In particular, any characteristic subgroup P_2 of P lies above P_1 in the Thompson ordering.

In a well-known sequel to [209], Thompson obtained (implicitly) the first in a long series of so-called "factorization theorems" [211]. Essentially he showed that for every p-group P there exists a nonidentity characteristic subgroup $J(P)$ such that if P is a Sylow p-subgroup of a group L, where $F^*(L)$ is a p-group and $SL(2,p)$ is not involved in L, then

(1.2) $L = C_L(Z(P))N_L(J(P))$.

Thompson's definition of $J(P)$ was slightly different from that given in the previous chapter; however, using the definition given there, Glauberman [79] later showed that, for odd p, (1.2) can be improved to

(1.2′) $L = N_L(Z(J(P)))$.

Since $Z(J(P))$ is characteristic in P, the preceding discussion shows that, in a group G of characteristic p type, every p-local subgroup of index divisible by p which does not involve $SL(2,p)$ can be pushed-up. Indeed, there exists at most one maximal p-local which does not involve $SL(2,p)$, namely $N_G(Z(J(P)))$ where $P \in Syl_p(G)$.

Unfortunately, examples show that no "single term" factorization such as (1.2′) can hold for $p = 2$ in many interesting situations—for example, even in the symmetric group S_4. As a substitute, one usually tries to obtain a "three against two" factorization theorem. This concept was also introduced by Thompson ([213], [214: I]). In its modern form, it consists of constructing three characteristic subgroups of a 2-group T such that whenever T is a Sylow subgroup of a group L in which $F^*(L)$ is a 2-group, and certain minimal configurations do not arise, then at least two of the three characteristic subgroups are normal in L. For example, Glauberman has obtained such a theorem when S_4 is not involved in L and, as a consequence, has obtained a spectacular simplification of Thompson's classification of 3′-groups [84].

Turning now to more recent work, we first generalize the notion of pushing-up as follows. We shall say that a family $\{L_i\}$ of subgroups of a group G is *p-coherent* if there exists a p-subgroup P such that $P \cap L_i \in Syl_p(L_i)$ for all i, and that a p-coherent family $\{L_i\}$ can be *pushed-up* if there exists a nonidentity p-subgroup Q which is normal in each L_i. The earlier notion is essentially the special case of the family $\{N(P), N(Q)\}$ where $P \in Syl_p(N(Q))$. However, from the point of view of the Thompson ordering, it is sufficient to push-up a family of the type $\{P_1, N(Q)\}$ where P_1 is any p-subgroup properly containing P. Thus there is the possibility of obtaining pushing-up theorems without a factorization theorem. This possibility was first realized by Sims [177] who observed that some graph-theoretic results of Tutte [226] were directly applicable to the case $p = 2$ and $N(Q)/Q \cong S_3$. This idea was later generalized by Glauberman [81]. The Sims–Tutte theorem states, with the above notation, that if Q is a 2-group such that

$$N(Q)/Q \cong S_3, \quad P \in Syl_2(N(Q)) \quad \text{and} \quad |P_1 : P| = 2,$$

then the pair $\{P_1, N(Q)\}$ can be pushed-up unless $|P| \leqslant 2^4$. In a subsequent refinement of this result, Glauberman [82] essentially obtained the following "weak factorization theorem".

(1.3) *If* $L/O_2(L) \cong S_3$, $P \in Syl_2(L)$ *and no characteristic subgroup of* P *is normal in* L, *then* L *has at most one noncentral* 2-*chief factor.*

The conclusion of (1.3) can be improved to obtain a classification of the "pushing-up obstructions" (see (2.14) of [93]). The difference between (1.3) and (1.2′) is an important one—in (1.3) the characteristic subgroup which is normal in L may depend on L (cf. discussion in Section 2 of Chapter 9), whereas in (1.2′) it depends only on P. Thus, for example, suppose that G is a group and L_1, L_2 are subgroups of G with $L_1 \cap L_2 \in Syl_2(L_i)$ and

$$L_i/O_2(L_i) \cong S_3 \quad (i = 1,2).$$

A strong factorization theorem such as (1.2′) would push-up the pair $\{L_1,L_2\}$, but (1.3) does not. Actually, a recent result of Goldschmidt [93] classifies all obstructions in this case and determines the pairs $\{L_1,L_2\}$ for which pushing-up will fail; there are 15 such pairs and, in particular, $\{L_1,L_2\}$ *can* be pushed-up unless L_i has at most two noncentral chief factors and order dividing 3.2^7 $(i = 1,2)$. The ideas behind the proof will be discussed briefly in the next section (cf. Chapter 9, Theorem 1.3).

In the foregoing results, it is best to think of S_3 as $SL(2,2)$. From this point of view, the following generalization of (1.3) obtained independently by Baumann [24] and Niles [155] is natural.

(1.3′) *If L is a group with $O^{2'}(L/O_2(L)) \cong SL(2,2^n)$, $P \in Syl_2(L)$, and no characteristic subgroup of P is normal in L, then L has at most one noncentral 2-chief factor, and $O^{2'}(L/O_2(L))$ induces the standard module for $SL(2,2^n)$ on it.*

Quite recently, Niles [156] has obtained a striking pushing-up theorem for certain families of size three or more. To describe his result, we shall say that a group L is of *rank r parabolic type* (over the field F) if $O^{p'}(L/O_p(L))$ is a group of Lie type over F of rank r. Niles considers a family $\{L_1,\ldots,L_r\}$ of subgroups of a group G which are of rank 1 parabolic type over fields F_i, and sets

$$L_{ij} = \langle L_i,L_j\rangle \quad (1 \leqslant i,j \leqslant r).$$

He then proves that if, for some fixed prime p,

(a) L_i and L_{ij} have a common Sylow p-subgroup $(1 \leqslant i,j < r)$,
(b) each field F_i has characteristic p and cardinality at least 5, and
(c) L_{ij} is of rank 2 parabolic type $(1 \leqslant i,j \leqslant r)$,

then $\langle L_1,\ldots,L_r\rangle$ is of rank r parabolic type. In particular, the family $\{L_1,\ldots,L_r\}$ can be pushed-up unless $O^{p'}(\langle L_1,\ldots,L_r\rangle)$ is a rank r group of Lie type.

The exceptions for small fields are given more precisely. Finally, Chermak [41] has obtained a rank 3 pushing-up theorem over $GF(2)$ where the quotients $L_{ij}/O_2(L_{ij})$ are assumed to be isomorphic to either $SL(3,2)$ or $SL(2,2) \times SL(2,2)$.

2. Local Geometries

Before returning to the result of Goldschmidt mentioned in the previous section, we look at some more general considerations.

Following Serre [175], define an *amalgam* as a triple of groups (P_1,B,P_2) together with monomorphisms $\varphi_i : B \to P_i (i = 1,2)$. We shall represent this diagramatically as

$$P_1 \xleftarrow{\varphi_1} B \xrightarrow{\varphi_2} P_2.$$

Then two amalgams (P_1,B,P_2) and $(\tilde{P}_1,\tilde{B},\tilde{P}_2)$ are *isomorphic* if there exist isomorphisms α_1, θ, and α_2 such that the obvious diagram

$$\begin{array}{ccccc} P_1 & \xleftarrow{\varphi_1} & B & \xrightarrow{\varphi_2} & P_2 \\ \alpha_1 \updownarrow & & \theta \updownarrow & & \alpha_2 \updownarrow \\ \tilde{P}_1 & \xleftarrow{\tilde{\varphi}_1} & \tilde{B} & \xrightarrow{\tilde{\varphi}_2} & \tilde{P}_2 \end{array}$$

commutes. In practice, we do not distinguish the amalgam

$$P_1 \xleftarrow{\varphi_1} B \xrightarrow{\varphi_2} P_2$$

from the amalgam

$$P_2 \xleftarrow{\varphi_2} B \xrightarrow{\varphi_1} P_1.$$

Given an amalgam

$$P_1 \xleftarrow{\varphi_1} B \xrightarrow{\varphi_2} P_2,$$

a *completion* is a group G together with homomorphisms

$$\psi_i : P_i \to G \ (i = 1,2)$$

such that

$$G = \langle P_1\psi_1, P_2\psi_2 \rangle$$

and the diagram

$$\begin{array}{ccccc} P_1 & \xleftarrow{\varphi_1} & B & \xrightarrow{\varphi_2} & P_2 \\ & \searrow^{\psi_1} & & \swarrow^{\psi_2} & \\ & & G & & \end{array}$$

commutes. It is well known that a *universal* completion exists, namely $P_1 *_B P_2$, the *free product with amalgamation* which was first studied by Schreier.

Suppose that P_1 and P_2 are subgroups of a group H. Setting $G = \langle P_1, P_2 \rangle$ and $B = P_1 \cap P_2$, we see that G may be regarded as a completion of the amalgam $P_1 \supseteq B \subseteq P_2$. Now construct a bipartite graph $\mathcal{G}$ as follows. The vertices are given by the union of cosets

$$G/P_1 \cup G/P_2$$

and the edges defined by the joins

$$P_1x \sim P_2y \Leftrightarrow P_1x \cap P_2y \neq \emptyset.$$

Then the edges correspond precisely to the cosets G/B. In general, the graph $\mathcal{G}$ may have circuits. However, these will correspond to relations between the elements of P_1 and P_2, and it can be shown that G is isomorphic to the universal completion of (P_1, B, P_2) if and only if $\mathcal{G}$ is a tree. In any case, G acts on $\mathcal{G}$ in a natural way, preserving the bipartite partition of the vertices.

Suppose that B contains no nonidentity normal subgroup of G. Since $G = \langle P_1, P_2 \rangle$, this happens if and only if P_1 and P_2 have no common nonidentity normal subgroup. So this is a property of the amalgam $P_1 \supseteq B \subseteq P_2$ alone. For technical reasons, we consider a slightly more restrictive property called *primitivity*. In particular, a completion of a primitive amalgam will act faithfully on the associated graph.

We are interested in the case where $\{P_1, P_2\}$ is the pair $\{L_1, L_2\}$ of the first section, $\{L_1, L_2\}$ cannot be pushed-up, and

$$|P_i : B| = 3 \quad (i = 1,2),$$

which we describe by saying that the amalgam has *index* (3,3). It turns out that $P_1 \supseteq B \subseteq P_2$ is primitive in this situation. Our main theorem is the following [93].

(2.1) *There are exactly* 15 *isomorphism classes of primitive amalgams* $P_1 \supseteq B \subseteq P_2$ *of index* (3,3) *in which B is finite. In particular,* $|B| \leqslant 2^7$.

This classification is obtained by studying universal completions. Given $P_1 \supseteq B \subseteq P_2$ and a universal completion G, the associated graph $\mathcal{G}$ is a cubic tree—i.e., every vertex is joined to precisely three others. (In fact, this is the Bruhat–Tits affine building associated with the arithmetic group $SL(2, \mathbf{Q}_2)$.) Thus we can determine the possible amalgams by studying a single combinatorial object. Tutte had earlier studied this tree, and Sims' theorem considered the special case in which it is assumed that P_1 and P_2 are conjugate in a (finite) completion G.

It is not possible to give an outline of the proof of (2.1) here, so we shall restrict ourselves to making a few combinatorial points. The subgroups P_1 and P_2 are represented by adjacent vertices α, β. Define

$$\Delta(\alpha) = \{\omega \in \mathcal{G} \mid \omega \sim \alpha\}.$$

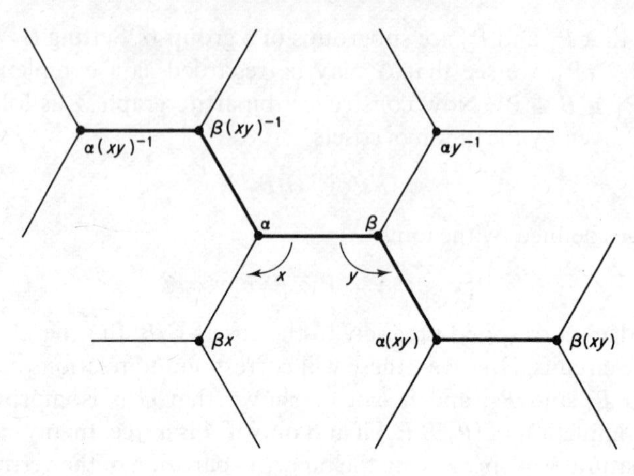

FIG. 1.

The stabilizer of α is just P_1, and hence P_1 acts on $\Delta(\alpha)$. Since $\beta \in \Delta(\alpha)$ and

$$|P_1 : P_1 \cap P_2| = 3,$$

it follows that this action is transitive. Also, since $\mathscr{G}$ is a cubic tree, the kernel of the action of P_1 on $\Delta(\alpha)$ is a 2-group; in fact, it is $O_2(P_1)$. Thus, in particular, B is a 2-group. Similar considerations apply to P_2.

Now choose $x \in P_1 - B$ and $y \in P_2 - B$. Then it is easily shown that if $\tau = xy$, then the images of α and β under the cyclic group $\langle \tau \rangle$ form an infinite chain T on which τ acts as a translation of length 2. (See Fig. 1.) We call (T,τ) a *track*. The study of tracks is central to the analysis; in particular, it is necessary to consider arcs contained in T. The reader who studies the proof in [93] is advised to consider graphical illustrations of the arguments.

References for Part A

Books

B. Burnside, W., "Theory of Groups of Finite Order", 2nd edn., Cambridge University Press, 1911; reprinted, Dover, New York, 1955.

C. Carter, R. W., "Simple Groups of Lie Type", Wiley, London and New York, 1972.

Dc. Dickson, L. E., "Linear Groups with an Exposition of the Galois Field Theory", Teubner, Leipzig, 1901; reprinted, Dover, New York, 1958.

De. Dieudonné, J., "La Géométrie des Groupes Classiques, 2nd edn., Springer-Verlag, Berlin, 1963.

FSG. Powell, M. B. and Higman, G. (eds.), "Finite Simple Groups", Academic Press, London, 1971.

G. Gorenstein, D., "Finite Groups", Harper and Row, New York, 1968.

H. Huppert, B., "Endliche Gruppen I", Springer-Verlag, Berlin, 1967.

S. Suzuki, M., "Group Theory I, II" (2 vols, Japanese), Iwanami, Tokyo, 1977, 1978.

T. Tits, J., "Buildings of Spherical Type and Finite (B,N)-Pairs" (Lecture Notes 386), Springer-Verlag, Berlin, 1974.

Articles

1. Alperin, J. L., Sylow intersections and fusion, *J. Algebra*, **6** (1967) 222–241.
2. Alperin, J. L., Brauer, R. and Gorenstein, D., Finite groups with quasi-dihedral and wreathed Sylow 2-subgroups, *Trans. Amer. Math. Soc.*, **151**(1970) 1–261.
3. Alperin, J. L., Brauer, R. and Gorenstein, D., Finite simple groups of 2-rank two, *Scripta Math.*, **29**(1973) 191–214.
4. Alperin, J. L. and Broué, M., Local methods in block theory, *Ann. of Math.*, **110**(1979), 143–157.
5. Alperin, J. L. and Gorenstein, D., The multiplicators of certain simple groups, *Proc. Amer. Math. Soc.*, **17**(1966) 515–519.
6. Alward, L., "Standard subgroups of type $O^-(8,2)'$", Ph.D. thesis, University of Oregon, 1979.
7. Andrilli, S., Ph.D. thesis, Rutgers, 1980.
8. Aschbacher, M., Finite groups generated by odd transpositions, I–IV, *Math. Z.*, **127**(1972) 45–56; *J. Algebra*, **26**(1973) 451–459; 460–478; 479–491.
9. Aschbacher, M., Finite groups with a proper 2-generated core, *Trans. Amer. Math. Soc.*, **197**(1974) 87–112.
10. Aschbacher, M., On finite groups of component type, *Illinois J. Math.*, **19**(1975) 87–115.
11. Aschbacher, M., Tightly embedded subgroups of finite groups, *J. Algebra*, **42**(1976) 85–101.

12. Aschbacher, M., Standard subgroups of alternating type centralized by a 4-group (to appear).
13. Aschbacher, M., Finite groups in which the generalized Fitting group of the centralizer of some involution is symplectic but not extraspecial, *Comm. Algebra*, **4**(1976) 595–616.
14. Aschbacher, M., On finite groups in which the generalized Fitting group of the centralizer of some involution is extraspecial, *Illinois J. Math.*, **21**(1977) 347–364.
15. Aschbacher, M., A characterization of Chevalley groups over finite fields of odd order, *Ann. of Math.*, **106**(1977) 353–398; 399–468; correction, **111**(1980) 411–414.
16. Aschbacher, M., Thin finite simple groups, *J. Algebra*, **54**(1978) 50–152.
17a. Aschbacher, M., On the failure of the Thompson factorization in 2-constrained groups, *Proc. London Math. Soc.* (to appear).
17b. Aschbacher, M., A factorization theorem for 2-constrained groups, *Proc. London Math. Soc.* (to appear).
18. Aschbacher, M., *GF*(2)-representations of finite groups (in preparation).
19a. Aschbacher, M., Finite groups of rank 3, I, II (to appear).
19b. Aschbacher, M., The uniqueness case for finite groups (to appear).
20. Aschbacher, M., Gorenstein, D. and Lyons, R., The embedding of 2-locals in finite groups of characteristic 2-type (to appear).
21. Aschbacher, M. and Seitz, G. M., Involutions in Chevalley groups over fields of even order, *Nagoya Math. J.*, **63**(1976) 1–91; **72**(1978) 135–136.
22. Aschbacher, M. and Seitz, G. M., On finite groups with a standard component of known type, *Osaka J. Math.*, **13**(1976) 439–482; II (to appear).
23. Baumann, B., Endliche nichtauflösbare Gruppen mit einer nilpotenten maximalen Untergruppen, *J. Algebra*, **38**(1976) 119–135.
24. Baumann, B., Über endliche Gruppen mit einer $L_2(2^n)$ isomorphen Faktorgruppe, *Proc. Amer. Math. Soc.*, **74**(1979) 215–222.
25. Beisiegel, B., Semi-extraspezielle *p*-Gruppen, *Math. Z.*, **156**(1977) 247–254.
26. Beisiegel, B., Über einfache endliche Gruppen mit Sylow 2-Gruppen der Ordnung höchstens 2^{10}, *Comm. Algebra*, **5**(1977) 113–170.
27. Bender, H., On the uniqueness theorem, *Illinois J. Math.*, **14**(1970) 376–384.
28. Bender, H., On groups with abelian Sylow 2-subgroups, *Math. Z.*, **117**(1970) 164–176.
29. Bender, H., Transitive Gruppen gerader Ordnung, in denen jede Involution genau einen Punkt festläßt, *J. Algebra*, **17**(1971) 527–554.
30. Bender, H., Goldschmidt's 2-signalizer functor theorem, *Israel J. Math.*, **22**(1975) 208–213.
31. Bender, H., On finite groups with dihedral Sylow 2-subgroups, *J. Algebra* (to appear).
32. Bender, H. and Glauberman, G., Characters of finite groups with dihedral Sylow 2-subgroups, *J. Algebra* (to appear).
33. Brauer, R., "On the structure of groups of finite order", Proc. Internat. Congr. Math. Amsterdam 1954, I, 209–217 (Noordhoff/N. Holland, 1957).
34. Brauer, R. and Suzuki, M., On finite groups of even order whose 2-Sylow subgroup is a quaternion group, *Proc. Nat. Acad. Sci. U.S.A.*, **45**(1959) 1757–1759.
35. Brauer, R., Suzuki, M. and Wall, G. E., A characterization of the one-dimensional unimodular groups over finite fields, *Illinois J. Math.*, **2**(1958) 718–745.
36. Burgoyne, N., Finite groups with Chevalley-type components, *Pacific J. Math.*, **72**(1977) 341–350.
37. Burgoyne, N. and Fong, P., The Schur multipliers of the Mathieu groups, *Nagoya Math. J.*, **27**(1966) 733–745; correction, **31**(1968) 297–304.

38. Campbell, N., "Pushing up in finite groups", Ph.D. thesis, California Institute of Technology, 1979.
39. Carter, R. W., Simple groups and simple Lie algebras, *J. London Math. Soc.*, **40**(1965) 193–240.
40. Chermak, A., Ph.D. thesis, Rutgers, 1975.
41. Chermak, A., On certain subgroups with parabolic-type subgroups over Z_2, *J. London Math. Soc.* (to appear).
42. Chevalley, C., Sur certains groupes simples, *Tôhoku Math. J.*, **7**(1955) 14–66.
43. Conway, J., A group of order 8,315,553,613,086,720,000, *Bull. London Math. Soc.*, **1**(1969) 79–88.
44. Conway, J. and Wales, D. B., Construction of the Rudvalis group of order 145,926,144,000, *J. Algebra*, **27**(1973) 538–548.
45. Cooperstein, B., Mason, G. and McClurg, P., Some questions concerning the 2-modular representations of Chevalley groups (in preparation).
46. Curtis, C. W., Central extensions of groups of Lie type, *J. Reine Ang. Math.*, **220**(1965) 174–185.
47. Dade, E. C., Lifting group characters, *Ann. of Math.*, **79**(1964) 590–596.
48. Dempwolff, U. and Wong, S. K., On finite groups whose centralizer of an involution has normal extraspecial and abelian subgroups, I, II, *J. Algebra*, **45**(1977) 247–253; **52**(1978) 210–217.
49. Egawa, Y., Ph.D. thesis, Ohio State University, 1980.
50. Feit, W., "The current situation in the theory of finite simple groups", Act. Cong. Internat. Math. Nice 1970, I, 55–93 (Gauthier-Villars, Paris, 1971).
51. Feit, W., Hall, M. Jr. and Thompson, J. G., Finite groups in which the centralizer of any nonidentity element is nilpotent, *Math. Z.*, **74**(1960) 1–17.
52. Feit, W. and Thompson, J. G., Finite groups which contain a self-centralizing subgroup of order 3, *Nagoya Math. J.*, **21**(1962) 185–197.
53. Feit, W. and Thompson, J. G., Solvability of groups of odd order, *Pacific J. Math.*, **13**(1963) 775–1029.
54. Finkelstein, L., Finite groups with a standard component of type Janko–Ree, *J. Algebra*, **36**(1975) 416–426.
55. Finkelstein, L., Finite groups with a standard component isomorphic to M_{23}, *J. Algebra*, **40**(1976) 541–555.
56. Finkelstein, L., Finite groups with a standard component of type HJ or HJM, *J. Algebra*, **43**(1976) 61–114.
57. Finkelstein, L., Finite groups with a standard component isomorphic to M_{22}, *J. Algebra*, **44**(1977) 558–572.
58. Finkelstein, L., Finite groups with a standard component whose centralizer has cyclic Sylow 2-subgroups, *Proc. Amer. Math. Soc.*, **62**(1977) 237–241.
59. Finkelstein, L., Finite groups with a standard component of type J_4, *Pacific J. Math.*, **71**(1977) 41–55.
60. Finkelstein, L. and Frohardt, D., A 3-local characterization of $L_7(2)$, *Trans. Amer. Math. Soc.*, **250**(1979) 181–194.
61. Finkelstein, L. and Frohardt, D., Simple groups with a standard 3-component of type $A_n(2)$, $n \geqslant 5$, *Proc. London Math. Soc.* (to appear).
62. Finkelstein, L. and Frohardt, D., Standard 3-subgroups of type $Sp(6.2)$, *Trans. Amer. Math. Soc.* (to appear).
63. Finkelstein, L. and Frohardt, D., Standard 3-subgroups of type $A_3(2)$ (in preparation).
64. Finkelstein, L. and Solomon, R., Standard components of type M_{12} and .3, *Osaka J. Math.*, **16**(1979) 759–774.
65. Finkelstein, L. and Solomon, R., Finite simple groups with a standard 3-component of type $Sp(2n,2)$, $n \geqslant 4$, *J. Algebra*, **59**(1979) 466–480.

66. Finkelstein, L. and Solomon, R., Finite simple groups with a standard 3-component of type $\Omega^{\pm}(2n,2)$ (to appear).
67. Fischer, B., Groups generated by 3-transpositions, *Invent. Math.*, **13**(1971) 232–246; University of Warwick (mimeo.).
68. Fletcher, L. R., Stellmacher, B. and Stewart, W. B., Endliche Gruppen, die kein Element der Ordnung 6 enthalten, *Quart. J. Math. Oxford* (2), **28**(1977) 143–154.
69. Fong, P. and Seitz, G. M., Groups with a (B,N)-pair of rank 1, I, II, *Invent. Math.*, **21**(1973) 1–57; **24**(1974) 191–237.
70. Foote, R., Finite groups with components of 2-rank 1, I, II, *J. Algebra*, **41**(1976) 16–46; 47–57.
71. Foote, R., Finite groups with maximal 2-components of type $L_2(q)$, q odd, *Proc. London Math. Soc.* (3), **37**(1978) 422–458.
72. Fritz, F., On centralizers of involutions with components of 2-rank two, I, II, *J. Algebra*, **47**(1977) 323–374; 375–399.
73. Gilman, R., Components of finite groups, *Comm. Algebra*, **4**(1976) 1133–1198.
74. Gilman, R. and Gorenstein, D., Finite groups with Sylow 2-subgroups of class two, I, II, *Trans. Amer. Math. Soc.*, **207**(1975) 1–101; 103–125.
75. Gilman, R. and Griess, R. L., A characterization of finite groups of Lie type in characteristic 2 (in preparation).
76. Gilman, R. and Solomon, R., Finite groups with small unbalancing 2-components, *Pacific J. Math.*, **83**(1979) 55–106.
77. Glauberman, G., Central elements in core-free groups, *J. Algebra*, **4**(1966) 403–420.
78. Glauberman, G., Subgroups of finite groups, *Bull. Amer. Math. Soc.*, **73**(1967) 1–12.
79. Glauberman, G., A characteristic subgroup of a p-stable group, *Canad. J. Math.*, **20**(1968) 1101–1135.
80. Glauberman, G., Prime-power factor groups of finite groups, *Math. Z.*, **107**(1968) 159–172.
81. Glauberman, G., Normalizers of p-subgroups in finite groups, *Pacific J. Math.*, **29**(1969) 137–144.
82. Glauberman, G., Isomorphic subgroups of finite groups II, *Canad. J. Math.*, **23**(1971) 1023–1039.
83. Glauberman, G., On solvable signalizer functors in finite groups, *Proc. London Math. Soc.* (3), **33**(1976) 1–27.
84. Glauberman, G., "Factorizations in local subgroups of finite groups", CBMS 33, Amer. Math. Soc., Providence, 1977.
85. Glauberman, G., Control of transfer for $p = 3$ (in preparation).
86. Glauberman, G. and Niles, R., A pair of characteristic subgroups for pushing-up (in preparation).
87. Goldschmidt, D. M., A conjugation family for finite groups, *J. Algebra*, **16**(1970) 138–142.
88. Goldschmidt, D. M., Solvable signalizer functors on finite groups, *J. Algebra*, **21**(1972) 137–148.
89. Goldschmidt, D. M., 2-signalizer functors on finite groups, *J. Algebra*, **21**(1972) 321–340.
90. Goldschmidt, D. M., Weakly embedded 2-local subgroups of finite groups, *J. Algebra*, **21**(1972) 341–351.
91. Goldschmidt, D. M., 2-fusion in finite groups, *Ann. of Math.*, **99**(1974) 70–117.
92. Goldschmidt, D. M., Strongly closed 2-subgroups of finite groups, *Ann. of Math.*, **102**(1975) 475–489.

93. Goldschmidt, D. M., Automorphisms of trivalent graphs, *Ann. of Math.*, **111**(1980) 377–406.
94. Gomi, K., Standard subgroups isomorphic to $Sp(4,2^n)$, *Japanese J. Math.*, **4**(1978) 1–76.
95. Gomi, K., Finite groups with a standard subgroup isomorphic to $PSU(4,2)$, *Pacific J. Math.*, **79**(1978) 399–462.
96. Gomi, K., Standard subgroups of type $Sp_6(2)$, I, II, *J. Fac. Sci. Tokyo*, **27**(1980) 87–107; 109–156.
97. Gorenstein, D., Finite simple groups and their classification, *Israel J. Math.*, **19**(1974) 5–66.
98. Gorenstein, D., The classification of finite simple groups, *Bull. Amer. Math. Soc.* (*New Series*), **1**(1979), 46–199.
99. Gorenstein, D. and Harada, K., Finite groups whose 2-subgroups are generated by at most 4 elements, *Mem. Amer. Math. Soc.*, **147**(1974) 1–464.
100. Gorenstein, D. and Lyons, R., Nonsolvable finite groups with solvable 2-local subgroups, *J. Algebra*, **38**(1976) 453–522.
101. Gorenstein, D. and Lyons, R., Non-solvable signalizer functors on finite groups, *Proc. London Math. Soc.* (3), **35**(1977) 1–33.
102. Gorenstein, D. and Lyons, R., The internal structure of finite groups of characteristic 2 type (to appear).
103. Gorenstein, D. and Walter, J. H., The characterization of finite groups with dihedral Sylow 2-subgroups, I, II, III, *J. Algebra*, **2**(1965) 85–151; 218–270; 354–393.
104. Gorenstein, D. and Walter, J. H., Centralizers of involutions in balanced groups, *J. Algebra*, **20**(1972) 284–319.
105. Gorenstein, D. and Walter, J. H., Balance and generation in finite groups, *J. Algebra*, **33**(1975) 224–287.
106. Griess, R. L., Schur multipliers of the known finite simple groups, *Bull. Amer. Math. Soc.*, **78**(1972) 68–71.
107. Griess, R. L., Schur multipliers of finite simple groups of Lie type, *Trans. Amer. Math. Soc.*, **183**(1973) 355–421.
108. Griess, R. L., Schur multipliers of some sporadic simple groups, *J. Algebra*, **32**(1974) 445–466.
109. Griess, R. L., "The structure of the 'Monster' simple group", Proceedings of a conference on finite groups, 113–118, eds. Scott, W. R. and Gross, F. (Academic Press, New York, 1976).
110. Griess, R. L., Mason, D. and Seitz, G. M., Bender groups as standard subgroups, *Trans. Amer. Math. Soc.*, **238**(1978) 179–211.
111. Griess, R. L. and Solomon, R., Finite groups with unbalancing 2-components of $\{L_3(4), He\}$-type, *J. Algebra*, **60**(1979) 96–125.
112. Guterman, M., A characterization of $F_4(4^n)$ as a group with a standard 3-component $B_3(4^n)$, *Comm. Algebra*, **7**(1979) 1079–1102.
113. Hall, M., Jr. and Wales, D. B., The simple group of order 604,800, *J. Algebra*, **9**(1968) 417–450.
114. Hall, P. and Higman, G., The *p*-length of a *p*-soluble group and reduction theorems for Burnside's problem, *Proc. London Math. Soc.* (3), **6**(1956) 1–42.
115. Harada, K., On finite groups having self-centralizing 2-subgroups of small order, *J. Algebra*, **33**(1975) 144–160.
116. Harada, K., "On the simple group F of order $2^{14} . 3^6 . 5^6 . 7 . 11 . 19$", Proceedings of a conference on finite groups, 119–276, eds. Scott, W. R. and Gross, F. (Academic Press, New York, 1976).

117. Harada, K., The automorphism group and the Schur multiplier of the simple group of order $2^{14}.3^6.5^6.7.11.19$, *Osaka J. Math.*, **15**(1978) 633–636.
118. Harris, M. E., Finite groups having an involution centralizer with a 2-component of dihedral type, II, *Illinois J. Math.*, **21**(1977) 621–647.
119. Harris, M. E., Finite groups containing an intrinsic 2-component of Chevalley type over a field of odd order (to appear).
120. Harris, M. E., $PSL(2,q)$-type 2-components and the unbalanced group conjecture, *J. Algebra* (to appear).
121. Harris, M. E. and Solomon, R., Finite groups having an involution centralizer with a 2-component of dihedral type I, *Illinois J. Math.*, **21**(1977) 575–620.
122. Hayashi, M., 2-factorization in finite groups, *Pacific J. Math.*, **84**(1979) 97–142.
123. Held, D., The simple groups related to M_{24}, *J. Algebra*, **13**(1969) 253–296.
124. Hering, C., Kantor, W. and Seitz, G. M., Finite groups with a split (B,N)-pair of rank 1, *J. Algebra*, **20**(1972) 435–475.
125. Higman, D. and Sims, C. C., A simple group of order 44,352,000, *Math. Z.*, **105**(1968) 110–113.
126. Higman, G., "Odd characterizations of finite simple groups" (mimeographed notes), University of Michigan, 1968.
127. Higman, G. and McKay, J., On Janko's simple group of order 50,232,960, *Bull. London Math. Soc.*, **1**(1969) 89–94; 219.
128. Ho, C.-Y., Quadratic pairs for odd primes, *Bull. Amer. Math. Soc.*, **82**(1976) 941–943.
129. Holt, D. F., Transitive permutation groups in which an involution central in a Sylow 2-subgroup fixes a unique point, *Proc. London Math. Soc.*(3), **37**(1978) 165–192.
130. Hunt, D. C., A characterization of the finite group $M(23)$, *J. Algebra*, **21**(1972) 103–112.
131. Ito, N., "Frobenius and Zassenhaus groups" (mimeographed notes), University of Illinois at Chicago Circle, 1969.
132. Janko, Z., A new finite simple group with abelian Sylow 2-subgroups and its characterization, *J. Algebra*, **3**(1966), 147–186.
133. Janko, Z., "Some new simple groups of finite order I", Ist. Naz. Alta Math., Symposia Mathematica 1, Odensi, Gubbio (1968), 25–64.
134. Janko, Z., Nonsolvable finite groups all of whose 2-local subgroups are solvable I, *J. Algebra*, **21**(1972) 458–517.
135. Janko, Z., A new finite simple group of order 86,775,571,046,077,562,880..., *J. Algebra*, **42**(1976) 564–596.
136. Klinger, K. and Mason, G., Centralizers of p-groups in groups of characteristic 2,p-type, *J. Algebra*, **37**(1975) 362–375.
137. Konvisser, M., Embedding of abelian subgroups in p-groups, *Trans. Amer. Math. Soc.*, **153**(1971) 469–481.
138. Landrock, P. and Michler, G., The character tables of groups of Ree type (unpublished).
139. Leon, J. and Sims, C. C., The existence and uniqueness of a simple group generated by $\{3,4\}$-transpositions, *Bull. Amer. Math. Soc.*, **83**(1977) 1039–1040.
140. Lyons, R., Evidence for a new simple group, *J. Algebra*, **20**(1972) 540–569.
141. MacWilliams, A., On 2-groups with no normal abelian subgroups of rank 3 and their occurrence as Sylow 2-subgroups of finite simple groups, *Trans. Amer. Math. Soc.*, **150**(1970) 345–408.
142. Manferdelli, J., "Standard components of type .2", Ph.D. thesis, University of California, Berkeley, 1979.
143. Mason, G., "Finite simple groups of characteristic 2,3-type", Proceedings of a

conference on finite groups, eds. Scott, W. R. and Gross, F., 37–45 (Academic Press, New York, 1976).

144. McBride, P., Nonsolvable signalizer functors on finite groups (in preparation).
145. McBride, P., A fusion theorem for groups of characteristic 2 type (in preparation).
146. McKay, J., "Computing with finite simple groups", Proc. 2nd Internat. Conf. Theory of Groups, Canberra, 1973, 448–452 (Lecture Notes 372, Springer-Verlag, Berlin, 1974).
147. McKay, J. and Wales, D. B., The multiplier of the Higman–Sims group, *Bull. London Math. Soc.*, **3**(1971) 283–285.
148. McKay, J. and Wales, D. B., The multipliers of the simple groups of order 604,800 and 50,232,960, *J. Algebra*, **17**(1971) 262–272.
149. McLaughlin, J., Some groups generated by transvections, *Arch. Math.*, **18**(1967) 364–368.
150. McLaughlin, J., A simple group of order 898,128,000, "Theory of Finite Groups", eds. Brauer, R. and Sah C.-H., 109–111 (Benjamin, New York, 1969).
151. Miller, R., Ph.D. thesis, Rutgers, 1974.
152. Miyamoto, I., Finite groups with a standard subgroup isomorphic to $U_4(2^n)$, $n \geqslant 2$, *Japanese J. Math.*, **5**(1979) 209–244.
153. Miyamoto, I., Finite groups with a standard subgroup of type $U_5(2^n)$, $n > 1$, *J. Algebra*, **64**(1980) 430–459.
154. Nah, C., "Uber endlichen einfach Gruppen die eine standard Untergruppe A besitzen derart, das $A/Z(A)$ zu $L_3(4)$ isomorph ist", Dissertation, Mainz, 1975.
155. Niles, R., Pushing up in finite groups, *J. Algebra*, **57**(1979) 26–63.
156. Niles, R., *BN*-pairs and finite groups with parabolic-type subgroups (to appear).
157. Norton, S., Ph.D. thesis, Cambridge, 1975.
158. O'Nan, M., Some evidence for the existence of a new simple group, *Proc. London Math. Soc.* (3), **32**(1976) 421–479.
59. O'Nan, M., "Some characterizations by centralizers of elements of order 3", Proceedings of a conference on finite groups, eds. Scott, W. R. and Gross, F., 79–84 (Academic Press, New York, 1976).
160. Parrott, D., On Thompson's simple group, *J. Algebra*, **46**(1977) 389–404.
161. Parrott, D., Characterizations of the Fischer groups III, *Trans. Amer. Math. Soc.* (to appear).
162. Patterson, N., Ph.D. thesis, Cambridge, 1972.
163. Patterson, N. and Wong, S. K., A characterization of the Suzuki sporadic simple group of order 448,345,497,600, *J. Algebra*, **39**(1976) 277–282.
164. Ree, R., A family of simple groups associated with the simple Lie algebra F_4, *Amer. J. Math.*, **83**(1961) 401–420.
165. Ree, R., A family of simple groups associated with the simple Lie algebra G_2, *Amer. J. Math.*, **83**(1961) 432–462.
166. Reifart, A., A characterization of the simple group $D_4^2(2^3)$, *J. Algebra*, **50**(1978) 63–68.
167. Reifart, A., On finite groups with large extraspecial subgroups, I, II, *J. Algebra*, **53**(1978) 452–470; **54**(1978) 273–289.
168. Reifart, A. and Stroth, G., Some simple groups with 2-local 3-rank at most 3, *J. Algebra*, **64**(1980) 102–139.
169. Rickman, B., Groups which admit a fixed-point-free automorphism of order p^2, *J. Algebra*, **59**(1979) 77–171.
170. Schur, I., Über die Darstellungen der symmetrischen und alternierenden Gruppen durch gebrochene lineare Substitutionen, *J. Reine Ang. Math.*, **139**(1911) 155–260.

171. Seitz, G. M., Standard subgroups of type $L_n(2^a)$, *J. Algebra*, **48**(1977) 417–438.
172. Seitz, G. M., Chevalley groups as standard subgroup I, *Illinois J. Math.*, **90**(1979) 36–57; II, III (to appear).
173. Seitz, G. M., Some standard subgroups (to appear).
174. Seitz, G. M., Generation of finite groups of Lie type, *Trans. Amer. Math. Soc.* (to appear).
175. Serre, J.-P., "Arbres, amalgams, et SL_2", Astérisque 46, Soc. Math. de France, 1977.
176. Sibley, D., Coherence in finite groups containing a Frobenius section, *Illinois J. Math.*, **20**(1976) 434–442.
177. Sims, C. C., Graphs and finite permutation groups, *Math. Z.*, **95**(1967) 76–86.
178. Sims, C. C., The existence and uniqueness of Lyons' group, "Finite Groups '72", eds. Gagen, T., Hale, M. and Shult, E., 138–141 (North-Holland, Amsterdam, 1973).
179. Smith, F., Finite simple groups all of whose 2-local subgroups are solvable, *J. Algebra*, **34**(1975) 481–520.
180. Smith, F., On finite groups with large extraspecial subgroups, *J. Algebra*, **44**(1977) 477–487.
181. Smith, F., On transitive permutation groups in which a 2-central involution fixes a unique point, *Comm. Algebra*, **7**(1979) 203–218.
182. Smith, S. D., A characterization of orthogonal groups over $GF(2)$, *J. Algebra*, **62**(1980) 39–60.
183. Smith, S. D., A characterization of finite Chevalley and twisted groups of type E over $GF(2)$, *J. Algebra*, **62**(1980) 101–117.
184. Smith, S. D., Large extraspecial groups of width 4 and 6, *J. Algebra*, **58**(1979) 251–281.
185. Solomon, R., Finite groups with Sylow 2-subgroups of type .3, *J. Algebra*, **28**(1974) 182–198.
186. Solomon, R., Finite groups with intrinsic 2-components of type A_n, *J. Algebra*, **33**(1975) 498–522.
187. Solomon, R., Maximal 2-components in finite groups, *Comm. Algebra*, **4**(1976) 561–594.
188. Solomon, R., Standard components of alternating type, I, II, *J. Algebra*, **41**(1976) 496–514; **47**(1977) 162–179.
189. Solomon, R., 2-signalizers in finite groups of alternating type, *Comm. Algebra*, **6**(1978) 529–549.
190. Solomon, R., Some sporadic components of sporadic type, *J. Algebra*, **53**(1978) 93–124.
191. Solomon, R., On certain 2-local blocks, *Proc. London Math. Soc.* (to appear).
192. Solomon, R. and Timmesfeld, F., A note on tightly embedded subgroups, *Arch. Math.*, **31**(1978) 217–223.
193. Stafford, R., A characterization of Janko's simple group J_4 by centralizers of elements of order 3, *J. Algebra*, **57**(1979) 555–566.
194. Steinberg, R., Variations on a theme of Chevalley, *Pacific J. Math.*, **9**(1959) 875–891.
195. Steinberg, R., "Générateurs, relations, et revétements de groupes algébriques", Colloque sur la théorie des groups algébriques, CBRM, Brussels, 113–127 (Libraire Universitaire, Louvain; Gauthier-Villars, Paris, 1962).
196. Steinberg, R., Representations of algebraic groups, *Nagoya Math. J.*, **22**(1963) 33–56.
197. Steinberg, R., "Lectures on Chevalley groups" (mimeographed notes), Yale, 1967.

198. Steinberg, R., Endomorphisms of linear algebraic groups, *Mem. Amer. Math. Soc.*, **80**(1968).
199. Stellmacher, B., Einfache Gruppen die von einer Konjugiertenklasse von Elementen der Ordnung 3 erzeugt werden, *J. Algebra*, **30**(1974) 320–356.
200. Stingl, V., Endliche, einfache Component-type-Gruppen, deren Ordnung nicht durch 2^{11} getielt wird", Dissertation, Mainz, 1976.
201. Stroth, G., A characterization of Fischer's sporadic simple group of order $2^{41}.3^{13}.5^6.7^2.11.13.17.19.23.31.47$, *J. Algebra*, **40**(1976) 499–531.
202. Stroth, G., Einige Gruppen vom Charakteristik 2-Typ, *J. Algebra*, **51**(1978) 107–144.
203. Stroth, G., Endliche Gruppen, die eine Maximale 2-lokale Untergruppe besitzen, so daß $Z(F^*(M))$eine TI-Menge in G ist, *J. Algebra*, **64**(1980) 460–528.
204. Suzuki, M., A characterization of simple groups $LF(2,p)$, *J. Fac. Sci. Univ. Tokyo* (*I*), **6**(1951) 259–293.
205. Suzuki, M., The nonexistence of a certain type of simple group of odd order, *Proc. Amer. Math. Soc.*, **8**(1957) 686–695.
206. Suzuki, M., On a class of doubly transitive groups, *Ann. of Math.*, **75**(1962) 105–145.
207. Suzuki, M., Finite groups in which the centralizer of any element of order 2 is 2-closed, *Ann. of Math.*, **82**(1965) 191–212.
208. Suzuki, M., A simple group of order 448,345,497,600, "Finite Groups", eds. Brauer, R. and Sah, C.-H., 113–119 (Benjamin, New York, 1969).
209. Thompson, J. G., Ph.D. thesis, University of Chicago, 1959.
210. Thompson, J. G., Normal p-complements for finite groups, *Math. Z.*, **72**(1960) 332–354.
211. Thompson, J. G., Normal p-complements for finite groups, *J. Algebra*, **1**(1964) 43–46.
212. Thompson, J. G., Fixed point of p-groups acting on p-groups, *Math. Z.*, **86**(1964) 12–13.
213. Thompson, J. G., Factorizations of p-solvable groups, *Pacific J. Math.*, **16**(1966) 371–372.
214. Thompson, J. G., Nonsolvable finite groups all of whose local subgroups are solvable, I–VI, *Bull. Amer. Math. Soc.*, **74**(1968) 383–437; *Pacific J. Math.*, **33**(1970) 431–536; **39**(1971) 483–534; **48**(1973) 511–592; **50**(1974) 215–297; **51**(1974) 573–630.
215. Thompson, J. G., "Quadratic pairs", Act. Cong. Internat. Math. Nice 1970, I, 375–376 (Gauthier-Villars, Paris, 1971).
216. Thompson, J. G., A simple subgroup of $E_8(3)$, "Finite Groups", ed. Iwahori, N. (Japan Soc. for Promotion of Science, Tokyo, 1976).
217. Timmesfeld, F., A characterization of the Chevalley and Steinberg groups over F_2, *Geom. Ded.*, **1**(1973) 269–323.
218. Timmesfeld, F., Groups generated by root-involutions, I, II, *J. Algebra*, **33**(1975) 75–135; **35**(1975) 367–441.
219. Timmesfeld, F., Groups with weakly closed TI-subgroups, *Math. Z.*, **143**(1975) 243–278.
220. Timmesfeld, F., On elementary abelian TI-subgroups, *J. Algebra*, **44**(1977) 457–476.
221. Timmesfeld, F., Finite simple groups in which the generalized Fitting group of the centralizer of some involution is extraspecial, *Ann. of Math.*, **107**(1978) 297–369.
222. Timmesfeld, F., On the structure of 2-local subgroups in finite groups, *Math. Z.*, **161**(1978) 119–136.

223. Timmesfeld, F., A condition for the existence of a weakly closed TI-set, *J. Algebra*, **60**(1979) 472–484.
224. Timmesfeld, F., On finite groups in which a maximal abelian normal subgroup of some maximal 2-local subgroup is a *T.I.*-set, *Proc. London Math. Soc.* (to appear).
225. Timmesfeld, F., A note on 2-groups of $GF(2^n)$-type, *Arch. Math.*, **32**(1979) 101–108.
226. Tutte, W., On the symmetry of cubic graphs, *Canad. J. Math.*, **11**(1959) 621–624.
227. Wagner, A., Determination of the finite primitive reflection groups over an arbitrary field of characteristic not two, I, II, III (to appear).
228. Walter, J. H., The characterization of finite groups with abelian Sylow 2-groups, *Ann. of Math.*, **89**(1969) 405–514.
229. Walter, J. H., A characterization of Chevalley groups, I, "Finite Groups", ed. Iwahori, N. (Japan Soc. for Promotion of Science, Tokyo, 1976).
230. Walter, J. H., Characterizations of Chevalley groups (to appear).
231. Walter, J. H., B-conjecture; 2-components in finite groups (to appear).
232. Ward, H. N., On Ree's series of simple groups, *Trans. Amer. Math. Soc.*, **121**(1966) 62–89.
233. Wong, W. J., Generators and relations for classical groups, *J. Algebra*, **32**(1974) 529–553.
234. Yamada, H., Finite groups with a standard subgroup isomorphic to $G_2(2^n)$, *J. Fac. Sci. Tokyo*, **26**(1979) 1–52.
235. Yamada, H., Finite groups with a standard subgroup isomorphic to ${}^3D_4(2^n)$, *J. Fac. Sci. Tokyo* (to appear).
236. Yamada, H., Finite groups with a standard subgroup isomorphic to $U_5(2)$, *J. Algebra*, **58**(1979) 527–562.
237. Yamada, H., Standard subgroups isomorphic to $PSU(6,2)$ or $SU(6,2)$, *J. Algebra*, **61**(1979) 82–111.
238. Yoshida, T., A characterization of Conway's group C_3, *Hokkaido Math. J.*, **3**(1974) 232–242.
239. Bombieri, E., Thompson's problem ($\sigma^2 = 3$), *Inventiones Math.*, **58**(1980) 77–100.
240. Lempken, W., The Schur multiplier of J_4 is trivial, *Arch. Math.*, **30**(1978) 267–270.
241. Thompson, J. G., Towards a characterization of $E_2^*(q)$, I, II, III, *J. Algebra*, **7**(1967) 406–414; **20**(1972) 610–621; **49**(1977) 162–166.
242. Thompson, J. G., Uniqueness of the Fischer–Griess group, *Bull. London Math. Soc.*, **11**(1979) 340–346.
243. McLaughlin, J., Some subgroups of $GL(\boldsymbol{F}_2)$, *Illinois J. Math.*, **13**(1969) 108–115.

Part B

Representation Theory of Groups of Lie Type

11

Complex Representation Theory of Finite Groups of Lie Type

R. W. CARTER

Fundamental progress has been made in recent years in determining the irreducible complex representations of the finite groups of Lie type. The key paper on this subject is that on "Representations of reductive groups over finite fields" by Deligne and Lusztig [4], and the results of this paper have been supplemented by more recent work of Lusztig [10], [11]. In this article we give a brief summary, without proofs, of the main results of the Deligne–Lusztig theory.

1. The Groups

The finite simple groups of Lie type can most conveniently be thought of as finite subgroups of simple algebraic groups over an algebraically closed field of prime characteristic. Let G be a simple algebraic group of adjoint type over K, the algebraic closure of the field F_p with p elements. Thus G is one of the following groups:

$$A_\ell(K), B_\ell(K), C_\ell(K), D_\ell(K), G_2(K), F_4(K), E_6(K), E_7(K), E_8(K).$$

Let $\sigma : G \to G$ be a surjective homomorphism of G to itself which has the property that the fixed point set $G_\sigma = \{g \in G \mid g^\sigma = g\}$ is finite. Then the commutator subgroup G'_σ of G_σ is usually simple, apart from a few exceptions in small characteristic.

For example let $G = A_\ell(K) = PGL_{\ell+1}(K)$ and σ be the map of G into itself which raises each matrix coefficient to the qth power where q is a power of p. Then $G_\sigma = PGL_{\ell+1}(q)$ and $G'_\sigma = PSL_{\ell+1}(q)$.

In general, the groups G_σ arising in this way are the Chevalley groups, the twisted groups of Steinberg and Tits, the Suzuki groups and the Ree groups. Their orders are given by the formula

$$|G_\sigma| = q^N(q^{d_1} - \varepsilon_1)(q^{d_2} - \varepsilon_2)\dots(q^{d_\ell} - \varepsilon_\ell)$$

where q is a power of p (possibly a half-integral power in the case of Suzuki and Ree groups), $d_1, d_2 \ldots d_\ell$ are certain positive integers, $\varepsilon_1, \varepsilon_2 \ldots \varepsilon_\ell$ are certain roots of unity, and N is given by

$$N = (d_1 - 1) + (d_2 - 1) + \ldots + (d_\ell - 1).$$

Further information about these groups can be found, for example, in the book of Carter [2] or the notes of Steinberg [14].

We shall consider representations of the groups G_σ rather than of their simple subgroups G'_σ, as the representation theory of G_σ turns out to be more elegant than that of G'_σ in a number of respects, although not differing too markedly from it.

2. Lang's Map

We need a fundamental theorem on algebraic groups over finite fields due to Lang [9]. Let $L : G \to G$ be defined by $L(g) = g^{-1}g^\sigma$. L is called *Lang's map*. It was shown by Lang that L is surjective whenever G_σ is an algebraic group over a finite field, and this result was extended by Steinberg [15] to the more general situation of an arbitrary surjective homomorphism σ for which G_σ is finite.

The algebraic group G contains a maximal torus T which is invariant under σ. Let

$$T_\sigma = \{t \in T \mid t^\sigma = t\}.$$

T_σ is a subgroup of G_σ which is abelian of order prime to p. T_σ is called a maximal torus of G_σ. The group G contains a Borel subgroup B containing T. B decomposes into a semidirect product

$$B = UT, \quad U \cap T = 1,$$

where U is the unipotent radical of B. The concepts from the theory of algebraic groups being used here are introduced and expounded, for example, in the books of Borel [1], or Humphreys [8].

Let $\tilde{X}$ be the subset $L^{-1}(U)$ of G defined by

$$\tilde{X} = L^{-1}(U) = \{g \in G \mid L(g) \in U\}.$$

$\tilde{X}$ is an algebraic subset of G, although not a subgroup. Thus $\tilde{X}$ is an affine algebraic variety.

Now G_σ acts on $\tilde{X}$ by left multiplication since, if $x \in \tilde{X}$, $g \in G_\sigma$, we have

$$L(gx) = (gx)^{-1}(gx)^\sigma = x^{-1}g^{-1}g^\sigma x^\sigma = x^{-1}x^\sigma \in U;$$

hence $gx \in \tilde{X}$. Also T_0 acts on $\tilde{X}$ by right multiplication since, if $x \in \tilde{X}$, $t \in T$, we have

$$L(xt) = (xt)^{-1}(xt)^\sigma = t^{-1}x^{-1}x^\sigma t^\sigma = t^{-1}(x^{-1}x^\sigma)t \in U;$$

hence $xt \in \tilde{X}$. Thus we see that $G_\sigma \times T_\sigma$ acts on the algebraic variety $\tilde{X}$ as a group of automorphisms. We wish to define a linear representation of $G_\sigma \times T_\sigma$ from this algebraic representation, and this is done as follows.

3. ℓ-Adic Cohomology Modules

Let X be an algebraic variety over $K = \bar{F}_p$ and let ℓ be a prime different from p. It has been shown by Grothendieck how one can define ℓ-adic cohomology groups with compact support $H_c^i(X,Q_\ell)$ where Q_ℓ is the field of ℓ-adic numbers. The method of construction of these groups is elaborate, and a recent exposition of this construction has been given by Deligne [3].‡ The $H_c^i(X,Q_\ell)$ have formal properties similar to those of the usual singular homology groups of X over the complex field. The formal properties which are needed in the work we are describing are listed in [10] and [11]. In particular, if Γ is a finite group of automorphisms of X, then Γ acts also on $H_c^i(X,Q_\ell)$ as a group of nonsingular linear maps. Thus $H_c^i(X,Q_\ell)$ becomes a Γ-module over a field of characteristic 0. The trace function

$$g \to tr\,(g,H^i(X,Q_\ell))$$

is therefore a character of Γ, and so takes values which are algebraic integers. Let $\mathscr{L}(g,X)$ be defined by

$$\mathscr{L}(g,X) = \sum_{i \geqslant 0} (-1)^i \operatorname{tr}(g,H^i(X,Q_\ell)), \qquad g \in \Gamma.$$

$\mathscr{L}(g,X)$ is called the *Lefschetz number* of g on X. It can be shown that $\mathscr{L}(g,X)$ is a rational integer. The values $\mathscr{L}(g,X)$ can be calculated without the use of ℓ-adic cohomology by making use of an analogue due to Grothendieck of the Lefschetz fixed point theorem. The map $g \to \mathscr{L}(g,X)$ is a generalized character of Γ.

We now specialize to the situation discussed earlier, replacing X by $\tilde{X}$ and Γ by $G_\sigma \times T_\sigma$. Thus the map

$$(g,t) \to \mathscr{L}((g,t),\tilde{X})$$

is a generalized character of $G_\sigma \times T_\sigma$. Let $\theta : T_\sigma \to C^*$ be an irreducible complex character of T_σ. It is then easy to verify that the map

$$g \to \frac{1}{|T_\sigma|} \sum_{t \in T_0} \mathscr{L}((g,t),\tilde{X})\theta(t)$$

is a generalized character of G_σ. We denote this generalized character by $R_{T,\theta,U}$, since its construction depends on the choice of the σ-stable maximal torus T of G, the complex character θ of T_σ, and the unipotent radical U of a Borel subgroup of G containing T. Properties of the ℓ-adic cohomology modules show that this generalized character is independent of the prime ℓ.

‡ Another exposition of this construction has recently been given by Srinivasan [16].

4. Orthogonality Relations

Now one knows from the theory of algebraic groups that any two maximal tori of G are conjugate. Let T,T' be maximal tori of G and let $N(T,T')$ be defined by

$$N(T,T') = \{g \in G \mid g^{-1}Tg = T'\}.$$

$N(T,T')$ is a union of cosets of T in G and we define $W(T,T')$ to be the corresponding set of cosets

$$W(T,T') = T \backslash N(T,T').$$

In particular, if $T = T'$, then we set $W(T,T') = W(T)$. All the $W(T)$ are isomorphic to the Weyl group W of G.

If T,T' are σ-stable, then σ will act on $N(T,T')$ and $W(T,T')$. In these circumstances one can prove the following formula giving the scalar product of two of the generalized characters $R_{T,\theta,U}$ and $R_{T',\theta',U'}$ of G_σ.

THEOREM [4]. *$(R_{T,\theta,U}, R_{T',\theta',U'})$ is the number of σ-stable $w \in W(T,T')_\sigma$ making the following diagram commute:*

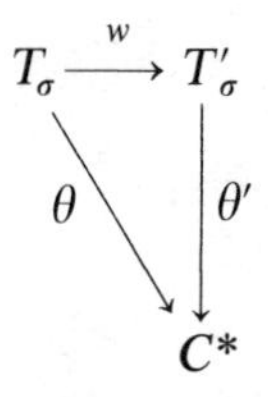

It follows from this formula that $R_{T,\theta,U}$ is independent of U. For

$$(R_{T,\theta,U}, R_{T,\theta,U}) = (R_{T,\theta,U}, R_{T,\theta,U'}) = (R_{T,\theta,U'}, R_{T,\theta,U'}),$$

and so we have

$$(R_{T,\theta,U} - R_{T,\theta,U'},\ R_{T,\theta,U} - R_{T,\theta,U'}) = 0$$

from which it follows that $R_{T,\theta,U} = R_{T,\theta,U'}$. We shall therefore subsequently write $R_{T,\theta,U}$ as $R_{T,\theta}$.

We can also deduce from the above formula the following orthogonality relation:

PROPOSITION. *If T_σ, T'_σ are not conjugate in G_σ and if θ,θ' are complex characters of T_σ, T'_σ respectively, then $(R_{T,\theta}, R_{T',\theta'}) = 0$.*

We note that if the pairs (T,θ) and (T',θ') are conjugate under G_σ, then $R_{T,\theta} = R_{T',\theta'}$. We shall say that the character θ of T_σ is in *general position* if the only element of $W(T)_\sigma$ fixing θ is the identity. If θ is in general position then we see from the above formula that $(R_{T,\theta}, R_{T,\theta}) = 1$. Hence we have:

PROPOSITION. *If θ is in general position then $R_{T,\theta}$ is, to within sign, an irreducible character of G_σ.*

5. Character Formulae

It is possible to give a formula which expresses the values of the generalized character $R_{T,\theta}$ in terms of those on the unipotent elements of G_σ. If $u \in G_\sigma$ is unipotent, we define $Q_{T,u}$ by

$$Q_{T,u} = R_{T,1}(u).$$

$Q_{T,u}$ is a function of a maximal torus and a unipotent conjugacy class in G_σ. It is called a *Green function*, following work of Green on these functions in the case when $G = GL_n$ [6].

THEOREM [4]. *Let $g \in G_\sigma$ and $g = su = us$ be the Jordan decomposition of g, where s is semisimple and u is unipotent. Then*

$$R_{T,\theta}(g) = \frac{1}{|Z^0(s)_\sigma|} \sum_{\substack{x \in G_\sigma \\ xsx^{-1} \in T}} Q^{Z^0(s)}_{x^{-1}Tx,u} \theta(xsx^{-1}).$$

Here $Z(s)$ is the centralizer of s in G and $Z^0(s)$ its connected component containing 1. Since $s \in x^{-1}Tx$, $x^{-1}Tx$ is a maximal torus of $Z^0(s)$ and u is a unipotent element of $Z^0(s)$. $Z^0(s)$ is a reductive group, i.e. has unipotent radical equal to the identity. The theory which we have outlined for simple groups holds more generally for reductive groups and so the Green functions for $Z^0(s)$ can be defined. This gives meaning to the expression $Q^{Z^0(s)}_{x^{-1}Tx,u}$ in the above formula.

The values of the Green functions appear to be very complicated in general. However it is possible to determine their values at the identity.

THEOREM [4]. *Let $\alpha(G)$ be the dimension of a maximal split torus in G_σ and let $\varepsilon_G = (-1)^{\alpha(G)}$. Then*

$$Q_{T,1} = \varepsilon_T \varepsilon_G \frac{|G_\sigma|_{p'}}{|T_\sigma|}.$$

$|G_\sigma|_{p'}$ denotes the part of $|G_\sigma|$ of order prime to p. We recall that $|T_\sigma|$ is always prime to p.

Using this degree formula for $R_{T,1}$ and the above character formula for $R_{T,\theta}$, we obtain the values of the generalized characters $R_{T,\theta}$ on the semisimple elements of G_σ.

PROPOSITION. *Let s be a semisimple element of G_σ. Then*

$$R_{T,\theta}(s) = \frac{\varepsilon_T \varepsilon_{Z^0(s)}}{|T_\sigma| \, |Z^0(s)_\sigma|_p} \sum_{\substack{x \in G_\sigma \\ xsx^{-1} \in T}} \theta(xsx^{-1}).$$

Here $|Z^0(s)_\sigma|_p$ denotes the part of $|Z^0(s)_\sigma|$ which is a power of p.

COROLLARY.

$$R_{T,\theta}(1) = \varepsilon_T \varepsilon_G \frac{|G_\sigma|_{p'}}{|T_\sigma|}$$

Thus $R_{T,\theta}(1)$ is independent of θ.

COROLLARY. *If θ is in general position then $\varepsilon_T \varepsilon_G R_{T,\theta}$ is an irreducible character of G_σ.*

There is a well known representation of G_σ, the Steinberg representation, whose character St is given by

$$St(su) = 0 \text{ unless } u = 1,$$
$$St(s) = \varepsilon_G \varepsilon_{Z^0(s)} |Z^0(s)_\sigma|_p.$$

By multiplying the character values of St by those of $\varepsilon_T \varepsilon_G R_{T,\theta}$ we find:

PROPOSITION.

$$\varepsilon_T \varepsilon_G R_{T,\theta} \otimes St = (\theta_{T_\sigma})^{G_\sigma}$$

where $(\theta_{T_\sigma})^{G_\sigma}$ denotes the induced representation of θ.

Thus $\varepsilon_T \varepsilon_G R_{T,\theta}$, although having very complicated character values, becomes rather simple when tensored with the Steinberg character.

We next form the sum of the generalized characters $\varepsilon_T \varepsilon_G R_{T,\theta}$ over all irreducible complex characters θ of T and over all σ-stable maximal tori of G.

THEOREM.

$$\frac{1}{|G_\sigma|_p} \sum_{\substack{T \\ T^\sigma = T}} \sum_\theta \varepsilon_T \varepsilon_G R_{T,\theta} \text{ is the character of the regular representation of } G_\sigma.$$

It follows from this result that every irreducible character of G_σ occurs as component of some generalized character $R_{T,\theta}$.

6. Geometric Conjugacy Classes

Let $\chi_1, \chi_2, \ldots, \chi_r$ be the irreducible characters of G_σ. We say that χ_i is *equivalent* to χ_j if there exist characters $\chi_{k_1}, \ldots, \chi_{k_t}$ with $\chi_i = \chi_{k_1}$, $\chi_{k_t} = \chi_j$ and such that, for each i, $\chi_{k_{i-1}}$ and χ_{k_i} are components of a common $R_{T,\theta}$. This divides the set of irreducible characters of G_σ into equivalence classes.

Similarly we can divide the set of pairs (T,θ) into equivalence classes, where T is a σ-stable maximal torus of G and θ an irreducible character of T_σ. We say (T,θ) *is equivalent to* (T',θ') if there exist $(T_{k_1},\theta_{k_1}) \ldots (T_{k_t},\theta_{k_t})$ with

$$(T,\theta) = (T_{k_1},\theta_{k_1}), \quad (T_{k_t},\theta_{k_t}) = (T',\theta'),$$

and such that, for each i, $R_{T_{k_{i-1}},\theta_{k_{i-1}}}$ and $R_{T_{k_i},\theta_{k_i}}$ contain a common irreducible component.

The equivalence classes of χ_i and (T,θ) are in natural bijection. These equivalence classes are called *geometric conjugacy classes.* We observe that, if θ is in general position, (T,θ) is in a geometric conjugacy class by itself. (We identify pairs (T,θ) related by conjugacy in G_σ, since they give rise to the same $R_{T,\theta}$).

THEOREM. *If G_σ is a Chevalley group defined from the map $\sigma : G \to G$ which raises each matrix coefficient to the qth power, then the number of geometric conjugacy classes of G_σ is q^ℓ where ℓ is the rank of G.*

Note. This result depends upon the choice of G as a simple algebraic group of adjoint type. It can be generalized suitably when G is a reductive group with connected centre and $\sigma : G \to G$ is an arbitrary surjective homomorphism for which G_σ is finite. However if the centre of G is not connected there is no simple formula for the number of geometric conjugacy classes of G_σ.

7. Duality

We describe a duality relation on the class of reductive groups. Let G be reductive over an algebraically closed field K and T be a maximal torus of G. T is isomorphic to $K^* \times \ldots \times K^*$ (ℓ factors) where K^* is the multiplicative group of K and ℓ is the rank of G. Let $X = Hom(T,K^*)$ and $Y = Hom(K^*,T)$ be the groups of algebraic homomorphisms from T to K^* and K^* to T respectively. X and Y are both free abelian groups of rank ℓ. X is called the *lattice of weights* and Y the *lattice of coweights.* G determines a finite set of elements of X, the *roots*, and a finite set of elements of Y, the *coroots.*

Let G,G^* be two reductive groups over K. We pick maximal tori T,T^* of G,G^* and define X,X^* and Y,Y^* as above. Then G,G^* are called *dual* if there is an isomorphism from X to Y^* taking the roots of G to the coroots of G^*. There will then be an isomorphism from Y to X^* taking the coroots of G to the roots of G^*. Each reductive group G has a unique dual G^*. Further, the dual of G^* is then equal to G.

The duality operation on reductive groups has the effect of both dualizing the root system (viz. interchanging long and short roots) and also dualizing the isogeny type (for example, interchanging adjoint and simply connected groups).

We now suppose that $K = \bar{F}_p$ and that $\sigma : G \to G$ and $\sigma^* : G^* \to G^*$ are surjective homomorphisms such that G_σ and $G^*_{\sigma^*}$ are finite. We say the pairs (G,σ) and (G^*,σ^*) are *dual* if G,G^* are dual and it is possible to choose a σ-

stable maximal torus T of G and a σ^*-stable maximal torus T^* of G^* such that there is an isomorphism from X to Y^* taking the roots of G to the coroots of G^* and making the following diagram commute:

$$\begin{array}{ccc} X & \longrightarrow & Y^* \\ \sigma \downarrow & & \downarrow \sigma^* \\ X & \longrightarrow & Y^* \end{array}$$

Now given G over $K = \bar{F}_p$ and a σ-stable maximal torus T, σ acts naturally on the associated lattices X and Y and one can recover both T_σ and its character group $\hat{T}_\sigma$ from the σ-actions on X and Y. In fact $\hat{T}_\sigma$ is naturally isomorphic to $X/(\sigma-1)X$ and T_0 is naturally isomorphic to $Y/(\sigma-1)Y$. For elements of X yield characters of T and the elements of $(\sigma-1)X$ are those which act trivially on T_σ. So

$$\hat{T}_\sigma \cong X/(\sigma-1)X.$$

In order to establish the isomorphism $T_\sigma \cong Y/(\sigma-1)Y$, we observe that

$$(\sigma-1)^{-1} : Y \otimes \boldsymbol{Q} \to Y \otimes \boldsymbol{Q}$$

is a nonsingular linear map of the rational vector space $Y \otimes \boldsymbol{Q}$ into itself. On restriction to Y, the image of $(\sigma-1)^{-1}$ lies in $Y \otimes \boldsymbol{Q}_{p'}$ where $\boldsymbol{Q}_{p'}$ consists of rationals with denominator prime to p. Taking the elements of $\boldsymbol{Q}_{p'}$ modulo $\boldsymbol{Z}$ we obtain a map

$$(\sigma-1)^{-1} : Y \to Y \otimes \boldsymbol{Q}_{p'}/\boldsymbol{Z}.$$

Now $\boldsymbol{Q}_{p'}/\boldsymbol{Z}$ is isomorphic to K^* since $K = \bar{F}_p$ and $Y \otimes K^*$ is isomorphic to T. Thus we have a map

$$(\sigma-1)^{-1} : Y \to T$$

whose image can be shown to be T_σ and whose kernel is $(\sigma-1)Y$. Hence $Y/(\sigma-1)Y$ is isomorphic to T_σ.

Now suppose we have two dual pairs (G,σ) and (G^*,σ^*). Then there is an isomorphism between X and Y^* consistent with the σ-action on X and the σ^*-action on Y^*. This gives an isomorphism between $X/(\sigma-1)X$ and $Y^*/(\sigma^*-1)Y^*$ which identifies characters of T_σ with elements of $T^*_{\sigma^*}$. This gives an isomorphism between the set of characters of the maximal torus T_σ of G_σ and the set of elements of the dual torus $T^*_{\sigma^*}$ of $G^*_{\sigma^*}$. Let this isomorphism take θ to θ^*.

We now return to our previous situation in which geometric conjugacy classes had been defined, and have the following result.

PROPOSITION [4]. *Let (G,σ), (G^*,σ^*) be dual pairs and T,T' be σ-stable maximal tori of G with corresponding σ^*-stable maximal tori T^*, $(T')^*$ of G^*.*

Then (T,θ) and (T',θ') are geometrically conjugate in G if and only if the corresponding elements $\theta^* \in T^*_{\sigma^*}$ and $(\theta')^* \in (T')^*_{\sigma^*}$ are conjugate in $G^*_{\sigma^*}$.

This proposition shows that a geometric conjugacy class of pairs (T,θ) in G determines a conjugacy class of semisimple elements containing θ^* in $G^*_{\sigma^*}$.

8. Semisimple Characters and Regular Characters

We now wish to obtain a cross-section of the geometric conjugacy classes of irreducible characters of G_σ. Let $\mathscr{C}$ be a geometric conjugacy class. $\mathscr{C}$ determines a corresponding set of pairs (T,θ). Then we have:

THEOREM [4].

$$\sum_{(T,\theta)\in\mathscr{C}} \frac{R_{T,\theta}}{(R_{T,\theta}, R_{T,\theta})}$$

is, to within sign, an irreducible character $\chi_{\mathscr{C}}$ of G_σ. (The sum extends over one pair (T,θ) in each G_σ-conjugacy class in $\mathscr{C}$.) $\chi_{\mathscr{C}}$ lies in the geometric conjugacy class $\mathscr{C}$. Let s^* be an element in the semisimple class corresponding to $\mathscr{C}$ of the dual group $G^*_{\sigma^*}$. Then the degree of $\chi_{\mathscr{C}}$ is given by

$$\chi_{\mathscr{C}}(1) = \frac{|G^*_{\sigma^*}|_{p'}}{|Z^0_{G^*}(s^*)_{\sigma^*}|_{p'}}.$$

The characters $\chi_{\mathscr{C}}$ form a cross-section of the geometric conjugacy classes of irreducible characters of G_σ. They are called the *semisimple characters* of G_σ. (*Note*. Since G has been assumed adjoint, its dual G^* will be simply connected and so $Z_{G^*}(s^*)$ will be connected. Thus $Z^0_{G^*}(s^*) = Z_{G^*}(s^*)$ in the above degree formula.)

In the special case when $G_\sigma = PGL_{\ell+1}(q)$, we have q^ℓ semisimple characters of G_σ defined in this way. Their degrees are polynomials in q which have non-zero constant term. In fact they are characterized by this property since all the remaining irreducible characters of $PGL_{\ell+1}(q)$ have degrees which are polynomials in q with constant term zero.

A second cross-section of the geometric conjugacy classes of irreducible representations of G_σ may be obtained as follows.

THEOREM [4].

$$\sum_{(T,\theta)\in\mathscr{C}} \frac{\varepsilon_T\varepsilon_G R_{T,\theta}}{(R_{T,\theta}, R_{T,\theta})}$$

is an irreducible character $\chi'_{\mathscr{C}}$ of G_σ. (The sum extends as before over one pair (T,θ) in each G_σ-conjugacy class in $\mathscr{C}$.) $\chi'_{\mathscr{C}}$ lies in the geometric conjugacy class $\mathscr{C}$. Its degree is given by

$$\chi'_{\mathscr{C}}(1) = \frac{|G^*_{\sigma^*}|_{p'}}{|Z^0_{G^*}(s^*)_{\sigma^*}|_{p'}} |Z^0_{G^*}(s^*)_{\sigma^*}|_p.$$

The characters $\chi'_{\mathscr{C}}$ also form a cross-section of the geometric conjugacy classes. They may be called the *regular characters* of G_σ.

In the special case when $G_\sigma = PGL_{\ell+1}(q)$, we have q^ℓ regular characters. Their degrees are polynomials in q which, as polynomials, all have the same degree. All the other irreducible characters of $PGL_{\ell+1}(q)$ have degrees which are polynomials in q of strictly smaller polynomial degree than that of the regular characters.

Note. The regular characters as defined here are irreducible and are not to be confused with the character of the regular representation of G_σ. They can be thought of as being dual to the regular conjugacy classes in $G^*_{\sigma^*}$.

9. The Brauer Complex

A geometrical system called the Brauer complex was introduced by Humphreys [7] which, in the case when G_σ is a Chevalley group, seems to be relevant when considering the relationship between the irreducible complex representations of G_σ and the irreducible p-modular representations.

We construct the Brauer complex by considering the semisimple classes in the dual group $G^*_{\sigma^*}$. Let T^* be a σ^*-stable maximal torus of G^*. Every semisimple element of G^* is conjugate to an element of T^* and two elements of T^* are conjugate in G^* if and only if they are equivalent under the action of the Weyl group W. Let $Y^* = \text{Hom}(K^*, T^*)$. Since G is assumed adjoint, G^* will be simply connected and this means that the coroots of G^* generate the lattice Y^*. We consider the rational vector space $Y^* \otimes \boldsymbol{Q}$. This contains the subset $Y^* \otimes \boldsymbol{Q}_{p'}$ of all points whose coordinates in terms of the fundamental coroots have denominators not divisible by p. We recall that $\boldsymbol{Q}_{p'}/\boldsymbol{Z}$ is isomorphic to K^* and that $Y^* \otimes K^*$ is isomorphic to T^*. Thus we have a map

$$Y^* \otimes \boldsymbol{Q}_{p'} \to Y^* \otimes \boldsymbol{Q}_{p'}/\boldsymbol{Z} \cong T^*$$

which has as kernel $Y^* \otimes \boldsymbol{Z} = Y^*$. This map provides a bijection between elements of T^* and elements of $Y^* \otimes \boldsymbol{Q}_{p'}/Y^*$. The Weyl group W also acts on Y^*, and the equivalence classes of T^* under W are in bijective correspondence with the equivalence classes of $Y^* \otimes \boldsymbol{Q}_{p'}$ under the group W_a generated by W and the translations by elements of Y^*. W_a is called the *affine Weyl group*. A fundamental region for $Y^* \otimes \boldsymbol{Q}_{p'}$ under W_a is given by the set A_0 which we shall now define.

We have a map

$$X^* \times Y^* \to Z$$
$$(x,y) \to \langle x,y \rangle$$

defined by saying that y followed by x is the homomorphism $t \to t^{\langle x,y\rangle}$ of K^* into K^*. Let $r_1, r_2, \ldots, r_\ell \in X^*$ be the fundamental roots of G^* and $R \in X^*$ be the highest root of G^*. Then the set A_0 defined by

$$A_0 = \{y \in Y^* \otimes \boldsymbol{Q}_{p'} \mid \langle r_i, y\rangle \geqslant 0, \langle R, y\rangle \leqslant 1\}$$

is a fundamental region for $Y^* \otimes \boldsymbol{Q}_{p'}$ under the action of W_a. Thus there is a bijective correspondence between semisimple conjugacy classes in G^* and elements of A_0. A_0 is called the *fundamental alcove.*

Now σ^* operates on the semisimple conjugacy classes of G^* and, assuming that we have a Chevalley group, there are q^ℓ σ^*-stable semisimple classes. In the same way σ^* can be made to act on A_0, and A_0 will then contain exactly q^ℓ σ^*-stable points.

We now construct the Brauer complex. Let B_0 be the subset of A_0 defined by

$$B_0 = \{y \in Y^* \otimes \boldsymbol{Q}_{p'} \mid \langle r_i, y\rangle \geqslant 0, \quad \langle R, y\rangle \leqslant 1/q\}.$$

Then B_0 is an alcove of the same shape as A_0 but of volume $1/q^\ell$ times as great. By reflecting B_0 in its wall $\langle R, y\rangle = 1/q$ and continuing to reflect the images in their walls, we will eventually cover A_0 by q^ℓ small alcoves obtained from B_0 by successive reflections. The complex formed in this way has q^ℓ simplices of maximum dimension ℓ, and is called the *Brauer complex.*

It was conjectured by Humphreys [7] that there is a natural bijection between the simplices of maximal dimension in the Brauer complex and the set of geometric conjugacy classes of G_σ. This is in fact so and is a consequence of the following result.

THEOREM [5]. *Let G_σ be a Chevalley group. Then for each simplex of maximal dimension in the Brauer complex, the closure of this simplex contains exactly one σ^*-stable point of A_0.*

Since the σ^*-stable points of A_0 are in bijective correspondence with the semisimple conjugacy classes of $G^*_{\sigma^*}$ we deduce the following.

THEOREM. *Let G_σ be a Chevalley group of adjoint type. Then each of the following sets is in natural bijective correspondence with each of the others:*

(i) *Simplices of maximal dimension in the Brauer complex;*
(ii) *Semisimple conjugacy classes of $G^*_{\sigma^*}$;*
(iii) *Semisimple representations of G_σ; and*
(iv) *Geometric conjugacy classes of G_σ.*

One can also find more detailed relations between the position of a given σ^*-stable point in the closure of the corresponding simplex and the degree of the associated semisimple representation of G_σ. (See Deriziotis [5].)

10. Unipotent Characters

We now wish to give a definition of what it means for a character of G_σ to be unipotent.

THEOREM [4]. *The following conditions on an irreducible character χ of G_σ are equivalent:*

(i) *χ is geometrically conjugate to the principal character* 1, *and*
(ii) *χ occurs as component in some $R_{T,1}$.*

The characters satisfying the conditions of this theorem are called *unipotent*. The only character which is both semisimple and unipotent is the principal character. The only character which is both regular and unipotent is the Steinberg character.

The unipotent characters of G_σ have been investigated by Lusztig [11]. The method of determining their number and degrees makes use of the process of induction from proper parabolic subgroups due to Harish–Chandra, in an adaptation to finite groups due to Springer.

We choose a σ-stable Borel subgroup B of G containing T and consider the σ-stable parabolic subgroups of G containing B. Each such parabolic subgroup has a semi-direct product decomposition $P = U_p L$ where U_p is the unipotent radical of P and L is a σ-stable reductive subgroup, called a Levi subgroup of P.

Let ρ be an irreducible representation of G_σ. ρ is called *cuspidal* if, for all proper σ-stable parabolic subgroups P, $(\rho_{(U_p)_\sigma}, 1_{(U_p)_\sigma}) = 0$.

THEOREM [11]. *Let ρ be an irreducible representation of G_σ. Then there is a σ-stable parabolic subgroup P and a cuspidal irreducible representation ρ' of L_σ such that*

$$((\rho'_{P_\sigma})^{G_\sigma}, \rho) \neq 0.$$

L_σ is determined to within conjugacy in G_σ.

Thus we obtain all irreducible representations of G_0 without repetition by taking all cuspidal representations of L_σ (one in each conjugacy class in G_σ), lifting to P_σ, inducing to G_σ, and decomposing into irreducible components.

We now restrict attention to the unipotent characters and have the following result.

THEOREM [11]. *Let ρ be an irreducible unipotent representation of G_σ. Then there is a σ-stable parabolic subgroup P and a cuspidal unipotent representation ρ' of L_σ such that*

$$(\rho'_{P_\sigma}{}^{G_\sigma}, \rho) \neq 0.$$

Furthermore all the irreducible components of $(\rho'_{P_\sigma})^{G_\sigma}$ are unipotent.

This result reduces the problem of determining the unipotent characters to the following two problems:

(i) Determine the cuspidal unipotent characters;

(ii) Decompose the induced character $(\rho'_{P_\sigma})^{G_\sigma}$ where ρ'_{P_σ} is an irreducible cuspidal unipotent character of L_σ lifted to P_σ.

11. Cuspidal Unipotent Characters

We first discuss the decomposition of the induced character $(\rho'_{P_\sigma})^{G_\sigma}$. We do this by considering the algebra of endomorphisms of this induced module.

Let Π be a set of simple reflections in a Coxeter group W and $\phi : \Pi \to C$ be a map which has the property that conjugate reflections have the same image. Let $\mathscr{H}_{W,\phi}$ be the associative algebra defined as follows by generators and relations. $\mathscr{H}_{W,\phi}$ is generated by elements T_i, $i \in \Pi$, subject to relations

$$T_i T_j T_i \ldots = T_j T_i T_j \ldots, \quad i \neq j,$$

(n_{ij} terms on each side where n_{ij} is the order of $w_i w_j$), and

$$(T_i + 1)(T_i - \phi(i)1) = 0.$$

The algebra $\mathscr{H}_{W,\phi}$ has dimension $|W|$. Algebras of the form $\mathscr{H}_{W,\phi}$ were first studied by Iwahori in the case when $\phi(i) = q$ for all i. Iwahori showed that if G_σ is a Chevalley group then the algebra $\mathrm{End}_{G_\sigma}(1_{B_\sigma}{}^{G_\sigma})$ of endomorphisms of the module induced from the principal module for B_σ is isomorphic to $\mathscr{H}_{W,\phi}$ for ϕ as above. More generally Matsumoto [12] showed that $\mathrm{End}_{G_\sigma}(1_{B_\sigma}{}^{G_\sigma})$ is always isomorphic to $\mathscr{H}_{W,\phi}$ for suitable W and ϕ.

We now consider the more general situation where we have the algebra $\mathrm{End}_{G_\sigma}(\rho'_{P_\sigma}{}^{G_\sigma})$ where ρ' is a cuspidal irreducible unipotent representation of L_σ.

Theorem [11]. *Let ρ' be an irreducible cuspidal unipotent representation of L_σ. Then $\mathrm{End}_{G_\sigma}(\rho'_{P_\sigma}{}^{G_\sigma})$ is isomorphic to $\mathscr{H}_{W,\phi}$ for suitable W and ϕ.*

We note that the Coxeter group W which appears here need not be the Weyl group of G. In fact W behaves like a "quotient Coxeter group" of G with respect to L.

Corollary. *The irreducible components of $\mathrm{End}_{G_\sigma}(\rho'_{P_\sigma}{}^{G_\sigma})$ are in bijective correspondence with the irreducible representations of W.*

This result shows that the induced representations $(\rho'_{P_\sigma})^{G_\sigma}$ decompose in a well-behaved manner into components in bijective correspondence with irreducible representations of an appropriate Coxeter group.

The fact that these induced representations decompose in a transparent way means that the problem of determining the unipotent representations of G_σ can be reduced to that of finding the cuspidal unipotent representations.

The number of cuspidal unipotent representations and their degrees have recently been determined by Lusztig, at least when q is not too small. The results are as follows.

THEOREM [11]. *The number of cuspidal unipotent characters of G_σ is as follows:*

A_ℓ	0
${}^2A_\ell$	$\begin{cases} 1 \text{ if } \ell+1 = (k^2+k)/2 \text{ for some } k \\ 0 \text{ otherwise} \end{cases}$
B_ℓ	$\begin{cases} 1 \text{ if } \ell = k^2+k \text{ for some } k \\ 0 \text{ otherwise} \end{cases}$
C_ℓ	$\begin{cases} 1 \text{ if } \ell = k^2+k \text{ for some } k \\ 0 \text{ otherwise} \end{cases}$
D_ℓ	$\begin{cases} 1 \text{ if } \ell = k^2 \text{ for some even } k \\ 0 \text{ otherwise} \end{cases}$
${}^2D_\ell$	$\begin{cases} 1 \text{ if } \ell = k^2 \text{ for some odd } k \\ 0 \text{ otherwise} \end{cases}$
G_2	4
F_4	7
E_6	2
${}^2E_6$	3
E_7	2
E_8	13
${}^3D_4$	2
${}^2B_2$	2
${}^2G_2$	6
${}^2F_4$	10

(In some cases this is known only if q is sufficiently large.)

We conclude this survey with a conjecture. Having defined semisimple characters and unipotent characters one would like to have an analogue of the Jordan decomposition of an arbitrary element into the product of commuting semisimple and unipotent elements. Such a result is not yet known in general. However one may make the following conjecture, which is consistent with the evidence available so far.

CONJECTURE.‡ *Let χ_s be a semisimple character of G_σ and s^* be an element in the corresponding semisimple class of the dual group $G^*_{\sigma^*}$. Form the dual group of the centralizer $Z_{G^*}(s^*)$, and consider the corresponding finite subgroup $((Z_{G^*}(s^*))^*)_\sigma$ of this dual group. Then there is a bijection between the irreducible characters χ of G_σ in the geometric conjugacy class containing χ_s and the unipotent characters χ_u of $((Z_{G^*}(s^*))^*)_\sigma$. This bijection $\chi \to \chi_u$ has the property that*

$$\chi(1) = \chi_s(1)\chi_u(1).$$

References

1. Borel, A., "Linear Algebraic Groups", Benjamin, New York, 1969.
2. Carter, R. W., "Simple Groups of Lie Type", Wiley, London, 1972.
3. Deligne, P., "Séminaire de Géometrie Algébrique (S.G.A. $4\frac{1}{2}$)", Lecture Notes in Mathematics 569, Springer-Verlag, Berlin, 1977.
4. Deligne, P. and Lusztig, G., Representations of reductive groups over finite fields, *Annals of Math.*, **103**(1976) 103–161.
5. Deriziotis, D., Ph.D. thesis, University of Warwick, 1977.
6. Green, J. A., The characters of the finite general linear groups, *Trans. Amer. Math. Soc.*, **80**(1955) 402–447.
7. Humphreys, J. E., "Ordinary and modular representations of Chevall y groups", Lecture Notes in Mathematics 528, Springer-Verlag, Berlin, 1976.
8. Humphreys, J. E., "Linear algebraic groups", Graduate Texts in Mathematics 21. Springer-Verlag, Berlin, 1975.
9. Lang, S., Algebraic groups over finite fields, *Amer. Jour. of Math.*, **78**(1956) 555–563.
10. Lusztig, G., "Representations of finite classical groups", Notes of Madison Conference (1977) by B. Srinivasan.
11. Lusztig, G., Representations of finite Chevalley groups, C.B.M.S. Regional Conference Series in Mathematics **39,** A.M.S. Providence, Rhode Island.
12. Matsumoto, H., Générateurs et relations des groupes de Weyl généralisés, *C.R. Acad. Sci. Paris*, **258**(1964) 3419–3422.
13. Springer, T. A., "Cusp forms for finite groups", Lecture Notes in Mathematics 131. Springer-Verlag, Berlin, 1970.
14. Steinberg, R., "Lectures on Chevalley groups", Yale University, 1968.
15. Steinberg, R., Endomorphisms of linear algebraic groups, *Mem. Amer. Math. Soc.*, **80**(1968).
16. Srinivasan, B., "Representations of finite Chevalley groups", Lecture Notes in Mathematics 764, Springer-Verlag, Berlin, 1979.

‡ This conjecture has recently been proved by Lusztig when q is sufficiently large.

12

Modular Representations of Finite Groups of Lie Type

J. E. HUMPHREYS

Introduction

Over C the main problems of representation theory for a finite group of Lie type are reasonably clear: construct the irreducible representations in some "natural" way and compute their characters. Over a field of characteristic dividing the group order, it is less clear how much one ought to expect, especially because there will usually exist infinitely many nonisomorphic indecomposable modules. We shall follow (more or less) the programme laid out by Brauer: study the irreducible modules and projective modules, then see how to decompose ordinary characters modulo p. For p we always take the prime defining the group of Lie type, since for other primes (except possibly 2) we cannot hope to say much about an entire Lie family. The organization of this chapter follows that of [23], but we hope to correct some of the loose reasoning used in Part II of those notes. Though we omit proofs, we shall give many explicit references to the literature.

One apology is required right at the outset. We have devoted much space to a study of certain modules for the ambient algebraic group G, rather than sticking closely to the finite groups. At the present stage this seems inevitable, since the deepest results about the modular representations of the finite groups of Lie type rely on algebraic groups to some extent. Possibly a more self-contained treatment will be found in the future.

The general framework of this chapter resembles that of Chapter 11, but we depart somewhat from that notation, and we use simply connected rather than adjoint groups for convenience. (Of course, in cases such as G_2 there is no distinction. But we want to use B_2 as an illustration, where both the ordinary and the modular theories are quite well developed.)

1. Irreducible Modules

1.1. First we establish some notation. For convenience we shall deal with a simply connected, simple algebraic group G over an algebraically closed field

K of characteristic $p > 0$, e.g. $SL(n,K)$ or $Sp(2n,K)$. We assume that G is defined and split over the prime field $\boldsymbol{F}_p$. Simple connectedness of G just means that the character group $X = X(T)$ of a maximal torus T of G is as "large" as possible, i.e. can be identified with the full lattice of weights associated with the root system of G. Denote by X_r the $\boldsymbol{Z}$-span of the roots in X; the index of connection $f = [X : X_r]$ is the determinant of the Cartan matrix of the root system. Let $W = N_G(T)/T$ be the Weyl group and w_0 its longest element. Finally, choose a Borel subgroup $B = T.U$ containing T or, equivalently, choose a set of positive roots. This determines the fundamental dominant weights $\omega_1, \ldots, \omega_l$, a $\boldsymbol{Z}$-basis of X, and the set $X^+ = \Sigma \boldsymbol{Z}^+ \omega_i$ of dominant weights. Write $\delta = \Sigma \omega_i$ (= half the sum of positive roots).

By *G-module* we mean a finite dimensional rational G-module M. It is a basic fact that T acts completely reducibly on M, giving rise to a weight space decomposition $M = \oplus M_\mu$ where, for $\mu \in X$,

$$M_\mu = \{v \in M \mid t.v = \mu(t)v \quad \text{for all} \quad t \in T\}.$$

We call μ a *weight* of the G-module M only when $M_\mu \neq 0$. Associated with M is its formal character,

$$\mathrm{ch}(M) = \sum (\dim M_\mu) e(\mu),$$

where the $e(\mu)$ are canonical basis elements of the group ring $\boldsymbol{Z}[X]$. Another basic fact is that $\mathrm{ch}(M)$ lies in the subring $\boldsymbol{Z}[X]^W$ of W-invariants: this comes from the observation that any representative of $w \in W$ in $N_G(T)$ maps M_μ onto $M_{w\mu}$. Clearly the formal character of M is the sum of the formal characters of the composition factors of M. If K is the algebraic closure of $\boldsymbol{F}_p$, we can embed K^* in $\boldsymbol{C}^*$ and pass in the obvious way from the formal character to a Brauer character, defined on semisimple elements of G (those having a conjugate in T).

1.2. Up to a point, the study of irreducible G-modules is the same as in characteristic 0 (cf. [22, Section 31]). Call $v \in M$ a *maximal vector* if $0 \neq v \in M_\lambda$ for some λ and U fixes v. Any nonzero G-module contains such a vector, by the Lie–Kolchin Theorem.

THEOREM (Chevalley).

(a) *If M is an irreducible G-module, then M possesses a maximal vector, unique up to scalar multiples, and its weight λ is dominant.*

(b) *All other weights μ of M satisfy $\mu \leqslant \lambda$ (i.e., $\lambda - \mu$ is a sum of positive roots), so λ is the highest weight.*

(c) *Two irreducible modules are isomorphic if and only if they have the same highest weight.*

(d) *If $\lambda \in X^+$, there exists an irreducible G-module $M(\lambda)$ of highest weight λ (G being simply connected).*

All of this imitates closely the classical theory of Cartan over C. One can even give a uniform construction of irreducible modules as subspaces of the ring of polynomial functions on G. But this is not very explicit. In general (1.5 below), $M(\lambda)$ turns out to have smaller dimension than the analogous module $V(\lambda)_C$ over C. Write $\mathrm{ch}(M(\lambda)) = p\text{-ch}(\lambda)$, to distinguish it from the well known $\mathrm{ch}(V(\lambda)_C) = \mathrm{ch}(\lambda)$ given by Weyl's formula. Each is a formal sum of W-orbits of weights, the orbit of λ occurring just once, and each set

$$\{p\text{-ch}(\lambda) \mid \lambda \in X^+\}, \quad \{\mathrm{ch}(\lambda) \mid \lambda \in X^+\},$$

serves equally well as a $\mathbf{Z}$-basis for $\mathbf{Z}[X]^W$.

The explicit determination of the $p\text{-ch}(\lambda)$ is the main unsolved problem about irreducible G-modules. (See [56] for a precise conjecture.) We shall sketch below what is known, and how it relates to the finite groups of Lie type.

1.3. Since we are working in characteristic p, we can exploit the Frobenius twist: by raising matrix entries to the pth power, we obtain from a G-module M a new module $M^{(p)}$, which is irreducible if M is. In particular, it is easy to see that $M(\lambda)^{(p)} \cong M(p\lambda)$. Denote by X_{p^n} the set of all $\lambda = \Sigma c_i\omega_i$ for which $0 \leqslant c_i < p^n$. Then each dominant weight λ lies in X_{p^n} for large enough n and has a p-adic expansion

$$\lambda = \lambda_0 + \lambda_1 p + \ldots + \lambda_{n-1}p^{n-1}, \quad \lambda_i \in X_p.$$

(If we increase n further, we just add 0 terms.) We call the weights in X_p *restricted*; they lie in a parallelopiped in the dominant Weyl chamber.

THEOREM (Steinberg). *With λ written as above,*

$$M(\lambda) \cong M(\lambda_0) \otimes M(\lambda_1)^{(p)} \otimes \ldots \otimes M(\lambda_{n-1})^{(p^{n-1})}.$$

It follows that a knowledge of the $M(\lambda)$ for λ restricted will suffice. In case $G = SL(2,K)$, the picture is easily completed: identifying a weight with its single integral coordinate (relative to ω_1), $M(\lambda)$ for $0 \leqslant \lambda < p$ is the familiar $(\lambda+1)$-dimensional module consisting of homogeneous polynomials of degree λ in two variables. (This case was already known by Brauer and Nesbitt.) We remark that Sullivan and Donkin have (independently) been able to prove the irreducibility of the twisted tensor product modules by more conceptual methods than that of Steinberg in [37], [4]. (See [50] for a new proof of the theorem.)

1.4. Now we consider how the preceding results on irreducible G-modules adapt to the finite Chevalley groups $\Gamma_n = G(\boldsymbol{F}_{p^n})$ (just write Γ when $n = 1$). The notation Γ is somewhat artificial, serving mainly to emphasize the distinction between G and its finite subgroups. Eventually we would like to use Steinberg's uniform set-up [38], in which any finite group of Lie type (split or not) can be regarded as the fixed points G_σ of a surjective endomorphism of G. But for the moment it seems preferable to focus on the split groups, commenting separately on the twisted types.

THEOREM (Curtis, Steinberg [4],[12],[37]). *The irreducible G-modules* $M(\lambda)$, $\lambda \in X_{p^n}$, *remain irreducible and inequivalent on restriction to* Γ_n, *and exhaust the irreducible* $K\Gamma_n$*-modules.*

For twisted groups, there are natural adaptations due to Steinberg [37] and formulated elegantly in [38]; e.g. $SL(m,q)$ and $SU(m,q)$ have essentially the same irreducible modules, coming by restriction from certain $SL(m,q^2)$-modules.

Remarks

(a) The number of irreducible modules for Γ_n (=number of semisimple classes) is $p^{n\ell}$, where ℓ = rank of G. For groups not of simply connected type, the answer is more complicated; however, it is not difficult to sort out which $M(\lambda)$ yield modules for (say) the associated simple group by looking at the action of the centre of Γ_n. For example, when p is odd, the irreducible modules for $PSL(2,p)$ are the $M(\lambda)$, $\lambda = 0,2,4,\ldots,p-1$.

(b) It is possible to classify and construct the irreducible modules for finite groups of Lie type entirely within the framework of split BN-pairs, using a suitable notion of "highest weight" (cf. [12], [34], [8]). However, this does not yield the refinement in 1.3, nor does it seem to lead very far in the direction of finding explicit formulas for Brauer characters, degrees, etc.

(c) Steinberg's work shows that $\boldsymbol{F}_{p^n}$ is already a splitting field for Γ_n. In case one wants to study irreducible modules over the prime field $\boldsymbol{F}_p$, further adaptation of the above results is needed, using the standard techniques of restriction and extension of scalars.

1.5. Returning to the G-modules $M(\lambda)$, we ask what is actually known about their dimensions and formal characters. There is as yet no "natural" model for $M(\lambda)$ from which such information could easily be read off, but one approach to proving the existence of $M(\lambda)$ does give some insight. Start with the irreducible module $V(\lambda)_{\boldsymbol{C}}$ for a simple Lie algebra (or simply connected Lie group) over $\boldsymbol{C}$ having the same root system as G. Fix a maximal vector v^+ (cf. 1.2) and apply the Kostant $\boldsymbol{Z}$-form $U_{\boldsymbol{Z}}$ of the universal enveloping algebra (cf. [39], [4]) to obtain an "admissible" lattice $V(\lambda)_{\boldsymbol{Z}}$. In turn, $V(\lambda) = V(\lambda)_{\boldsymbol{Z}} \otimes \boldsymbol{K}$ acquires the structure of a G-module, generated by the maximal vector $v^+ \otimes 1$ (of weight λ relative to T). Following [7] we call $V(\lambda)$ a *Weyl module:* $\mathrm{ch}(\lambda) = \mathrm{ch}(V(\lambda))$ is given by Weyl's formula. Now $V(\lambda)$ is easily seen to have a unique maximal submodule, the irreducible quotient being $M(\lambda)$. In particular, the construction shows that

$$\dim M(\lambda)_\mu \leqslant \dim V(\lambda)_\mu \quad \text{for all} \quad \mu \in X.$$

It is now natural to look for the composition factors of $V(\lambda)$; if we knew these (and their multiplicities), we could recursively obtain the weight space dimensions for $M(\lambda)$ by working upwards through the partial ordering of dominant weights. The starting point for such a recursion is, of course, the fact that we must have $V(\lambda) = M(\lambda)$ if λ is minimal in the dominant chamber; e.g., $\lambda = 0$ yields the trivial 1-dimensional module for any p. What we need to find are the coefficients $a_{\lambda\mu} \in \boldsymbol{Z}^+$ occurring in the formal character equations

$$\text{ch}(\lambda) = \sum a_{\lambda\mu} p\text{-ch}(\mu).$$

Here $a_{\lambda\lambda} = 1$, and the sum is taken over dominant weights $\mu \leqslant \lambda$. As indicated in 1.2, we could equally well write

$$p\text{-ch}(\lambda) = \sum b_{\lambda\mu} \text{ch}(\mu),$$

where now $b_{\lambda\mu} \in \boldsymbol{Z}$, $b_{\lambda\lambda} = 1$, and the sum is again over $\{\mu | \mu \leqslant \lambda\}$. So the problem is to determine either the *a*'s or the *b*'s. This has been solved completely only in a few cases: A_1, A_2, A_3, B_2, G_2 (except when $p = 5$), C_3 (almost finished). The methods used are due mainly to Jantzen [28], [29] (see 1.9 below as well as his contribution to this volume). Thanks to 1.3, it is only necessary to determine the composition factors of Weyl modules having restricted highest weights, but some of these composition factors may have highest weights outside X_p, which complicates matters.

1.6. The construction of Weyl modules sketched in 1.5 depends on reduction modulo p, starting with known representations in characteristic 0, but there is an intrinsic characterization as well.

THEOREM. *$V(\lambda)$ is the "universal" highest weight module of weight λ for G; i.e., given any G-module M generated by a maximal vector of weight λ, there exists an epimorphism (unique up to scalars) $V(\lambda) \to M$.*

This is proved in [29, Satz 1], following suggestions of the author. The proof uses Kempf's vanishing theorem for cohomology of line bundles on G/B: in effect, $V(-w_0\lambda)$ is dual to $H^0(G/B, \mathscr{L}(\lambda))$. (The Appendix to this chapter discusses further aspects of this interpretation.) A more direct proof of the theorem would of course be desirable (cf. [28, Satz 12] for a treatment of the case $G = SL(n,K)$).

The theorem shows that Weyl modules are analogous to "Verma modules" for a simple Lie algebra over $\boldsymbol{C}$. Indeed, the analogy is strong enough to permit a parallel treatment of the two theories in [28].

1.7. A first step in analysing the possible composition factors of a Weyl module is suggested by the early work of Harish–Chandra on universal highest weight modules (Verma modules) for a simple Lie algebra over **C**. There the idea was to study the scalars by which elements in the centre of the universal enveloping algebra act on highest weight modules. This approach via "central characters" is formally analogous to the classification of indecomposable modules into blocks for the modular group algebra of a finite group.

Motivated by Harish–Chandra's work, Verma proposed in 1967 the following criterion, which the author has called the "linkage principle". Call $\lambda, \mu \in X_p$ *linked* if there exists $w \in W$ such that

$$\lambda + \delta \equiv w(\mu + \delta) \,(\mathrm{mod}\, pX);$$

more generally, call $\lambda, \mu \in X^+$ linked if λ_0 and μ_0 are linked (where $\lambda = \lambda_0 + p\lambda_1 + \ldots, \mu = \mu_0 + p\mu_1 + \ldots$ are the p-adic expansions as in 1.3).

THEOREM. *Let V be an indecomposable G-module, having $M(\lambda)$ and $M(\mu)$ as composition factors. Then λ and μ are linked.*

This was proved for $p > h$ (=Coxeter number of W) by the author [17], using an argument which imitates one proof of Harish–Chandra's theorem. By imitating another (less constructive) proof, Kac and Weisfeiler [33] were able to remove the restriction on p. Meanwhile, Carter and Lusztig [7] had given a fairly direct proof in the case of $SL(n,K)$, stated just for V a Weyl module. These treatments all emphasize the Lie algebra of G, or its restricted universal enveloping algebra u; linkage is really a condition on weights in the restricted region, and the $M(\lambda)$, $\lambda \in X_p$, are precisely the irreducible modules for u (cf. the exposition of Curtis' work in [4]).

To say that $\lambda, \mu \in X^+$ are linked is to say that they are conjugate under the transformation group $\tilde{W}$ generated by W along with translations by elements of pX, where now W acts by the twisted rule

$$w \,.\, \lambda = w(\lambda + \delta) - \delta$$

(i.e. the origin is moved to $-\delta$). But the weights of an indecomposable G-module must all lie in the same coset of X_r in X. If $p \nmid f = [X : X_r]$, it follows that the highest weights of composition factors of an indecomposable G-module must be conjugate under the subgroup W_a generated by W and the translations by pX_r; this is called the *affine Weyl group* (relative to p). When $p \mid f$, this is still conjectured to be true; it holds for type A_ℓ thanks to [7] and has been verified by Jantzen [29] in low rank cases such as B_2. (See [41] for a detailed discussion of the role of affine Weyl groups.) Very recent work of Andersen [46] proves the conjecture for all p.

The affine hyperplanes relative to pX_r in $\mathbf{R} \otimes X$ define *alcoves*: these are the connected components of the complement of the union of hyperplanes.

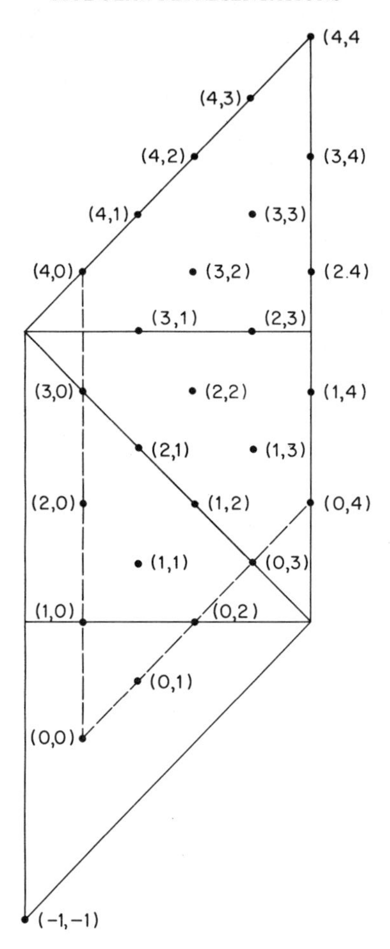

FIG. 1. Restricted weights for B_2: $(r,s) = r\omega_1 + s\omega_2$ $(p = 5)$.

The closure of any alcove is then a fundamental domain for W_a. The restricted weights X_p lie in $|W|/f$ alcoves, as pictured in Fig. 1 for B_2 when $p = 5$. Among these restricted alcoves, there is always a unique highest and a unique lowest with respect to the partial order of weights.

Remark. The analogy with block theory mentioned above can be pushed considerably further, using work of Green [16]. Green defines "blocks" for G, each involving certain indecomposable injective modules occurring as summands of the ring $K[G]$ of polynomial functions on G. Although the injectives are infinite dimensional, they are unions of finite dimensional submodules, so we can assign the $M(\lambda)$ to blocks in a reasonable way; thus blocks of G correspond to a partition of X^+. It is known that there are infinitely many blocks. But only finitely many of them involve *p-regular* weights (i.e. weights lying inside alcoves for W_a). The blocks have been determined by Donkin; in the p-regular case, two weights lie in the same block if and only if they are linked and lie in the same coset of X_r in X. See [11], [14], [26], [42], [51] for further details.

1.8. One immediate consequence of the linkage principle is an irreducibility criterion for Weyl modules.

THEOREM. *Let $\lambda \in X^+$. If no dominant weight $\mu < \lambda$ is linked to λ, then $V(\lambda) = M(\lambda)$.*

This criterion applies in particular when λ is a p-regular weight located in the lowest alcove (or a weight lying in the closure of this alcove). For $SL(2,K)$ there are no other restricted weights, and we recover the well known situation described in 1.3.

The criterion applies also to the weight $(p-1)\delta$, which is minimal in its linkage class. The irreducible Weyl module belonging to this highest weight is denoted by St and called the *Steinberg module* (for u or Γ); its dimension is p^m (m = number of positive roots), the order of a Sylow p-subgroup of Γ. It follows readily from Weyl's dimension formula and 1.3 that, for $\lambda = (p^n-1)\delta$, we also have $V(\lambda) = M(\lambda)$. This module is denoted by St_n and called the Steinberg module for Γ_n; its dimension is $(p^n)^m$, again the order of a Sylow p-subgroup.

It is not yet completely determined when $V(\lambda) = M(\lambda)$. Jantzen [26] has observed that it will suffice to answer this question for $\lambda \in X_p$, and has written down [27,II.8] a precise answer for $G = SL(n,K)$; cf. also [7]. His results on composition factors [29, Satz 10] do provide one general criterion.

THEOREM. *Let λ be p-regular. Then $V(\lambda) = M(\lambda)$ if and only if λ lies in the lowest alcove.*

We remark that James and Murphy [53] obtain a criterion for ordinary characters of the symmetric group S_n to remain irreducible modulo p, which formally resembles the criteria just discussed. The ordinary characters are associated with Specht modules (analogous to Weyl modules), and the affine Weyl group enters as in the set-up of [7].

1.9. Next we outline briefly several of the techniques which have been introduced to study composition factors of Weyl modules. The case B_2 will serve as an illustration of what is known.

(a) Following the lead of Verma, one can ask what are the possible G-module homomorphisms from $V(\mu)$ to $V(\lambda)$. For most p, Jantzen ([28], [29]) has verified the necessary condition for existence conjectured by Verma [41]:

μ must be obtainable from λ by applying a sequence of reflections from W_a in such a way that successive weights are dominant and lower in the partial order. On the other hand, the explicit construction of homomorphisms seems to be very difficult. For $SL(n,K)$, Carter and Lusztig [7] carried out such constructions, starting with key elements of $U_{\mathbf{Z}}$. Their result, roughly, is that one gets a nonzero map $V(\mu) \to V(\lambda)$ provided μ is obtainable from λ by one reflection but by no longer sequence of reflections, and then the space of maps is 1-dimensional. More general results for $SL(n,K)$ will appear in the Warwick thesis of Payne. For $SL(3,K)$ there is now emerging a clear picture of the various images and kernels of maps between Weyl modules having highest weights in "general" position (cf. the Appendix); the situation is quite similar to an analogous one involving the finite group $SL(3,p)$ (cf. [8]) but the analogy is poorly understood.

(b) In his thesis and subsequent work ([27], [28], [29]), Jantzen has studied "contravariant" forms on highest weight modules. For Weyl modules the idea is to define a suitable symmetric bilinear form (over $\mathbf{Z}$) on $V(\lambda)_{\mathbf{Z}}$, relative to which different weight spaces are orthogonal. The form is nondegenerate over $\mathbf{Z}$, and in most cases its determinant can be calculated explicitly. Divisibility of the determinant by the prime p is then equivalent to having $V(\lambda) \neq M(\lambda)$. Moreover, a close study of the determinant gives good information about the possible composition factors of $V(\lambda)$. In [43] Wong had initiated the use of the contravariant form as a means of proving irreducibility criteria. (An analogous form is definable for Specht modules for the symmetric group S_n; calculation of the determinant leads to the result mentioned at the end of 1.8.)

(c) In [28] Jantzen obtained a "translation principle", which shows that the constants $a_{\lambda\mu}$, $b_{\lambda\mu}$ defined in 1.5 depend only on the alcoves to which λ,μ belong (in case these weights are p-regular). More precisely, if λ belongs to an alcove and λ' lies in the "upper closure" of this alcove, one can convert character formulas involving λ into formulas involving λ', and vice versa if λ' also lies in the alcove. As a result, one need only find p-ch(λ) for a single weight in each restricted alcove in order to solve completely the problem posed in 1.5. (However, when p is too small—smaller than the Coxeter number h—there are no p-regular weights, and one has to look more closely.) The translation principle is proved by systematically tensoring Weyl modules and extracting the appropriate G-summand corresponding to the linkage class of weights under consideration. For this one has to keep track of weights by using classical (characteristic 0) formulas for the decomposition of tensor products, some of which appropriately go back to the earliest work of Brauer.

These and other methods are described more fully by Jantzen in the next chapter. To illustrate, we give the final results for type B_2. Here there are 4 alcoves in the restricted region, which happen to be ordered linearly; number them 1 to 4 from top to bottom, and let λ_i be a typical weight in the ith alcove. Then, taking the λ_i to be linked, we have:

$$\begin{aligned}
p\text{-ch}(\lambda_1) &= \text{ch}(\lambda_1)-\text{ch}(\lambda_2)+\text{ch}(\lambda_3)-\text{ch}(\lambda_4),\\
p\text{-ch}(\lambda_2) &= \text{ch}(\lambda_2)-\text{ch}(\lambda_3)+\text{ch}(\lambda_4),\\
p\text{-ch}(\lambda_3) &= \text{ch}(\lambda_3)-\text{ch}(\lambda_4),\\
p\text{-ch}(\lambda_4) &= \text{ch}(\lambda_4).
\end{aligned}$$

Inverting, one finds that each Weyl module $V(\lambda_1)$, $V(\lambda_2)$, $V(\lambda_3)$ has two composition factors, involving the given weight and the adjacent one. If λ is not p-regular, we get $V(\lambda) = M(\lambda)$ with the following exceptions: if λ lies in the "vertical" wall of the top alcove (cf. Fig. 1), then $V(\lambda)$ has a second composition factor whose highest weight is linked and lies in the vertical wall of the alcove numbered 2; or if $p = 2$, we get

$$2\text{-ch}(1,0) = \text{ch}(1,0)-\text{ch}(0,0),$$

as pictured in [23, p.19].

1.10. As noted in 1.3, Steinberg's twisted tensor product theorem reduces the determination of p-ch(λ) to the case in which λ is restricted. But there is something to be gained by looking directly at Weyl modules having nonrestricted highest weight, because then certain regularities become more apparent. Consider the "p^2-alcoves" determined by an affine Weyl group whose translations are relative to p^2X_r rather than pX_r (later one can consider even higher powers of p). If λ is in sufficiently general position inside the lowest p^2-alcove, Jantzen [30] shows that the decomposition behaviour of $V(\lambda)$ is generic, i.e. independent of λ. (In his thesis, Winter independently

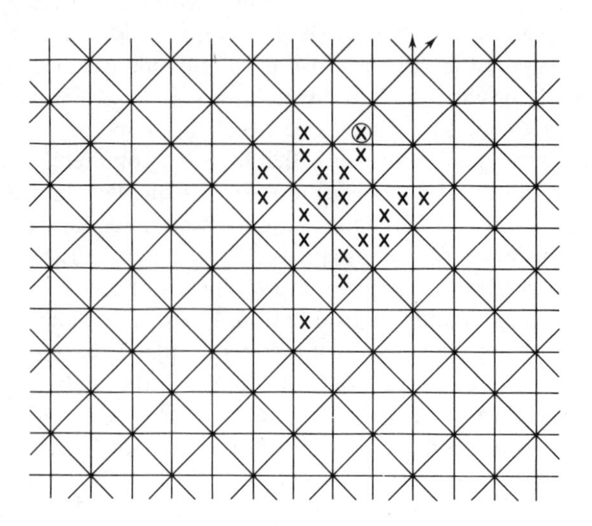

FIG. 2. A generic pattern for B_2.

observed this in ranks 1,2). For example, a generic Weyl module for G of type B_2 has 20 composition factors, each with multiplicity 1, and the alcoves (for W_a) in which their highest weights lie depend just on which of the four types of alcoves λ belongs to. Here the "type" of an alcove is determined by translating back to the restricted parallelogram. The resulting configuration is pictured in Fig. 2 for λ in an alcove of "type 1" (to use the numbering of 1.9 above). In [30] the case of an alcove of "type 2" is pictured.

The explicit working out of the generic decomposition patterns for B_2 required a knowledge of the behaviour of weights in the restricted region, as given in 1.9. However, it should be observed that the generic patterns imply the restricted results, as follows. If the highest weight λ in Fig. 2 is placed in the alcove numbered 1 in the restricted region, only it and the 3 weights just below it will remain in the dominant Weyl chamber; the other 16 small alcoves in Fig. 2 no longer contribute dominant weights. Those at a distance less than p from the walls of the dominant chamber can be disregarded, while the 2 leftmost alcoves are brought back to "cancel" the 2 lowest alcoves of the restricted region. So the Weyl module in question has just 2 composition factors, as stated in 1.9.

The generic patterns for B_2 do not look very symmetric, but a dual formulation reveals quite a bit of underlying symmetry. Instead of asking for weights μ such that $M(\mu)$ occurs as a composition factor of $V(\lambda)$, we can turn the question around and ask for all $V(\lambda)$ in which a given $M(\mu)$ occurs as a composition factor. (As above, we consider only weights in sufficiently general position inside the lowest p^2-alcove, ignoring the fact that $M(\mu)$ may occur as composition factor of Weyl modules having arbitrary "big" highest weight.) For fixed μ in each of the 4 types of alcove for W_a, we can fill in the alcoves where the permitted λ occur and get 4 pictures which are just like those occurring in [23, pp. 82–3]. These latter pictures are quite symmetric, but were discovered in an entirely different context; they involve respectively 8,16,24,32 alcoves.

Finally, we point out what happens for larger powers p^n. Here again, a Weyl module whose highest weight lies in general position in the lowest p^n-alcove will exhibit generic decomposition behaviour, but the number of composition factors will go up. For B_2 and p^3, there will be 400 composition factors, occurring as 20 configurations of 20 each (and so on!).

We have apologized at the outset for spending so much time on the modules for G. But some of the phenomena encountered here have remarkably sharp analogues for the finite groups. For example, the generic decomposition patterns just discussed seem to be imitated in the reduction modulo p of ordinary characters of Γ (or Γ_n, where the higher powers of p become relevant). For type B_2 (i.e. the finite group $Sp(4,p)$), we can confidently expect that a "typical" ordinary irreducible character will involve 20 composition factors modulo p, whose highest weights will be distributed among the 4 types of alcoves with the frequency predicted by Jantzen's patterns (2,4,6,8 for the alcoves numbered 1,2,3,4 in 1.9). Of course, it may be necessary to choose p fairly large in order that most characters become "typical" in this sense.

There is also a strange connection (see Appendix) between the generic decomposition patterns and the cohomology of line bundles on G/B. This is hardly understood as yet.

1.11. Before leaving the subject of Weyl modules, we should note some other aspects worth investigating. So far we have emphasized mainly the determination of formal characters (or composition factors), since this information is what we need to pin down the characters of the irreducible modules. But we may also ask for explicit structural properties of Weyl modules, e.g. minimal submodules or the full lattice of submodules. In general it is expected (though not yet proved) that Weyl modules will have infinitely many submodules, as happens usually for Verma modules; so a complete picture might be hopelessly complicated.

When $G = SL(2,K)$, there are only finitely many submodules, and their lattice structure has been described in detail by Carter and Cline [6] (with proofs yet to appear). Though not stated explicitly by them, it follows that there is always a unique minimal (as well as a unique maximal) submodule, whose highest weight is obtained from the highest weight of the Weyl module by using a reflection which involves the highest possible power of p. Andersen has recently looked at such phenomena from a cohomological point of view. (See the Appendix.)

In higher ranks not much is known about the submodule structure of Weyl modules, apart from the limited information that results from existence of nonzero maps between Weyl modules as in 1.9(a). There cannot always be a unique minimal submodule, as shown by the adjoint module for type D_{2n} when $p = 2$. But this is true "generically" [54].

1.12. Finally, we make explicit what has been implicit in many of the results mentioned: the essential phenomena connected with Weyl modules and irreducible modules (and other modules to be encountered below) seem to be essentially independent of the choice of p. There are, however, some special statements about small p or about primes dividing f, so one cannot make a definitive conjecture at this point about the precise sense in which "essentially" is to be understood.

2. Projective Modules

2.1. Although we did not emphasize it above, modules for the Lie algebra of G (or, equivalently, for its restricted universal enveloping algebra u) play an important role in the original study of irreducible modules by Curtis and Steinberg (cf. [4], [37]). The idea was to start with the irreducible u-modules $M(\lambda)$, $\lambda \in X_p$, "lift" them to G-modules (via associated projective representations of the adjoint group), then enlarge the collection of G-modules by forming twisted tensor products, and finally restrict the resulting modules to the finite groups. We will sketch now a similar approach to the PIM's, which was first worked out in the case $G = SL(2,K)$ by Jeyakumar [32].

To make the framework more balanced, it is useful to set things up as follows. Taking $U_{\mathbf{Z}}$ as before to be the Kostant $\mathbf{Z}$-form of the (complex) universal enveloping algebra, set $U_K = U_{\mathbf{Z}} \otimes K$ (the *hyperalgebra* of G). In a natural way, U_K-modules correspond to G-modules (cf. [11], [24]). Moreover, there is a nice filtration of U_K by finite dimensional subalgebras u_n, where $u_1 = u$ and $\dim u_n = (p^n)^{\dim G}$; note that this power of p is the leading term in the polynomial which gives the order of Γ_n. The algebras u_n arise as "Frobenius kernels", while Γ_n is the group of fixed points of a Frobenius operator. The representation theory of u_n parallels very closely that of $K\Gamma_n$, starting with the fact that the irreducible modules for the two algebras are the same (via G). But in some ways u_n is the easier algebra to work with, as we shall see below.

2.2. The first indecomposable projective modules one encounters are the Steinberg modules.

THEOREM. *St_n is a projective module for u_n and $K\Gamma_n$.*

For u_n this can be seen fairly directly or via the linkage principle (St_n is alone in its "block"); cf. [23, 5.5] for $n = 1$. For $K\Gamma_n$ it seems to require the existence of an irreducible complex character of Γ_n of degree $(p^n)^m$ (cf. [38]), along with Brauer's theory of blocks of defect 0. It is true for both u_n and $K\Gamma_n$ that the dimension of a PIM must be divisible by $(p^n)^m$. In particular, no irreducible module except St_n can also be projective. Current knowledge suggests that (to quote [9]), "As the structure of the irreducible Brauer character gets more complicated, so the structure of the corresponding indecomposable character gets simpler. For instance the Steinberg character is the most complicated irreducible Brauer character and the simplest projective character."

To construct further PIM's, it is reasonable to tensor St_n with various modules, say irreducible ones; for either u_n or $K\Gamma_n$, this procedure yields projective modules which can (in theory, at least) be decomposed into PIM's. To carry this out in practice is quite difficult, except for $SL(2,q)$, where it is straightforward ([32], [19]). Recently, Chastkofsky and Feit ([9], [10]) have had some success with groups of types C_2 and ${}^2C_2$ when $p = 2$. So far, all general results about PIM's (notably those of Ballard [2]) have relied heavily on tensoring with St_n.

2.3. Denote by $Q(\lambda)$, $\lambda \in X_p$, the PIM of u having $M(\lambda)$ as quotient. (Actually, since u_n and $K\Gamma_n$ are both symmetric algebras, their PIM's are also injective and have the same top and bottom composition factor.) It is known [17] that $\dim Q(\lambda) = a_\lambda d_\lambda p^m$, where a_λ is the cardinality of the linkage class of λ in X_p and where d_λ is the multiplicity of $M(\lambda)$ as a composition factor of the "universal" highest weight module $Z(\lambda)$ for u. Although the decomposition numbers d_λ are not yet known in general, Jantzen has shown that they remain constant on alcoves (and their upper closures) (cf. [30]); they are computable once the p-ch(v) are all known. For B_2 we find $d_\lambda =$ 1,2,3,4 for λ in the alcove numbered similarly in 1.9 (from top to bottom). For G_2, A_3, the results get more complicated (cf. [23, p. 24]). It does turn out in general that $d_\lambda = 1$ for λ in the top alcove of X_p, consistent with the quotation in 2.2, so we first examine this situation. (In 2.5 we shall return to the general case.)

In outline, our strategy is as follows: Given λ, choose μ in such a way that $Q(\lambda)$ occurs just once in any decomposition of $M(\mu) \otimes St$ as a direct sum of PIM's. Next, show (if possible) that a summand $Q(\lambda)$ is actually a G-summand; this will be the case, for example, if the natural summand of the tensor product belonging to the linkage class of λ has the same dimension as $Q(\lambda)$. *Its formal character is $s(\mu)ch(St)$, where $s(\mu) = \Sigma\, e(w\mu)$ is the formal* Now the restriction to Γ of the G-module $Q(\lambda)$ is projective and can be analyzed further.

The first step goes through easily for arbitrary $\lambda \in X_p$. Define $\hat{\lambda} = w_0\lambda + (p-1)\delta$. This pairing of X_p with itself satisfies $\hat{\hat{\lambda}} = \lambda$. Note that 0 gets paired with $(p-1)\delta$.

THEOREM [23, 8.2]. *Let $\lambda \in X_p$, $\mu = \hat{\lambda}$. Then $Q(\lambda)$ occurs precisely once as a u-summand in any decomposition of $M(\mu) \otimes St$ into PIM's; moreover, its irreducible submodule $M(\lambda)$ is G-stable.*

Next we look for a suitable G-summand. The only "natural" G-summands of the tensor product seem to be those corresponding to linkage classes (this is a "block" decomposition, or at least a first approximation to one; cf. 1.7). Usually these G-submodules will not be indecomposable.

THEOREM. *If λ lies in the top alcove of X_p, $\mu = \hat{\lambda}$, then the G-summand of $M(\mu) \otimes St$ corresponding to the linkage class of λ coincides with a u-summand $Q(\lambda)$. Its formal character is $s(\mu)ch(St)$, where $s(\mu) = \Sigma\, e(w\mu)$ is the formal orbit sum.*

This is stated in [23, 9.2] just for the case in which μ is assumed to lie in the bottom alcove. A more refined version is given in [30, Satz 5], yielding G-module structure on $Q(\lambda)$ for all λ in the closure of the top alcove. The proofs involve a study of the formal character of the tensor product; the contribution of the linkage class of λ turns out to be just $\Sigma\, \mathrm{ch}(w\mu + (p-1)\delta)$, which can be rewritten as $s(\mu)\mathrm{ch}(St)$. In particular, comparison with $\dim Q(\lambda)$ above shows that the linkage class summand equals $Q(\lambda)$ and that $d_\lambda = 1$. (A similar calculation was done independently by Ballard [2].)

It should be pointed out that when $G = SL(2,K)$, there is only one alcove in X_p, so this procedure yields G-structure on all PIM's of u. Except for St, the dimension of each is $2p$.

2.4. We consider now how to adapt the construction in 2.3 to Γ. For the moment, let $\lambda \in X_p$ be arbitrary, $\mu = \hat{\lambda}$. In case a u-summand $Q(\lambda)$ of $M(\mu) \otimes St$ is also a G-summand, $Q(\lambda)$ is also a projective $K\Gamma$-module. Since the u-submodule $M(\lambda)$ of $Q(\lambda)$ is G-stable (2.3), the $K\Gamma$-module $M(\lambda)$ occurs in the socle of $Q(\lambda)$, forcing the corresponding PIM $R(\lambda)$ to occur as a $K\Gamma$-summand of $Q(\lambda)$. It remains to be seen whether $Q(\lambda) = R(\lambda)$ and, if not, how $Q(\lambda)$ decomposes. In case λ lies in the top alcove, the existing evidence favours the conclusion that equality should hold. But for weights in the closure of that alcove, the situation is more complicated: e.g. for $SL(2,K)$, $Q(0) = R(0) + St$.

Ballard [2, Section 5] has obtained a lower bound for $\dim R(\lambda)$ which leads to a general criterion for $(\dim R(\lambda))/p^m$ to equal the cardinality $|W\mu|$ of the W-orbit of μ. This is based on the observation that certain PIM's for the Borel subgroup of Γ must occur in the restriction of $R(\lambda)$. These PIM's have dimension p^m and correspond to characters coming from the W-orbit of $\mu \bmod (p-1)X$ (i.e. characters of the finite split torus). In particular, when μ is suitably "small" it follows that $\dim R(\lambda) = \dim Q(\lambda)$ (cf. [2, Section 7]). Ballard's criteria are not yet optimal, but do support the conjecture that equality should hold here for all λ in the top alcove.

TABLE 1. *$Sp(4,5)$, principal block*

λ	$\dim Q(\lambda)/p^4$	$\dim R(\lambda)/p^4$
(3,2)	8	8
(2,2)	16	16
(2,0)	24	16
(0,0)	32	9
(2,4)	4	4
(1,4)	8	8
(1,0)	16	8
(3,4)	4	4
(0,4)	8	7
(0,2)	16	9
(4,0)	4	3
(3,0)	12	5
(4,2)	4	4
(1,2)	12	12

TABLE 2. (*Chastkofsky*) $Sp(4,p)$, $p \geqslant 5$

$Q(0,0) = R(0,0)+R(0,p-1)+R(p-1,0)+R(1,p-1)+R(p-1,2)+St$
$Q(0,p-1) = R(0,p-1)+St$
$Q(0,p-2) = R(0,p-2)+R(p-1,p-2)$
$Q(0,p-3) = R(0,p-3)+R(p-1,p-3)+3St$
$Q(0,s) = R(0,s)+R(p-1,s)+R(p-1,s+2),\ 1 \leqslant s < p-3$
$Q(p-1,0) = R(p-1,0)+St$
$Q(p-2,0) = R(p-2,0)+R(p-2,p-1)+3St$
$Q(r,0) = R(r,0)+R(r,p-1),\ r = (p-3)/2$
$Q(r,0) = R(r,0)+R(r,p-1)+R(r+1,p-1),\ 1 \leqslant r < p-2,\ r \neq (p-3)/2$

A few examples for type B_2 ($p = 5$) may help to clarify the range of possibilities (cf. Tables 1 and 2).

(a) For the weight $\lambda = (3,2)$ in the top alcove, $\mu = (1,2)$ lies on the wall between the second and third alcoves. We have $a_\lambda = 8$ (the linkage class is of full size $|W|$), and $\dim Q(\lambda) = 8p^4$. We also have $|W\mu| = 8$, but taken mod 4 the orbit only has size 4, so Ballard's criterion only forces $\dim R(\lambda) \geqslant 4p^4$. But in fact $R(\lambda) = Q(\lambda)$.
(b) Next consider the weight $\lambda = (4,2)$ in the closure of the top alcove, $\mu = (0,2)$. Here $a_\lambda = 4 = |W\mu|$, and the W-orbit of μ has the same size mod 4. So $\dim Q(\lambda) = 4p^4 = \dim R(\lambda)$.
(c) Take $\lambda = (4,0)$, $\mu = (0,4)$. Again $\dim Q(\lambda) = 4p^4$, and $|W\mu| = 4$, but mod 4 the orbit size is 1. It turns out in this case that $Q(\lambda) = R(\lambda) \oplus St$.

2.5. Having seen directly that certain $Q(\lambda)$ have the structure of G-modules in such a way that the restriction to $K\Gamma$ is projective, it is natural to ask whether this is true for arbitrary $\lambda \in X_p$. In [23, 8.2] the author gave an incomplete proof of this general assertion, which has since been completed by Ballard [3] under an unpleasant restriction on p which one hopes eventually to remove. Jantzen [54] has lowered the bound on p:

THEOREM. *Assume $p \geqslant 2h-2$ (h = Coxeter number). Let $\lambda \in X_p$, $\mu = \hat{\lambda}$. Then $Q(\lambda)$ is a G-summand of $M(\mu) \otimes St$; so its restriction to Γ is a projective $K\Gamma$-module having $R(\lambda)$ as summand.*

Ballard and Donkin have (independently) analysed some low rank cases when $p < 2h-2$, and have been able to verify that the theorem is still valid; in particular, it holds for B_2, for all p. It should be pointed out that, for arbitrary p, $Q(\lambda)$ (resp. $R(\lambda)$) occurs exactly once as a u-summand (resp. $K\Gamma$-summand) of the tensor product; cf. 2.3 above (resp. [2, Cor. 2.7]). But only via the G-module structure can we expect to relate $Q(\lambda)$ to $R(\lambda)$.

It should also be mentioned that for any G-module Q whose restriction to u is isomorphic to $Q(\lambda)$, the G-module structure and hence $\mathrm{ch}(Q)$ is well determined. This has been shown (independently) by Donkin and Jantzen.

What seems especially problematic is to characterize $Q(\lambda)$ intrinsically as a G-module, independently of its occurrence in a particular tensor product construction. One hopeful indication is the natural connection which exists (when the above theorem is valid) between $Q(\lambda)$ and the injective indecomposable G-module having $M(\lambda)$ as its socle: cf. 2.8 below.

2.6. As in the special situation of 2.4, it still has to be seen how $Q(\lambda)$ decomposes as a direct sum of PIM's for $K\Gamma$. Verma suggested that for regular λ we should have $Q(\lambda) = R(\lambda)$. Here a "regular" weight is one lying in an ordinary Weyl chamber. In particular, weights in the top alcove are regular. Some evidence supporting the regularity conjecture is summarized in [23,10.3], but Chastkofsky has now found some probable counterexamples.

Recently Chastkofsky has found a way to obtain explicit decompositions of the $Q(\lambda)$ in low rank cases such as A_2 (where he has rigorously justified the formulas of [23, p.53]), ${}^2A_2$ and B_2 (see Table 2 for $p \geqslant 5$). The formulas are rather uniform with respect to p, although they degenerate somewhat for small p. To count multiplicities of various $Q(\nu)$ or $R(\nu)$ in tensor products $M(\mu) \otimes St$, one reformulates the question (cf. [23, bottom of p.41]) and looks for the multiplicity of St as a composition factor (or summand) of certain tensor products.

Further insight into the behaviour of PIM's for $K\Gamma$ will probably require a comparison with ordinary characters; cf. Section 3 below.

2.7. What can be said about the Brauer character of PIM's of $K\Gamma$? The formal character of $Q(\lambda)$ found in the special setting of 2.3 is just the product of a W-orbit sum and the Steinberg character, each of which yields a Brauer character to be denoted by $s(\mu)$, st respectively ($\mu = \hat{\lambda}$). In his thesis, Ballard showed that products of this form yield a $\boldsymbol{Z}$-basis for the group generated by Brauer characters of PIM's. The precise statement is as follows [2, Theorem 2.5].

THEOREM. *The Brauer character of $R(\lambda)$, $\lambda \in X_p$, has the form*

$$s(\mu)st + \sum c(\nu)s(\nu)st,$$

where $c(\nu) \in \boldsymbol{Z}$ and the sum is over $\nu \in X_p$ for which $\nu < \mu$ and $\hat{\nu} > \hat{\mu} = \lambda$.

Here p can be replaced by $q = p^n$ and $R(\lambda)$ by the PIM to be denoted $R(\lambda,n)$ in 2.9 below. The theorem is in a sense dual to the description given earlier by Wong [43] of the Brauer character of $M(\lambda)$. A consequence of the theorem is the observation that the Steinberg character "divides" all characters of PIM's; this was proved independently by Lusztig in a slightly more general setting.

2.8. Once the u-modules $Q(\lambda)$ have been given compatible G-module structures, it makes sense to form twisted tensor products as was done in 1.3 for irreducible modules. Write $\lambda \in X^+$ as $\lambda_0 + p\lambda_1 + \ldots + p^{n-1}\lambda_{n-1}$ $(\lambda_i \in X_p)$, and define

$$Q(\lambda,n) = Q(\lambda_0) \otimes Q(\lambda_1)^{(p)} \otimes \ldots \otimes Q(\lambda_{n-1})^{(p^{n-1})}.$$

For example, if $\lambda \in X_p$, $Q(\lambda,1) = Q(\lambda)$. Here n has to be built into the notation since, for fixed λ and increasing n, we get larger modules, unlike the situation for $M(\lambda)$ where tensoring by $M(0)^{(p^k)}$ has no effect. If $\lambda \in X_{p^n}$, the G-module $Q(\lambda,n)$ is easily seen to be the PIM for u_n whose top (or bottom) composition factor is isomorphic to $M(\lambda)$ (cf. [3], [26]). We remark that, without assuming G-structure on $Q(\lambda)$, Donkin has shown that the dimension of the PIM of u_n corresponding to $\lambda \in X_{p^n}$ is $\Pi \dim Q(\lambda_i)$. This is a further indication that the G-structure ought to be there.

Since $Q(0)$, and hence $Q(0)^{(p^k)}$, has a G-submodule $M(0)$, we have natural embeddings

$$Q(\lambda,n) \subsetneq Q(\lambda,n+1) \subsetneq \ldots,$$

when $\lambda \in X_{p^n}$. So it makes sense to pass to the direct limit. Generalizing a result of Winter [42] for $SL(2,K)$, Donkin [14] and Ballard [3] have shown independently that this direct limit has intrinsic significance for G. (Cf. [54].)

THEOREM. *Assume $p \geqslant 2h-2$. Then*

$$\varinjlim_n Q(\lambda,n)$$

is an indecomposable injective G-module with socle $M(\lambda)$.

This may help to explain the close connection found by the authors of [11] between cohomology groups of u_n and U_K (or G). By contrast, their earlier work exhibits a more complicated connection between the cohomology of Γ_n and that of G.

2.9. The twisted tensor product construction in 2.8 yields immediately some information about PIM's of $K\Gamma_n$ (always under the assumption that PIM's of u have suitable G-module structure as in 2.5). If $R(\lambda,n)$ denotes the PIM having $M(\lambda)$, $\lambda \in X_{p^n}$, as top composition factor, then $R(\lambda,n)$ occurs as a $K\Gamma_n$-summand of $Q(\lambda,n)$. (In fact, it occurs just once; cf. [2, Cor. 2.7].)

For $SL(2,p^n)$, a simple dimension comparison (cf. [32], [19]) shows that $Q(\lambda,n) = R(\lambda,n)$ if $\lambda \neq 0$, whereas $Q(0,n) = R(0,n) \oplus St_n$. In particular, $\dim R(0,n) = (2^n - 1)p^n$, which makes it clear that in general we cannot expect all $R(\lambda,n)$ to be expressible in a sensible way as twisted tensor products. (Thus it may be even more difficult to study PIM's than to study irreducible modules wholly within the context of finite groups.)

The main problem now (as in 2.6) is to decompose $Q(\lambda,n)$ effectively as a direct sum of PIM's, in particular to decide when $Q(\lambda,n) = R(\lambda,n)$. Chastkofsky's approach may be successful here. His joint work with Feit for $p = 2$ (cf. [9], [10]) already exhibits some reasonable patterns for the dimensions.

In case $Q(\lambda,n) = R(\lambda,n)$, or in case $Q(\lambda,n)$ can be explicitly decomposed into PIM's for $K\Gamma_n$, the Cartan invariants of $K\Gamma_n$ can in principle be found once we know the irreducible modules; see [20] for a discussion of some crude computational techniques. However, none of this is very easy in practice, even in the case of $SL(2,p^n)$, where the Cartan invariants have recently been calculated by Alperin [1] for $p = 2$ and by Upadhyaya [40] for odd p. One problem is that the composition factors of the G-module $Q(\lambda,n)$ often have highest weights lying outside X_{p^n}. For example, when $n = 1$, $M(\mu) \otimes M(\nu)^{(p)}$ becomes an ordinary tensor product $M(\mu) \otimes M(\nu)$ on restriction to Γ, and composition factors are complicated to compute.

2.10. The methods outlined above for studying PIM's of a Chevalley group can be adapted readily to the twisted groups of types A, D, E_6. This is discussed in [23, Section 15]. The starting point is the fact that the twisted and untwisted group of a given type share the same collection of irreducible modules, with St_n being a PIM for each (if the groups are constructed over the field of order p^n). Accordingly, the same sort of tensor product construction of PIM's can be attempted in the twisted case. When $Q(\lambda)$, $\lambda \in X_p$, can be realized as a G-summand of $M(\mu) \otimes St$ as in 2.5, its restriction to the twisted group over $\boldsymbol{F}_p$ will be projective and involve the PIM corresponding to λ as a direct summand. Again one has the problem of decomposing $Q(\lambda)$ in irregular cases. And again, the twisted tensor products allow one to study the groups over $\boldsymbol{F}_{p^n}$.

So far there are not many precise results about the PIM's of twisted groups; for $p = 2$, Chastkofsky and Feit [10] have been able to compute the dimensions of PIM's for $SL(3,p^n)$ and $SU(3,p^n)$. The author has also studied $SU(3,5)$, finding the Cartan invariants and dimensions of PIM's, and Chastkofsky has worked out in general the way in which the $Q(\lambda)$ decompose for $SU(3,p)$. (The formulas in [23, p.98] are inaccurate. Instead one has

$$Q(0,0) = R'(0,0) + R'(p-1,1) + R'(1,p-1) + St,$$
$$Q(r,0) = R'(r,0) + R'(r+1,p-1) \quad \text{for} \quad 0 < r < p-2,$$

and similarly for $Q(0,s)$.)

The twisted groups of Ree and Suzuki have been less well understood from our point of view. Here the connection with the corresponding Chevalley group is more devious, as one sees already in the case of irreducible modules [37]. But for types C_2 and ${}^2C_2$ ($p = 2$), Chastkofsky and Feit [9] have been able to study the groups in tandem. Again, tensoring with a "Steinberg module" gives effective information about PIM's.

3. Relation with Ordinary Characters

3.1. Having surveyed what is known about the modular theory, we now ask how the results tie in with "ordinary" (characteristic 0) representation theory of finite groups of Lie type.

An irreducible module Z (over C) for a finite group can be reduced modulo p (in various ways) to yield a modular representation $\bar{Z}$, whose composition factors are independent of the particular reduction chosen. Since PIM's are also obtainable by reduction modulo p (via lifting of idempotents, etc.), each PIM involves certain of the $\bar{Z}$. According to Brauer's reciprocity principle, the number of times $\bar{Z}$ occurs in a PIM is equal to the number of times the top composition factor of that PIM occurs in a composition series of $\bar{Z}$. This is usually expressed in the form $C = {}^tD.D$, where C is the Cartan matrix of the modular group algebra and D is the decomposition matrix, recording the composition factor multiplicities of the various $\bar{Z}$.

For finite groups of Lie type, quite a bit is now known about the ordinary representations or their characters. (See Chapter 11.) To make a comparison with the modular theory, we again start with the split groups Γ, Γ_n. As in 2.1, there is a very close parallel between $K\Gamma_n$ and u_n, including a version of Brauer reciprocity for u_n (cf. [17], [31]). Here the "intermediate" modules between irreducibles and PIM's are the modules $Z(\lambda)$ or $Z(\lambda,n)$ mentioned in 2.3: the universal highest weight modules, analogous to Verma modules. Again one has an equation $C = {}^tD.D$. All evidence so far supports the hypothesis that the number of composition factors (and the distribution by types of alcoves) for a typical $Z(\lambda,n)$ and for a typical $\bar{Z}$ will coincide. The meaning of "typical" in each case has to be clarified, but the example $SL(2,p^n)$ indicates what is meant: the number of composition factors is usually 2^n in either case. When $n = 1$, this number is just the sum over a full linkage class of the numbers d_λ mentioned in 2.3; for general n, one takes the nth power.

The algebraic group G provides a connection between these two parallel theories, and it too may enjoy a sort of Brauer reciprocity (suggested by the author and recently worked out for $SL(2,K)$ by Upadhyaya). For example, when $n = 1$, the ingredients for such a principle are the G-modules $M(\lambda)$ and $Q(\lambda)$, $\lambda \in X_p$ (assuming G-structure has been defined on the latter), along with the "intermediate" modules $V(\lambda)$, where λ ranges over the lowest p^2-alcove; cf. the universal property of Weyl modules discussed in 1.6. When such a Weyl module exhibits generic decomposition behaviour (1.10), we expect that it will have the same number of composition factors as the generic number above for $Z(\lambda)$ or $\bar{Z}$ (cf. [30]). The number in question is 2 for $SL(2,K)$, 9 for $SL(3,K)$, 20 for $Sp(4,K)$,... ([23, p. 25]).

In this section, we take as our central problem the explicit determination of the decomposition matrix D for Γ_n. Knowledge of D implies, in turn, knowledge of C (though the latter might also be found directly in the framework of Section 2 above; cf. [20]). Since there are not yet many general results in this direction, we shall illustrate our ideas in the concrete case $\Gamma =$ Sp(4,5), taking advantage of the fact that both the ordinary and the modular theory are fairly well worked out for type B_2. (For discussion of type A_2, see [23].)

3.2. Decomposition of the modular group algebra of a finite group into blocks (indecomposable two-sided ideals) induces a corresponding block decomposition of the matrix D, but in our situation block theory is of little direct use.

THEOREM. *For a finite simple group of Lie type, there are only two p-blocks: the block* (*of defect* 0) *containing just the Steinberg module, and the principal block* (*of highest defect*).

In more detail, since we have started with a simply connected G, $K\Gamma_n$ has as many blocks of highest defect as the order of the centre of Γ_n, but no other blocks except for that of St_n. These results were first obtained by Dagger in special cases, then in general by the author [18]. It is not hard to distribute the irreducible modules $M(\lambda)$ into blocks; cf. [23, p. 59]. For example, for type B_2 and p odd, the principal block of $K\Gamma$ involves weights other than $(p-1,p-1)$ having even second coordinate relative to the fundamental dominant weights (taking the first simple root to be long); cf. Table 1 above for $p = 5$.

3.3. The work of Deligne and Lusztig [13] shows that a finite group of Lie type has families $R_{T,\theta}$ of virtual characters parametrized by certain classes of maximal tori T of G and characters θ of the corresponding finite tori. (We have been using T just to denote a split torus, but this new notation should cause no confusion.) Their theory is easier to describe when G is of adjoint type or, more generally, has connected centre (cf. Chapter 11); in particular, the character values are simpler in that case. But most of the theory is equally valid for the groups we are considering, such as $SL(n,K)$ and $Sp(2n,K)$.

In a split group, the classes of maximal tori in question are parametrized naturally by the classes of W; if T corresponds to w, $\pm R_{T,\theta}(1)$ equals the p'-part of the order of Γ_n divided by the order of the finite torus $T(\boldsymbol{F}_{p^n})$, which is computable in terms of w. This virtual degree is a polynomial in $q = p^n$ with leading term q^m. For example, when $\Gamma_n = SL(2,q)$, there are two families with respective degrees $q+1$, $q-1$. When $\Gamma_n = Sp(4,q)$ there are five families, which we label (purely for convenience) A–E, with the following degrees:

A	$(q+1)^2(q^2+1)$
B,C	q^4-1
D	$(q^2-1)^2$
E	$(q-1)^2(q^2+1)$.

Every irreducible character occurs as a constituent of some $R_{T,\theta}$, and, for θ in "general position", $\pm R_{T,\theta}$ is irreducible. The fraction of all irreducible characters of Γ_n coming from a given class of tori is approximately equal to the cardinality of the corresponding class of W divided by $|W|$; e.g. for $Sp(4,q)$ the families A–E contribute (respectively) approximately $\frac{1}{8}, \frac{1}{4}, \frac{1}{4}, \frac{1}{4}, \frac{1}{8}$ of all irreducible characters.

To obtain irreducible characters from $R_{T,\theta}$, one has to be able to decompose induced characters (these occur for $Sp(4,q)$ in the families A–C). A further complication (not well understood) may arise when G has a finite nontrivial centre: $R_{T,\theta}$ may decompose into a number of characters of equal

TABLE 3. *Characters of Sp(4,5) (principal block)* [36]

Family	*Character*	*Degree*	$q = 5$
A	θ_0	1	1
	θ_1	$q^2(q^2+1)/2$	325
	θ_2	same	
	θ_3	$(q^2+1)/2$	13
	θ_4	same	
	θ_9	$q(q+1)^2/2$	90
	θ_{11}	$q(q^2+1)/2$	65
	θ_{12}	same	
	$\chi_8(1)$	$(q+1)(q^2+1)$	156
	$\chi_9(1)$	$q(q+1)(q^2+1)$	780
	Φ_5	$(q+1)(q^2+1)/2$	78
	Φ_6	same	
	Φ_7	$q(q+1)(q^2+1)/2$	390
	Φ_8	same	
	Φ_9	$q(q^2+1)$	130
B	$\chi_5(1,1)$	q^4-1	624
	$\xi_1(2)$	$(q-1)(q^2+1)$	104
	$\xi_1'(2)$	$q(q-1)(q^2+1)$	520
	$\xi_{21}(2)$	$(q^4-1)/2$	312
	$\xi_{22}(2)$	same	
	$\xi_{41}'(1)$	same	
	$\xi_{42}'(1)$	same	
C	$\chi_2(2)$	q^4-1	624
	$\chi_6(1)$	$(q-1)(q^2+1)$	104
	$\chi_6(2)$	same	
	$\chi_7(1)$	$q(q-1)(q^2+1)$	520
	$\chi_7(2)$	same	
D	$\chi_1(2)$	$(q^2-1)^2$	576
	$\chi_1(4)$	same	
	$\chi_1(8)$	same	
E	$\xi_{21}'(1)$	$(q-1)^2(q^2+1)/2$	208
	$\xi_{22}'(1)$	same	
	θ_{10}	$q(q-1)^2/2$	40

degree, all taking the same values on semisimple elements; cf. the "exceptional" characters of $SL(2,q)$. We see this phenomenon clearly in the character table for $Sp(4,q)$, q odd, which was computed by Srinivasan [36] before the development of the general theory of [13]. (For even q, see [15].) Table 3 gives the degrees of the 33 characters in the principal block of $Sp(4,5)$, where a number of pairs (such as θ_1 and θ_2) behave in the way just described.

3.4. Thanks to Brauer reciprocity, the question of computing decomposition numbers is the same as the question of "decomposing" a PIM into ordinary characters. It is natural to look first at the "small" PIM's of dimension $|W|p^m$ for $K\Gamma$ considered in 2.3, 2.4. Here we assume that $Q(\lambda) = R(\lambda)$, though this is not yet established for all λ in the top alcove. A formal identity of Solomon [35, formula (5)] yields a helpful clue.

THEOREM. *$|W|p^m$ equals the sum of the various (positive) degrees $\pm R_{T,\theta}(1)$, each taken as often as the cardinality of the corresponding class of W.*

We are led by this (and by some experience with the Γ-composition factors of $Q(\lambda)$) to conjecture that, for λ in sufficiently general position in the top alcove, $R(\lambda)$ will involve $|W|$ irreducible characters of the types suggested by the theorem. Moreover, the characters θ involved should come from the restrictions to appropriate finite tori of the dual weight $\hat{\lambda}$ (the values in K^* being replaced by complex values in the usual way). This would be compatible with the character formula 2.3, at least on regular semisimple elements, where the Steinberg character takes values ± 1. All of this is well known, of course, for $SL(2,p)$, where $2p = (p+1)+(p-1)$. (*See* [52], [55].)

The meaning of "sufficiently general position" is yet to be specified, but study of an example for $Sp(4,5)$ illustrates what has to be avoided. The weight $\lambda = (3,2)$ lies inside the top alcove (Fig. 1), and $Q(\lambda) = R(\lambda)$ has dimension $8p^4$, $p = 5$. Nevertheless this PIM involves 12, rather than 8, ordinary characters (Table 4). The reason for this will show up clearly when we introduce the Brauer complex (3.6 below; see Fig. 3): (3,2) is not far enough from the walls of the alcove, nor from the "reflecting" line which bisects the top alcove.

3.5. For a weight not in the top alcove, but in sufficiently general position, we expect the corresponding PIM to have dimension equal to some multiple of $|W|p^m$ and to involve this multiple of $|W|$ ordinary characters in the proportions suggested by Theorem 3.4. At the moment we do not know enough about PIM's to make this more precise.

We can also ask directly for a way to list the composition factors of an ordinary representation taken modulo p. The number of these will, typically, be very large for Γ (and larger still for Γ_n), so some organizing principle is

TABLE 4. *Characters in $R(\lambda)$ for principal block of $Sp(4,5)$*

λ	A	B	C	D	E
(3,2)	$\chi_9(1), \chi_8(1)$	$\chi_5(1,1), \xi'_{41}(1), \xi'_{42}(1)$	$\chi_2(2), \chi_6(2), \chi_7(2)$	$\chi_1(2), \chi_1(8)$	$\xi'_{21}(1), \xi'_{22}(1)$
(2,2)	$2\chi_9(1), \Phi_5, \Phi_6, \Phi_7, \Phi_8$	$\chi_5(1,1), \xi_1(2), \xi'_1(2), \xi_{21}(2), \xi_{22}(2), \xi'_{41}(1), \xi'_{42}(1)$	$2\chi_2(2), 2\chi_7(2)$	$\chi_1(2), 2\chi_1(8), \chi_1(4)$	$\xi'_{21}(1), \xi'_{22}(1)$
(2,0)	$\chi_9(1), \Phi_9, 2\Phi_7, 2\Phi_8, \theta_3, \theta_4$	$2\xi'_1(2), \xi_{21}(2), \xi_{22}(2), \xi'_{41}(1), \xi'_{42}(1)$	$2\chi_2(2), 2\chi_7(2), \chi_6(1), \chi_6(2)$	$\chi_1(2), 2\chi_1(8), \chi_1(4)$	$\xi'_{21}(1), \xi'_{22}(1)$
(0,0)	$2\Phi_9, \Phi_7, \Phi_8, \theta_0$	$\xi_{21}(2), \xi_{22}(2), \xi_1(2), \xi'_1(2)$	$\chi_2(2), 2\chi_7(2), \chi_6(1)$	$\chi_1(2), \chi_1(8)$	$\xi'_{21}(1), \xi'_{22}(1)$
(2,4)	$\Phi_9, \theta_1, \theta_2$	$\xi_{21}(2), \xi_{22}(2)$	$\chi_7(2)$	$\chi_1(8)$	
(1,4)	$\chi_9(1), 2\theta_1, 2\theta_2$	$\xi_{21}(2), \xi_{22}(2), \xi'_{41}(1), \xi'_{42}(1)$	$\chi_7(2)$	$2\chi_1(8)$	
(1,0)	$\chi_9(1), \chi_8(1)$	$\chi_5(1,1), \xi'_{41}(1), \xi'_{42}(1)$	$\chi_2(2), \chi_7(2), \chi_6(2)$	$\chi_1(4), \chi_1(8)$	$\xi'_{21}(1), \xi'_{22}(1)$
(3,4)	$\chi_9(1)$	$\chi_5(1,1)$	$\chi_7(1)$	$\chi_1(4)$	
(0,4)	$\chi_9(1), \theta_9, \theta_{12}$	$\chi_5(1,1), \xi_1(2), \xi'_1(2)$	$\chi_7(1), \chi_7(2)$	$\chi_1(2), \chi_1(4)$	
(0,2)	$\chi_9(1), \Phi_5, \Phi_6, \theta_{11}$	$\chi_5(1,1), \xi'_1(2), \xi_{21}(2), \xi_{22}(2)$	$\chi_2(2), \chi_7(1), \chi_7(2)$	$\chi_1(8), \chi_1(4)$	θ_{10}
(4,0)	θ_9, θ_{11}	$\xi'_1(2)$	$\chi_7(1), \chi_6(1)$	$\chi_1(4)$	
(3,0)	$\chi_8(1), \theta_{12}$	$\chi_5(1,1), \xi'_1(2)$	$\chi_2(2), \chi_7(1)$	$\chi_1(2)$	θ_{10}
(4,2)	Φ_7, Φ_8	$\xi'_1(2)$	$\chi_2(2)$	$\chi_1(2)$	
(1,2)	$\chi_9(1), \chi_8(1), \Phi_7, \Phi_8$	$\chi_5(1,1), \xi'_1(2), \xi'_{41}(1), \xi'_{42}(1)$	$2\chi_2(2), \chi_6(1), \chi_7(1)$	$2\chi_1(2), \chi_1(4)$	$\xi'_{21}(1), \xi'_{22}(1)$

needed. Guided by the analogy with the u-modules $Z(\lambda)$ mentioned in 3.1, the author introduced in [21] a notion of "deformation" of linkage classes. The idea is to begin with the composition factors of a typical $Z(\lambda)$, taken with multiplicities d_μ, and then deform the highest weights slightly to obtain the composition factors of a typical $\bar{Z}$. For example, when $G = SL(2,K)$, the linkage class $\{\lambda, \; p-\lambda-2\}$ yields either $\{\lambda, \; p-\lambda-1\}$ or $\{\lambda, \; p-\lambda-3\}$ (corresponding to ordinary characters of degrees $p+1$ or $p-1$).

The rationale behind this can be described briefly as follows. A G-composition factor of $Q(\lambda)$ typically has the form $M(\mu) \otimes M(\nu)^{(p)}$, with $\mu \in X_p$ and ν "small". For u this yields (dim $M(\nu)$) copies of $M(\mu)$, and μ is linked to λ. But for $K\Gamma$ we get $M(\mu) \otimes M(\nu)$, whose Γ-composition factors are in general obtained by adding to μ the various W-conjugates of ν or (smaller) subweights.

Consider $Sp(4,K)$. Here a typical linkage class has 8 elements, 2 in each alcove, and the multiplicity of $M(\mu)$ in $Z(\lambda)$ is $d_\mu = 1,2,3$ or 4 for μ in the alcove numbered similarly in 1.9. (This is independent of the choice of λ in the given linkage class.) The resulting 20 weights are deformed according to a recipe given in [21], yielding in general 20 distinct weights distributed in the same way among alcoves. There is one deformation for each element (or class) of W, and it can be checked formally that the sum of 20 dimensions is in each case the generic character degree discussed in 3.3. This makes it overwhelmingly likely that the $\bar{Z}$ behave as expected, but as yet there is no rigorous proof for B_2. Of course, weights not in such general position will correspond to ordinary characters not of generic degree; cf. 3.6 below.

The above method, supplemented by *ad hoc* dimension comparisons and the like, has been used by the author to obtain the decomposition matrix (and Cartan matrix) for the principal block of $Sp(4,5)$; cf. Table 4. (In checking that the indicated sums of ordinary characters vanish except on p-regular classes, we found a few systematic sign errors in the character table [36], later confirmed by comparison with a list of errata prepared by Srinivasan.) The determinant of the Cartan matrix was checked by computer to be 5^{11} (as predicted by Brauer's theory).

3.6. As an aid in visualizing the rather complicated decomposition behaviour of ordinary characters, the author formulated in [23] a "Brauer complex" generalizing the Brauer tree of $SL(2,p)$. The idea is to divide the top alcove of X_p into p^ℓ small alcoves (ℓ = rank of G), each of which will correspond to the set of irreducible characters of Γ in a geometric conjugacy class. When Γ has more than one block of highest defect, there will be one such complex associated with each block, and most characters will be repeated a corresponding number of times in the complex (twice for $Sp(4,p)$, p odd). This shows up in Fig. 3, where the dotted line is a sort of mirror.

The vertices of the complex which are "special points" are labelled with weights from the top alcove belonging to the block in question. The alcoves surrounding such a vertex (of which there are generally $|W|$) should contain the ordinary characters which occur in the PIM attached to the corresponding weight (cf. 3.4). But when a geometric conjugacy class contains more than one character, these may or may not all occur in the PIM: this occurs for $Sp(4,5)$ along the outer walls of the complex and also along the "mirror", where pairs of characters of equal degree occur. (Two such pairs, of types B and E, occur in the PIM of (3,2) in Fig. 3. The characters of types A, C along the wall are induced characters which split into two pieces. So there are a total of 12 irreducible characters in this PIM.)

For Sp (4,p), we thus visualize a PIM for a typical weight in the top alcove as a square comprising 8 alcoves. For weights in lower alcoves, we expect respectively 16, 24, 32 small alcoves, arranged in the way shown in [23, p. 83]. These patterns are precisely the duals of Jantzen's generic decomposition patterns (1.10)! If p is large enough, such regular patterns can be expected to predominate, but for $p = 5$ they are not in evidence.

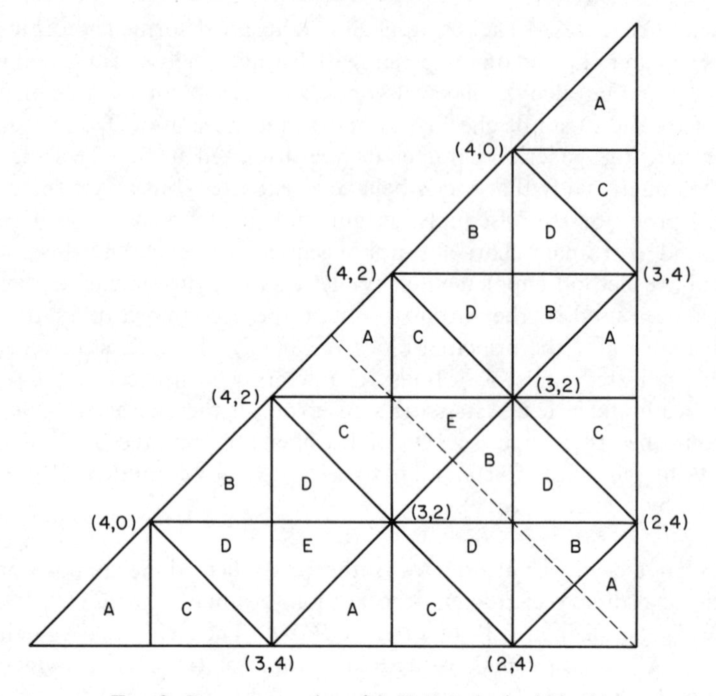

FIG. 3. Brauer complex of $Sp(4,5)$ (principal block).

For type B_2 we expect all nonzero decomposition numbers to be "generically" 1 (though 2 may sometimes occur, as in Table 4), but for other types, such as G_2, this will certainly not be true (cf. the diagrams in [30, pp. 457–8]). So the Brauer complex as pictured here must be supplemented with information about the PIM's for weights in lower alcoves, including multiplicities.

3.7. Recently Carter has suggested a way of looking at the Brauer complex which ties in more directly with the work of Deligne and Lusztig. This is more straightforward in the case that G has connected centre. Consider G_2. Here there are p^2 small alcoves in the complex, and also p^2 "semisimple" characters, one in each geometric conjugacy class. Carter observes that there is a natural way to assign the semisimple classes of G (here, its own dual group) to points of the fundamental alcove, so that the closure of each small alcove receives precisely one semisimple class of Γ. Moreover, the only such points not lying in the interior of an alcove lie in an outside wall. Using the correspondence between semisimple classes of a group and semisimple characters of the dual group, one has in this way another approach to assigning characters to small alcoves. (Carter's student Deriziotis has worked out this idea in detail in his 1977 Warwick thesis.)

3.8. At the opposite extreme from semisimple characters are the "unipotent" ones. (See Chapter 11.) What can be said about their reduction modulo p? This can be expected to be highly irregular, but on the other hand may involve relatively few weights (placed recognizably).

Consider $Sp\,(4,q)$, which has six unipotent characters (denoted as in [36]; cf. Table 3): $\theta_0, \theta_9, \theta_{10}, \theta_{11}, \theta_{12}, \theta_{13}$ (Steinberg character). All except θ_{10} are constituents of the character induced from the 1-character of the Borel subgroup; θ_{10} occurs in the "discrete series". For small primes, direct computation shows that the reduction modulo p of θ_{10} has as composition factors just the two irreducible Weyl modules $V\,(0,p-3)$ and $V(p-2,0)$ (contrary to the expectation voiced in [23, p. 86]!). This is likely to be true for all p. Similarly, dimension comparison suggests strongly that the reductions of other unipotent characters will involve these pairs of Weyl modules:

$$\theta_9 : V(p-1,0), \quad V(0,p-1),$$

$$\theta_{11} : V(p-1,0), \quad V(0,p-3),$$

$$\theta_{12} : V(p-2,0), \quad V(0,p-1).$$

For $Sp(4,q)$ it should suffice to replace p by q; but then the Weyl modules involved need not be irreducible. It may be possible to verify these conjectures by direct comparison of Brauer characters.

Lusztig has recently examined the unipotent characters of other groups of low rank and has found some similar candidates for composition factors.

3.9. Most of our discussion has centred on the decomposition behaviour of ordinary characters of Γ. For Γ_n it is quite likely that the number of composition factors will go up as the nth power of the generic number (20^n for $Sp(4,p^n)$). It is not yet clear, however, how to organize the vast numbers of weights involved (cf. 2.9 above). The case $\Gamma_n = SL(2,p^n)$ has been worked out in a character-theoretic way of Srinivasan, but even here it is difficult to get much insight. There are of course some regular patterns, related to the fact that the ordinary character degrees are expressible uniformly as polynomials in $q = p^n$.

3.10. Finally, we should mention the twisted groups, which have been neglected thus far. Both the ordinary and the modular representations of the twisted groups of types A, D, E_6 are closely related to the representations of the corresponding untwisted groups. The deformations of linkage classes discussed in 3.5 can be adapted (as indicated in [21]), so there is some hope for a unified treatment from this point of view. However, it seems less likely that a reasonable substitute for the Brauer complex can be found.

The Ree and Suzuki twisted groups pose a further challenge; at least in the case of the Suzuki groups, we now have sufficient information about the ordinary and the modular representations to be optimistic about computing the decomposition matrix. Here (as in our approaches above) we would like to see such a computation carried out using just "integer arithmetic", rather than using detailed character values (as in the work of Burkhardt [48] on the decomposition numbers of Suzuki groups), since the end result is to be just a matrix of integers.

Appendix

The cohomology groups associated with line bundles (locally free sheaves of rank 1) on G/B afford natural representations of G, but are not yet very well understood. Here we shall indicate briefly what is known, emphasizing the connection with Weyl modules. For further details see [44]–[47], [25].

A.1. Given an arbitrary weight $\lambda \in X$ (or the B-module it defines), the associated line bundle $\mathscr{L}(\lambda)$ is a sheaf on G/B whose space $H^0(G/B, \mathscr{L}(\lambda))$ of global sections consists of all polynomial functions $f: G \to K$ satisfying $f(xb) = \lambda(b)^{-1} f(x)$ for $x \in G$, $b \in B$. Since G/B is a projective variety, this space is finite dimensional over K. Moreover, G acts naturally on it (by left translations). Further, it has a unique B-stable line and hence a unique minimal submodule (of highest weight λ); in particular, $H^0(G/B, \mathscr{L}(\lambda))$ is nonzero if and only if λ is dominant. The first two of these three properties carry over to the right derived functors of the global section functor, the cohomology spaces $H^i(G/B, \mathscr{L}(\lambda))$. The largest value of i for which the cohomology could be nonzero is $m = \dim G/B =$ number of positive (or negative) roots. For $i > 0$ there is no concrete description of these spaces in general; thus it is difficult to say anything directly about their structure as G-modules.

One gets some idea of what to expect by recalling Bott's well-known theorem for the corresponding situation over C. Here G_C is a simple Lie group, B_C a Borel subgroup, while X, W, etc. play the same role as for G.

THEOREM (Bott).

(a) *If $\lambda + \delta$ is W-irregular (i.e., lies in a wall of some Weyl chamber), then all $H^i(G_C/B_C, \mathscr{L}(\lambda))$ are* 0.

(b) *If $\lambda + \delta$ is W-regular and dominant, then, for all $i > 0$,*

$$H^i(G_C/B_C, \mathscr{L}(\lambda)) = 0,$$

while $H^0(G_C/B_C, \mathscr{L}(\lambda))$ affords the irreducible G_C-module $V(\lambda)_C$ of highest weight λ.

(c) *If $\lambda + \delta$ is W-regular, and w is the (unique) element of W for which $w. \lambda = w(\lambda + \delta) - \delta$ is dominant, then $H^i(G_C/B_C, \mathscr{L}(\lambda)) = 0$ for $i \neq \ell(w)$, the length of w in W, while $H^{\ell(w)}(G_C/B_C, \mathscr{L}(\lambda))$ affords $V(w,\lambda)_C$.*

Of course, parts (b) and (c) could be combined, but we state them separately for reasons which will become clear below.

A.2. Part of Bott's theorem remains valid in characteristic p. Let us adopt the following terminology: $\lambda \in X$ has "standard vanishing behaviour" if either $\lambda+\delta$ is W-irregular and $H^i(G/B, \mathscr{L}(\lambda)) = 0$ for all i, or else $\lambda+\delta$ is W-regular and $H^i(G/B, \mathscr{L}(\lambda)) = 0$ for all $i \neq \ell(w)$, where w is the unique element of W for which $w.\lambda = w(\lambda+\delta)-\delta$ is dominant. Define Weyl chambers by letting all reflecting hyperplanes pass through $-\delta$. Call a Weyl chamber an "H^i-chamber" if $\ell(w) = i$ for the unique $w \in W$ which maps this chamber to the dominant one (via the dot action).

THEOREM (Kempf). *If $\lambda+\delta$ lies in the closure of the dominant Weyl chamber, then λ has standard vanishing behaviour, and the G-module $H^0(G/B, \mathscr{L}(\lambda))$ has formal character $ch(\lambda)$ (defined to be 0 if λ is not dominant).*

The proof of Kempf's theorem is quite difficult, but recently Andersen and Haboush (independently) have found a simpler approach.

The second part of the theorem is actually a consequence of the first part along with Bott's theorem: it is a general fact that the Euler characteristic (the alternating sum of the formal characters of the H^i) is invariant under change of base.

Using Kempf's theorem, one concludes easily that $V(\lambda)$, $\lambda \in X^+$, has the universal property stated in 1.6. The point is that the dual module of a highest weight module (of weight $\lambda^* = -w_0\lambda$) embeds naturally in $H^0(G/B, \mathscr{L}(\lambda))$, and by dimension comparison $V(\lambda^*)$ must be the largest possible highest weight module. As a result, $V(\lambda^*)$ is dual as a G-module to $H^0(G/B, \mathscr{L}(\lambda))$. Thanks to Serre duality, we also obtain

$$V(\lambda^*) = H^m(G/B, \mathscr{L}(w_0 . \lambda)).$$

A.3. Parts (a), (c) of Bott's theorem do not always remain true in characteristic p (unless $G = SL(2,K)$). In his 1976 Harvard thesis, Griffith worked out the exact pattern of nonvanishing for $SL(3,K)$. Here it happens that both H^1 and H^2 are nonzero for weights lying in certain alcoves near walls dividing H^1-chambers from H^2-chambers; more precisely, these alcoves are "p^n-alcoves", with n growing as one gets farther away from $-\delta$.

Recently Andersen has recovered these results by another method, based in part on Seshadri's formulation of Griffith's work, and has found similar phenomena in other cases such as B_2. He has also obtained some very general results, notably a precise necessary and sufficient condition for H^1 to be nonzero; qualitatively, it says that H^1 is nonzero only for weights in an H^1-chamber or for weights in certain alcoves "near" the wall separating a chamber from an adjacent H^1-chamber. At the same time, he shows that $H^1(G/B, \mathscr{L}(\lambda))$ has a unique minimal submodule, of a specified highest weight. He also finds all B-stable lines: their weights are of the form $\lambda + k\alpha$, where translation with respect to the root α takes λ into X^+. This shows in particular that even when a weight λ in an H^1-chamber has standard vanishing behaviour, so that $H^1(G/B, \mathscr{L}(\lambda))$ has formal character $\mathrm{ch}(s_\alpha . \lambda)$ (s_α = reflection with respect to α), the G-module structure will usually differ from that of $H^0(G/B, \mathscr{L}(s_\alpha . \lambda))$, which has a unique B-stable line. By the time one reaches H^m (a Weyl module), there are usually a large number of B-stable lines, corresponding to nonzero maps between Weyl modules.

Close study of some exact sequences has also enabled Andersen to describe precisely some images and kernels of maps between Weyl modules for $SL(3,K)$ and $Sp(4,K)$. (These results agree with independent findings of Jantzen based on different methods.) The situation here is remarkably similar to that described by Carter and Lusztig [8] for analogous induced modules for $SL(3,p)$: those induced from linear characters of the "Borel subgroup $B(\boldsymbol{F}_p)$.

A.4. As mentioned at the end of 1.10, the generic decomposition patterns for Weyl modules may be related to cohomology. This was suggested by the author [25] and has been partially confirmed for $SL(3,K)$ by Andersen [44]. The idea is that nonstandard vanishing behaviour should be correlated with cancellations occurring (as in 1.10) when some weights in the generic pattern lie far outside the dominant Weyl chamber. The weights to be cancelled should show up in the "unexpected" H^i, and the cancellation itself should be given by the Euler characteristic. As stated earlier, all of this is hardly understood as yet; but it may lead to new ways of looking at Weyl modules and hence irreducible modules for G.

References

1. Alperin, J. L., Projective modules for $SL(2,2^n)$, *J. Pure Appl. Algebra*, **15** (1979) 219–234.
2. Ballard, J. W., Projective modules for finite Chevalley groups, *Trans. Amer. Math. Soc.* **245** (1978) 221–249.
3. Ballard, J. W., Injective modules for restricted enveloping algebras, *Math. Z.*, **163** (1978) 57–63.
4. Borel, A., "Properties and linear representations of Chevalley groups", Lecture Notes in Mathematics 131, Springer-Verlag, Berlin, 1970.
5. Carter, R. W., "Simple Groups of Lie Type", J. Wiley & Sons, London, 1972.
6. Carter, R. and Cline, E., "The submodule structure of Weyl modules for groups of type A_1", Proc. Conf. Finite Groups, Univ. Utah, Park City, Utah, 1975, 303–311 (Academic Press, New York, 1976).
7. Carter, R. W. and Lusztig, G., On the modular representations of the general linear and symmetric groups, *Math. Z.*, **136** (1974) 193–242.
8. Carter, R. W. and Lusztig, G., Modular representations of finite groups of Lie type, *Proc. London Math. Soc.* (3), **32** (1976) 347–384.
9. Chastkofsky, L. and Feit, W., On the projective characters in characteristic 2 of the groups $Suz(2^m)$ and $Sp_4(2^n)$, *Inst. Hautes Etudes Sci. Publ. Math.*
10. Chastkofsky, L. and Feit, W., On the projective characters in characteristic 2 of the groups $SL_3(2^m)$ and $SU_3(2^m)$, *J. Algebra*, **63** (1980) 124–142.

11. Cline, E., Parshall, B. and Scott, L., Cohomology, hyperalgebras, and representations, *J. Algebra*, **63** (1980) 98–123.
12. Curtis, C. W., "Modular representations of finite groups with split BN-pairs", Lecture Notes in Mathematics 131, Springer-Verlag, Berlin, 1970.
13. Deligne, P. and Lusztig, G., Representations of reductive groups over finite fields, *Ann. of Math.*, **103** (1976) 103–161.
14. Donkin, S., Hopf complements and injective comodules for algebraic groups, *Proc. London Math. Soc.* (3), **40** (1980) 298–319.
15. Enomoto, H., The characters of the finite symplectic groups $Sp(4,q)$, $q = 2^f$, *Osaka J. Math.*, **9** (1972) 75–94.
16. Green, J. A., Locally finite representations, *J. Algebra*, **41** (1976) 137–171.
17. Humphreys, J. E., Modular representations of classical Lie algebras and semisimple groups, *J. Algebra*, **19** (1971) 51–79.
18. Humphreys, J. E., Defect groups for finite groups of Lie type, *Math. Z.*, **119** (1971) 149–152.
19. Humphreys, J. E., Projective modules for $SL(2,q)$, *J. Algebra*, **25** (1973) 513–518.
20. Humphreys, J. E., Some computations of Cartan invariants for finite groups of Lie type, Comm. *Pure Appl. Math.*, **26** (1973) 745–755.
21. Humphreys, J. E., Weyl groups, deformations of linkage classes, and character degrees for Chevalley groups, *Comm. Algebra*, **1** (1974) 475–490.
22. Humphreys, J. E., "Linear algebraic groups", Graduate Texts in Mathematics 21, Springer-Verlag, Berlin, 1975.
23. Humphreys, J. E., "Ordinary and modular representations of Chevalley groups", Lecture Notes in Mathematics, 528, Springer-Verlag, Berlin, 1976.
24. Humphreys, J. E., On the hyperalgebra of a semisimple algebraic group, "Contributions to Algebra: A Collection of Papers Dedicated to Ellis Kolchin", 203–210, Academic Press, New York, 1977.
25. Humphreys, J. E., Weyl modules and Bott's theorem in characteristic p, "Lie Theories and their Applications", Queen's Papers in Pure & Appl. Math. 48, 474–483, Kingston, Ontario, 1978.
26. Humphreys, J. E. and Jantzen, J. C., Blocks and indecomposable modules for semisimple algebraic groups, *J. Algebra*, **54** (1978) 494–503.
27. Jantzen, J. C., Darstellungen halbeinfacher algebraischer Gruppen und zugeordnete kontravariante Formen, *Bonner Math. Schriften*, **67** (1973).
28. Jantzen, J. C., Zur Charakterformel gewisser Darstellungen halbeinfacher Gruppen und Lie-Algebren, *Math. Z.*, **140** (1974) 127–149.
29. Jantzen, J. C., Darstellungen halbeinfacher Gruppen und kontravariante Formen, *J. Reine Angew. Math.*, **290** (1977) 117–141.
30. Jantzen, J. C., Über das Dekompositionsverhalten gewisser modularer Darstellungen halbeinfacher Gruppen und ihrer Lie-Algebren, *J. Algebra*, **49** (1977) 441–469.
31. Jantzen, J. C., Über Darstellungen höherer Frobenius-Kerne halbeinfacher algebraischer Gruppen, *Math. Z.*, **164** (1979) 271–292.
32. Jeyakumar, A. V., Principal indecomposable representations for the group $SL(2,q)$, *J. Algebra*, **30** (1974) 444–458.
33. Kac, V. and Weisfeiler, B., Coadjoint action of a semi-simple algebraic group and the centre of the enveloping algebra in characteristic p, *Indag. Math.*, **38** (1976) 136–151.
34. Richen, F. A., Modular representations of split BN-pairs, *Trans. Amer. Math. Soc.*, **140** (1969) 435–460.
35. Solomon, L. Invariants of finite reflection groups, *Nagoya Math. J.*, **22** (1963) 57–64.

36. Srinivasan, B., The characters of the finite symplectic group $Sp(4,q)$, *Trans. Amer. Math. Soc.*, **131** (1968) 488–525.
37. Steinberg, R., Representations of algebraic groups, *Nagoya Math. J.*, **22** (1963) 33–56.
38. Steinberg, R., Endomorphisms of linear algebraic groups, *Mem. Amer. Math. Soc.*, **80** (1968).
39. Steinberg, R., "Lectures on Chevalley Groups", Yale University Math. Dept., 1968.
40. Upadhyaya, B. Shreekantha, Composition factors of the principal indecomposable modules for the special linear group $SL(2,q)$, *J. London Math. Soc.* (2), **17** (1978) 437–445.
41. Verma, D. N., Role of affine Weyl groups in the representation theory of algebraic Chevalley groups and their Lie algebras, "Lie Groups and their Representations", 653–705, Halsted, New York, 1975.
42. Winter, P. W., On the modular representation theory of the two-dimensional special linear group over an algebraically closed field, *J. London Math. Soc.* (2) **16** (1977) 237–252.
43. Wong, W. J., Irreducible modular representations of finite Chevalley groups, *J. Algebra*, **20** (1972) 355–367.
44. Andersen, H. H., Cohomology of line bundles on G/B, *Ann. Sci. Ecole Norm. Sup.*, **12** (1979) 85–100.
45. Andersen, H. H., The first cohomology group of a line bundle on G/B, *Invent. Math.*, **51** (1979) 287–296.
46. Andersen, H. H., The strong linkage principle, *J. Reine Angew. Math.*, **315** (1980) 53–59.
47. Andersen, H. H., On the structure of Weyl modules, *Math. Z.*, **170** (1980) 1–14.
48. Burkhardt, R., Über die Zerlegungszahlen der Suzukigruppen $Sz(q)$, *J. Algebra*, **59** (1979) 421–433.
49. Burkhardt, R., Über die Zerlegungszahlen der unitären Gruppen $PSU(3,2^{2f})$, *J. Algebra*, **61** (1979) 548–581.
50. Cline, E., Parshall, B., Scott, L., On the tensor product theorem for algebraic groups, *J. Algebra*, **63** (1980) 264–267.
51. Donkin, S., The blocks of a semisimple algebraic group, *J. Algebra* (to appear).
52. Humphreys, J. E., Deligne-Lusztig characters and principal indecomposable modules, *J. Algebra*, **62** (1980) 299–303.
53. James, G. D., Murphy, G. E., The determinant of the Gram matrix for a Specht module, *J. Algebra*, **59** (1979) 222–235.
54. Jantzen, J. C., Darstellungen halbeinfacher Gruppen und ihrer Frobenius-Kerne, *J. Reine Angew. Math.* (to appear).
55. Jantzen, J. C., Zur Reduktion modulo p der Charaktere von Deligne und Lusztig (to appear).
56. Lusztig, G., Some problems in the representation theory of finite Chevalley groups, Proc. Symp. Pure Math. 37, Amer. Math. Soc. (to appear).

13.

Weyl Modules for Groups of Lie Type

JENS C. JANTZEN

1. Introduction

If one wants to study the representations of a group, the first thing to try to get hold of are the simple modules. When dealing with a finite group of Lie type defined over a field of characteristic p and its representations over an algebraically closed field of the same characteristic, the previous chapter makes it clear that one should first look at the ambient algebraic group G and at its simple modules. The best method hitherto known to attack these is to take the Weyl modules and to try to find out their structure, especially the multiplicity $[V(\lambda):M(\mu)]$ of a simple module $M(\mu)$ as a simple factor in a Jordan–Hölder series of a Weyl module $V(\lambda)$. This chapter should serve as an introduction to the known results about these multiplicities and to the techniques by which we get them.

In order to avoid repetitions we shall keep to the conventions in part 1 of Chapter 12 and shall use the notation $G,X,W,W_a,V(\lambda),M(\lambda),f,h,e(\mu)$ without any new definition.

2. Contravariant Forms

Let us remember the introduction of the Weyl modules (Chapter 12, 1.5). We started with a simple module $V(\lambda)_C$ for the simple Lie algebra $\mathfrak{g}_C$ over C having the same root system R as G and chose a maximal vector v^+, obtaining a lattice $V(\lambda)_Z = U_Z v^+$ by applying Kostant's Z-form U_Z. It will be useful to recall the definition of U_Z. One needs a Chevalley basis of $\mathfrak{g}_C$; i.e. one chooses root vectors X_α for $\alpha \in R$ such that $\alpha([X_\alpha,X_{-\alpha}]) = 2$ and such that there exists an involutory antiautomorphism $x \mapsto {}^t x$ of $\mathfrak{g}_C$ with ${}^t X_\alpha = X_{-\alpha}$ for all $\alpha \in R$. Now U_Z is the Z-subalgebra of the universal enveloping algebra of $\mathfrak{g}_C$ generated by all $X_\alpha^n/(n!)$ with $\alpha \in R$ and $n \in N$. In view of this construction it is obvious that $x \mapsto {}^t x$ induces involutory antiautomorphisms (also denoted by $u \mapsto {}^t u$) of U_Z as well as of the K-algebra $U_K = U_Z \otimes K$.

The Weyl module $V(\lambda)$ is now constructed as $V(\lambda)_Z \otimes K$; this vector space is first a U_K-module, and the image of G under the representation on $V(\lambda)$ is generated by operators of the form $\Sigma_{n\geqslant 0}(X_\alpha^n/(n!)) \otimes a^n$ with $a \in K$ and $\alpha \in R$. Therefore G- and U_K-submodules of $V(\lambda)$ coincide; it is usually more convenient to work with U_K. Denote $x \otimes 1$ by $\bar{x}$ for any $x \in V(\lambda)_Z$.

The $\mathfrak{g}_C$-module $V(\lambda)_C$ is a direct sum of weight spaces $V(\lambda)_{C,\mu}$ with $\mu \in X$ (cf. 1.1 of Chapter 12), as is $V(\lambda)_Z$ of the $V(\lambda)_{Z,\mu} = V(\lambda)_{C,\mu} \cap V(\lambda)_Z$, and we have $V(\lambda)_\mu = V(\lambda)_{Z,\mu} \otimes K$. In particular, we know $V(\lambda)_{Z,\lambda} = Zv^+$; hence $V(\lambda)_\lambda = K\overline{v^+}$, and $V(\lambda) = U_K\overline{v^+}$. Each submodule of $V(\lambda)$ is a direct sum of its weight spaces, so each proper submodule is contained in $\oplus_{\mu\neq\lambda} V(\lambda)_\mu$. This has to apply to the sum of all proper submodules, too, which therefore is the only maximal submodule of $V(\lambda)$, equal to rad $V(\lambda)$. Obviously it consists of those $x \in V(\lambda)$ with $U_K x \subset \oplus_{\mu\neq\lambda} V(\lambda)_\mu$. Equivalently we see that

$$\text{rad}\, V(\lambda) = K\overline{\{x \in V(\lambda)_Z | U_Z x \subset \oplus_{\mu\neq\lambda} V(\lambda)_{Z,\mu} \oplus pZv^+\}}.$$

Let $\pi_\lambda : V(\lambda)_Z \to Z$ be the map with

$$\pi_\lambda(v)v^+ - v \in \bigoplus_{\mu\neq\lambda} V(\lambda)_{Z,\mu} \qquad \text{for all } v \in V(\lambda)_Z.$$

Then we get

$$\text{rad}\, V(\lambda) = K\overline{\{x \in V(\lambda)_Z | \pi_\lambda(U_Z v) \subset Zp\}}.$$

We can interpret this in another way. It is easy to check that on $V(\lambda)_Z$ a symmetric bilinear form may be introduced by

$$(u_1 v^+, u_2 v^+) = \pi_\lambda({}^t u_1 u_2 v^+) \qquad \text{for all } u_1, u_2 \in U_Z.$$

This form has the property

$$(1) \qquad (uv_1, v_2) = (v_1, {}^t u v_2) \qquad \text{for all } v_1, v_2 \in V(\lambda)_Z,\ u \in U_Z.$$

Any bilinear form on a U_Z- or U_K-module satisfying (1) is called a *contravariant form* (the name going back to Wong [7]). We have shown that

$$(2) \qquad \text{rad}\, V(\lambda) = K\overline{\{x \in V(\lambda)_Z | (V(\lambda)_Z, x) \subset Zp\}}.$$

Any contravariant form has the property that weight spaces for different weights are orthogonal. If we denote the determinant of (,) with respect to a Z-basis of $V(\lambda)_{Z,\mu}$ by $D_\lambda(\mu)$, we may deduce from (2) that $V(\lambda)$ is simple if and only if no $D_\lambda(\mu)$ is divisible by p. The same arguments as above, applied to $V(\lambda)_C$ instead of $V(\lambda)$, show that $D_\lambda(\mu)$ is always an integer different from zero; it is in fact positive (cf. [1], Satz I.11). So a computation of the $D_\lambda(\mu)$ provides a criterion for the irreducibility of $V(\lambda)$ (cf. [1], Satz II.9, for the case $G = SL(n,K)$). We can deduce stronger information. Set

$$V(\lambda)_Z^i = \{x \in V(\lambda)_Z | (V(\lambda)_Z, x) \subset Zp^i\}$$

and

$$V(\lambda)^i = K\overline{V(\lambda)_Z^i} \qquad \text{for all } i \in N.$$

Then

$$V(\lambda)^0 = V(\lambda) \supset V(\lambda)^1 = \text{rad}\, V(\lambda) \supset V(\lambda)^2 \supset V(\lambda)^3 \supset \ldots$$

is a descending chain of submodules in $V(\lambda)$ with $V(\lambda)^i = 0$ for sufficiently large i, and we can show the following ([3], Lemma 3).

LEMMA. $$\sum_{i>0} \text{ch}\, V(\lambda)^i = \sum_{\mu \in X} v_p(D_\lambda(\mu))e(\mu).$$

Here v_p denotes the p-adic valuation of the integers; i.e., $v_p(p^n q) = n$ if $p \nmid q$.

In order to formulate our results on the $v_p(D_\lambda(\mu))$, we need some more notation. To each root $\alpha \in R$ there corresponds a reflection $s_\alpha = s_{\alpha,0} \in W$, which can be written in the form $s_\alpha \lambda = \lambda - \langle \lambda, \alpha^\vee \rangle \alpha$ for all $\lambda \in X$. Denote by $s_{\alpha,m}$, for all $m \in \mathbf{Z}$, the affine reflection

$$\lambda \mapsto s_\alpha \lambda + m\alpha = \lambda - (\langle \lambda, \alpha^\vee \rangle - m)\alpha \qquad \text{for all } \lambda \in X.$$

The $s_{\alpha,mp}$ with $m \in \mathbf{Z}$ are elements of the affine Weyl group W_a. Recall the convention $w \,.\, \lambda = w(\lambda + \delta) - \delta$, which we are going to extend to all $w \in W_a$.

For all $\mu \in X$ set

$$\text{ch}(\mu) = \left(\sum_{w \in W} \det(w) e(w \,.\, \mu) \right) \Big/ \left(\sum_{w \in W} \det(w) e(w \,.\, 0) \right).$$

Because of Weyl's character formula, this coincides for $\mu \in X^+$ with the definition of $\text{ch}(\mu)$ given in 1.2 of Chapter 12. Obviously we have $\text{ch}(w \,.\, \mu) = \det(w)\text{ch}(\mu)$ for all $\mu \in X$ and $w \in W$, so for all $\mu \in W \,.\, X^+$ we conclude that $\text{ch}(\mu)$ is, up to a sign, the character of a Weyl module. For $\mu \in X - W \,.\, X^+$, there is a root α with $s_\alpha \,.\, \mu = \mu$; hence $\text{ch}(\mu) = 0$. We see in particular that $\text{ch}(\mu)$ is in $\mathbf{Z}[X]$ for all $\mu \in X$. Now we can state ([4], Theorem 1 and Satz 4):

THEOREM, *Suppose $p \geqslant h$ (=Coxeter number of R) or that R is of type A_l. Then we have, for all $\lambda \in X^+$,*

$$\sum_{\mu \in X} v_p(D_\lambda(\mu))e(\mu) = \sum_{\alpha \in R^+} \sum_{0 < mp < \langle \lambda + \delta, \alpha^\vee \rangle} v_p(mp)\text{ch}(s_{\alpha,mp} \,.\, \lambda).$$

(Here, as well as later, R^+ denotes the set of positive roots.)

The formula is conjectured to be true for all p and R and has been checked for all p in a few additional small rank cases. The lemma, the theorem, and the preceding remarks imply that (for $p \geqslant h$, say) we can write $\Sigma_{i>0}\text{ch}\, V(\lambda)^i$ as an integral linear combination $\Sigma_{\mu \in X^+} a_\mu \text{ch}\, V(\mu)$, where $a_\mu \neq 0$ implies $\mu < \lambda$.

In fact we can show that μ satisfies a stronger condition which is mentioned in the proposition below. Using induction we can prove that the same condition is satisfied by every $\mu \in X^+$ with $[V(\lambda):M(\mu)] \neq 0$ because, for $\mu \neq \lambda$, we have equivalences

$$[V(\lambda):M(\mu)] \neq 0 \Leftrightarrow [V(\lambda)^1:M(\mu)] \neq 0 \Leftrightarrow \sum_{i>0} [V(\lambda)^i:M(\mu)] \neq 0,$$

and the last condition implies $[V(\mu'):M(\mu)] \neq 0$ for some $\mu' \in X^+$ with $a_{\mu'} \neq 0$. Thus we get ([4], Theorem 2)

PROPOSITION. *Suppose the theorem holds for R and p. Then for $\lambda,\mu \in X^+$ with $[V(\lambda):M(\mu)] \neq 0$ there exists a chain of weights $\lambda = \lambda_0 > \lambda_1 > \lambda_2 > \ldots > \lambda_r = \mu$ and of affine reflections $s_{\alpha_i,m_ip} \in W_a$ (with $\alpha_i \in R, m_i \in \mathbf{Z}$) such that $\lambda_i = s_{\alpha_i,m_ip} . \lambda_{i-1}$ for $1 \leqslant i \leqslant r$.*

One might want to deduce Theorem 1.7 of Chapter 12 as a corollary to this proposition; unfortunately Theorem 1.7 is used in the proof of the theorem above except for G of type A_1 and some low rank cases.

The converse of the proposition does not hold, as the example below will illustrate. Two situations where one does get positive multiplicities are described in [4], Satz 10 and Satz 11.

In order to see how much and how little we can deduce from these results, let us look at an example. In type A_1, A_2, or B_2, the theorem gives us the decomposition of $V(\lambda)$ for all $\lambda \in X_p$ which is all we need because of Steinberg's tensor product theorem (Chapter 12, 1.3). So let us look at $G = SL(4,K)$, i.e. R of type A_3. Denote the three simple roots by $\alpha_1,\alpha_2,\alpha_3$ in the usual ordering (i.e. $\langle \alpha_1,\alpha_3^\vee \rangle = 0$), write $s_i = s_{\alpha_i}$ $(i = 1,2,3)$ and $s_0 = s_{\alpha_0,p}$ where $\alpha_0 = \alpha_1+\alpha_2+\alpha_3$. Assume $p \geqslant 5$; then we can choose $\lambda \in X^+$ with $\langle \lambda+\delta,\alpha_0^\vee \rangle < p$. From results to be mentioned later on, it will be clear that we know all multiplicities if we know the $[V(w.\lambda):M(w'.\lambda)]$ with $w,w' \in W_a$ and $w.\lambda$, $w'.\lambda \in X^+$. As above we may restrict to those w with $w.\lambda \in X_p$. Exactly the following six elements of W_a satisfy this condition:

$$1, \quad w_1 = s_0, \quad w_2' = s_0s_3, \quad w_2'' = s_0s_1, \quad w_3 = s_0s_1s_3, \quad w_4 = s_0s_1s_3s_2.$$

For these w, there are eight elements $w' \in W_a$ for which $[V(w.\lambda):M(w'.\lambda)]$ may be nonzero in view of the proposition: the six elements mentioned already and $w_3' = w_2's_2$, $w_3'' = w_2''s_2$. The condition in the proposition defines, in general, an order relation on X^+. If we restrict it to the eight elements $w'.\lambda$, we get the following order diagram:

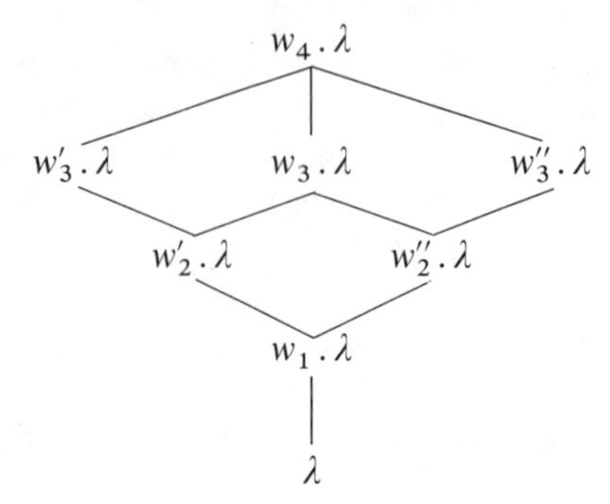

Now from the theorem and lemma we can deduce the following.

(A) $\Sigma_{i>0}\operatorname{ch} V(\lambda)^i = 0$; hence $V(\lambda)^i = 0$ for all $i > 0$, $\operatorname{rad} V(\lambda) = 0$ and $V(\lambda) = M(\lambda)$ is simple (cf. 1.8 of Chapter 12).

(B) For $w \in \{w_1, w'_2, w''_2, w'_3, w''_3\}$ there is exactly one $w' \in W_a$ such that $w'.\lambda$ is maximal among those elements in the order diagram below $w.\lambda$. Calculations and induction yield

$$\sum_{i>0} \operatorname{ch} V(w.\lambda)^i = \operatorname{ch} M(w'.\lambda).$$

This implies $V(w.\lambda)^1 \cong M(w'.\lambda)$ and $V(w.\lambda)^i = 0$ for $i \geqslant 2$. We have a non-split extension

$$0 \to M(w'.\lambda) \to V(w.\lambda) \to M(w.\lambda) \to 0.$$

(C) For w_3 we get, using (A) and (B),

$$\sum_{i>0} \operatorname{ch} V(w_3.\lambda)^i = \operatorname{ch} M(w'_2.\lambda) + \operatorname{ch} M(w''_2.\lambda) + 2\operatorname{ch} M(w_1.\lambda) + \operatorname{ch} M(\lambda).$$

Therefore we must have for $w \in \{w'_2, w''_2, 1\}$ that

$$[V(w_3.\lambda)^i : M(w.\lambda)] = \begin{cases} 1 & \text{for } i = 0,1 \\ 0 & \text{for } i \geqslant 2. \end{cases}$$

For $M(w_1.\lambda)$ there are two possibilities: either

(a) $$[V(w_3.\lambda)^i : M(w_1.\lambda)] = \begin{cases} 2 & \text{for } i = 0,1 \\ 0 & \text{for } i \geqslant 2 \end{cases} \quad \text{or}$$

(b) $$[V(w_3.\lambda)^i : M(w_1.\lambda)] = \begin{cases} 1 & \text{for } i = 0,1,2 \\ 0 & \text{for } i \geqslant 3. \end{cases}$$

(The second possibility would imply $V(w_3.\lambda)^2 \cong M(w_1.\lambda)$ and $V(w_3.\lambda)^3 = 0$; the first one $V(w_3.\lambda)^2 = 0$.) Whether (a) or (b) holds cannot be decided using only the results mentioned until now. In order to find out we shall develop other methods and return to the example afterwards.

3. Tensor Products

The affine Weyl group W_a is generated by the reflection s_α with α a simple root and $s_{\alpha_0,p}$ where α_0 is the greatest short root in R; denote this system of generators by S.

Set $C = \{\lambda \in X \mid \; 0 < \langle \lambda + \delta, \alpha^\vee \rangle < p \quad \text{for all } \alpha \in R^+\}$

and $\bar{C} = \{\lambda \in X \mid \; 0 \leqslant \langle \lambda + \delta, \alpha^\vee \rangle \leqslant p \quad \text{for all } \alpha \in R^+\}$.

Then C(resp. $\bar{C}$) is the intersection of X with an open alcove for the action of the reflection group W_a on $X \otimes \mathbf{R}$ (resp. with its closure), and S is the set of reflections with respect to the walls of this alcove. For any $\lambda \in \bar{C}$, therefore, the stabilizer of λ in W_a is generated by $S_\lambda = \{s \in S | s.\lambda = \lambda\}$. The set $\bar{C}$ is a fundamental domain for the operation of W_a on X.

If M is a G-module and $\lambda \in \bar{C}$, we say that M *belongs to* λ if all composition factors of M are of the form $M(w.\lambda)$ with $w \in W_a$. If $p \nmid f$ or if R is of type A_1, then we know that every indecomposable G-module belongs to some $\lambda \in \bar{C}$ (cf. 1.7 of Chapter 12). Assume now that our group has this property. We can decompose any G-module M into a direct sum $M = \bigoplus_{i=1}^n M_i$ of indecomposables. For every $\lambda \in \bar{C}$, set $p_\lambda(M)$ equal to the sum of the M_i belonging to λ. Then $p_\lambda(M)$ is independent of the choice of the decomposition, and we have $M = \bigoplus_{\lambda \in \bar{C}} p_\lambda(M)$.

Take $\lambda, \mu \in \bar{C}$ and let F be the simple G-module with highest weight in $W(\mu - \lambda)$. For each G-module M belonging to λ, set

$$T_\lambda^\mu M = p_\mu(M \otimes F).$$

This is a module belonging to μ, and T_λ^μ is an exact functor from the category of the G-modules belonging to λ to that of those belonging to μ. Furthermore, T_λ^μ and T_μ^λ are adjoint to each other.

In our following statements we assume that we start with a weight $\lambda \in C$. This is done partly in order to simplify the results and partly because some of the proofs work only in this case (e.g. in [2], Section 5). Let us remark that the existence of such a λ implies that $p \nmid f$ or that R is of type A_l. Therefore the assumption which we made before defining T_λ^μ is satisfied.

The effect of T_λ^μ on Weyl modules is easy to describe (combining [2], Satz 1, Satz 5 and [4], Satz 1):

LEMMA. *Suppose $\lambda \in C$, $\mu \in \bar{C}$ and $w \in W_a$ with $w.\lambda \in X^+$. Then:*

$$T_\lambda^\mu V(w.\lambda) = \begin{cases} V(w.\mu) & \text{if } w.\mu \in X^+ \\ 0 & \text{otherwise.} \end{cases}$$

(If $w.\mu$ is not dominant, we can deduce from our assumptions that $S_\mu \neq \emptyset$ and $w(\mu + \delta) \in X^+$; hence $\operatorname{ch}(w.\mu) = 0$.)

In order to formulate what T_λ^μ does to simple modules, we introduce the following notation for all $w \in W_a$:

$$\tau(w) = \{s \in S | ws.\lambda < w.\lambda\},$$

where λ is an arbitrary element of C. Then we get ([2], Theorem 1 and Theorem 2):

PROPOSITION. *Suppose λ, μ, w are as in the Lemma. Then:*

$$T_\lambda^\mu M(w.\lambda) = \begin{cases} M(w.\mu) & \text{if } S_\mu \cap \tau(w) = \emptyset, \\ 0 & \text{otherwise.} \end{cases}$$

The condition $S_\mu \cap \tau(w) = \emptyset$ can be interpreted in another way. For each $\alpha \in R^+$ there is an $m_\alpha \in \mathbf{N}$ with

$$m_\alpha p < \langle w(\lambda+\delta),\alpha^\vee \rangle < (m_\alpha+1)p.$$

Then $S_\mu \cap \tau(w) = \emptyset$ if and only if

$$m_\alpha p < \langle w(\mu+\delta),\alpha^\vee \rangle \leqslant (m_\alpha+1)p \qquad \text{for all } \alpha \in R^+.$$

In [2] we said that $w.\mu$ was in the upper closure of $w.C$ to describe this property.

Combining the lemma and the proposition with the exactness of T_λ^μ, we get ([5], Satz 7 and Lemma 11):

THEOREM (*Translation principle*). *Suppose* $\lambda \in C, w, w' \in W_a$ *with* $w.\lambda$, $w'.\lambda \in X^+$.

(a) *For all* $\mu \in \bar{C}$ *with* $\tau(w') \cap S_\mu = \emptyset$, *we have*

$$[V(w.\lambda):M(w'.\lambda)] = \begin{cases} [V(w.\mu):M(w'.\mu)] & \text{if } w.\mu \in X^+, \\ 0 & \text{otherwise.} \end{cases}$$

(b) *For all* $s \in S$ *with* $s \notin \tau(w')$, *we have*

$$[V(w.\lambda):M(w'.\lambda)] = \begin{cases} [V(ws.\lambda):M(w'.\lambda)] & \text{if } ws.\lambda \in X^+, \\ 0 & \text{otherwise.} \end{cases}$$

In order to show (b) one has to choose $\mu(s) \in \bar{C}$ with $S_{\mu(s)} = \{s\}$ (cf. [2], Satz 10) and apply (a).

Let us return to the example considered at the end of Section 2 and keep the notation used there. We have $S = \{s_0, s_1, s_2, s_3\}$ and, for example, $\tau(w_1) = \{s_0, s_2\}$; hence $s_1 \notin \tau(w_1)$. From part (b) of the theorem, we now get

$$[V(w_3.\lambda):M(w_1.\lambda)] = [V(w_3 s_1.\lambda = w_2'.\lambda):M(w_1.\lambda)] = 1.$$

This shows that, of the two possibilities discussed in Section 2, the second (b) is the right one.

If we look at w_4, the theorem of Section 2 tells us now that

$$\sum_{i>0} \operatorname{ch} V(w_4.\lambda)^i = \operatorname{ch} M(w_3.\lambda) + \operatorname{ch} M(w_3'.\lambda) + \operatorname{ch} M(w_3''.\lambda)$$
$$+ 2\operatorname{ch} M(w_2'.\lambda) + 2\operatorname{ch} M(w_2''.\lambda) + \operatorname{ch} M(w_1.\lambda).$$

Obviously, for $w \in \{w_3, w_3', w_3'', w_1\}$,

$$[V(w_4.\lambda)^i:M(w.\lambda)] = \begin{cases} 1 & \text{for } i = 0,1 \\ 0 & \text{for } i \geqslant 2. \end{cases}$$

Using part (b) of the theorem above in the same way as for w_3 we get, for $w \in \{w_2', w_2''\}$,

$$[V(w_4.\lambda)^i:M(w.\lambda)] = \begin{cases} 1 & \text{for } i = 0,1,2 \\ 0 & \text{for } i \geqslant 3. \end{cases}$$

In all cases considered in this example, all multiplicities not equal to 0 in the $V(w.\lambda)^i/V(w.\lambda)^{i+1}$ are 1. It follows easily from our construction of the $V(\mu)^i$ that there is always a nondegenerate contravariant form on $V(\mu)^i/V(\mu)^{i+1}$ for all μ and i. Now a G-module M admitting such a nondegenerate form is semisimple if all simple factors of M occur with multiplicity 1. In fact, take a simple submodule L of M and denote the orthogonal to L with respect to the form by $L^\perp$, which is a submodule because the form is contravariant. From the nondegeneracy of the form we get $\mathrm{ch}(L)+\mathrm{ch}(L^\perp) = \mathrm{ch}(M)$ and, from the assumption about the multiplicity, $[L^\perp : L] = 0$; hence $L \cap L^\perp = 0$ and $M = L \oplus L^\perp$. Induction over the length of M completes the proof.

In our example, therefore, the quotients $V(w.\lambda)^i/V(w.\lambda)^{i+1}$ which are neither 0 nor simple are

$$V(w_3.\lambda)^1/V(w_3.\lambda)^2 \cong M(\lambda) \oplus M(w_2'.\lambda) \oplus M(w_2''.\lambda),$$

$$V(w_4.\lambda)^1/V(w_4.\lambda)^2 \cong M(w_1.\lambda) \oplus M(w_3.\lambda) \oplus M(w_3'.\lambda) \oplus M(w_3''.\lambda),$$

and

$$V(w_4.\lambda)^2/V(w_4.\lambda)^3 \cong M(w_2'.\lambda) \oplus M(w_2''.\lambda).$$

The last formula shows that the socle of $V(w_4.\lambda)$ is not simple (cf. Chapter 12, 1.11).

In all the eight cases considered for type A_3 we could observe:

(1) All $V(w.\lambda)^i/V(w.\lambda)^{i+1}$ are semisimple.

(2) If $[V(w.\lambda)^i/V(w.\lambda)^{i+1} : M(w'.\lambda)] \neq 0$, then $l(w) - l(w') \equiv i \bmod 2$.

Here $l(w)$ is the length of w with respect to the system of generators S. It is an interesting question to ask how far these properties hold in general (for $\lambda \in C$). One will certainly have to make exceptions if $\langle w(\lambda+\delta), \alpha^\vee \rangle > p^2$ for some $\alpha \in R^+$.

We conclude with some remarks on T_μ^λ with $\lambda \in C$, $\mu \in \bar{C}$, and will restrict ourselves to the simplest case where $S_\mu = \{s\}$ consists of one reflection only. For each $w \in W_a$ with $w.\lambda \in X^+$ and $w.\lambda < ws.\lambda$ (i.e. $s \notin \tau(w)$) we get a (nonsplit) exact sequence

$$0 \to V(ws.\lambda) \to T_\mu^\lambda V(w.\mu) \to V(w.\lambda) \to 0.$$

This gives us the following information. Suppose we know (by induction, say) the multiplicities in the $V(w'.\lambda)$ with $w'.\lambda < ws.\lambda$. If we want to compute them for $V(ws.\lambda)$, we have to do so for $T_\mu^\lambda V(w.\mu)$. Now the theorem gives us the multiplicities in $V(w.\mu) = T_\lambda^\mu V(w.\lambda)$, so because of T_μ^λ being exact, we have to know the simple factors of the $T_\mu^\lambda M(w'.\mu)$ with $w'.\lambda \leqslant w.\lambda$. Again, by induction we conclude that we can compute the multiplicities in $V(ws.\lambda)$ if we can do so for $T_\mu^\lambda M(w.\mu)$. The following results about the structure of

$T_\mu^\lambda M(w\,.\,\mu)$ were proved mostly by Vogan in a different situation (for Harish–Chandra modules in [6]), but his arguments carry over.

Looking at $\mathrm{Hom}_G(M(w'\,.\,\lambda), T_\mu^\lambda M(w\,.\,\mu))$ and $\mathrm{Hom}_G(T_\mu^\lambda M(w\,.\,\mu), M(w'\,.\,\lambda))$ for all $w' \in W_a$ with $w'\,.\,\lambda \in X^+$, using the adjointness of T_μ^λ and T_λ^μ and the proposition above one proves that

$$\mathrm{soc}\; T_\mu^\lambda M(w\,.\,\mu) \cong M(w\,.\,\lambda)$$

and

$$T_\mu^\lambda M(w\,.\,\mu)/\mathrm{rad}\; T_\mu^\lambda M(w\,.\,\mu) \cong M(w\,.\,\lambda).$$

Set

$$s{*}M(w\,.\,\lambda) = \mathrm{rad}\; T_\mu^\lambda M(w\,.\,\mu)/\mathrm{soc}\; T_\mu^\lambda M(w\,.\,\mu).$$

Then (cf. [2], proof of Theorem 3 and [6], 3.9/10) the following hold.

PROPOSITION.

(a) $$[s{*}M(w\,.\,\lambda) : M(ws\,.\,\lambda)] = 1.$$

(b) *For all* $w' \in W_a$ *with* $w' \neq ws$ *and* $w'\,.\,\lambda \in X^+$,

$$[s{*}M(w\,.\,\lambda) : M(w'\,.\,\lambda)] \leqslant 2[V(w\,.\,\lambda) : M(w'\,.\,\lambda)].$$

(c) *For all* $w' \in W_a$ *with* $w'\,.\,\lambda \in X^+$,

$$[s{*}M(w\,.\,\lambda) : M(w'\,.\,\lambda)] \neq 0 \Rightarrow s \in \tau(w').$$

(d) *Let* $s' \in \tau(w)$ *and* $w' \in W_a$ *with* $w'\,.\,\lambda \in X^+$ *and* $[s{*}M(w\,.\,\lambda) : M(w'\,.\,\lambda)] \neq 0$.
If $(ss')^2 = 1$, *then* $s' \in \tau(w')$.
If $(ss')^3 = 1$, *then* $s' \in \tau(w')$ *except for* $w' \in \{ws, ws'\}$.

Let us be more explicit in the last case. There is exactly one $w' \in \{ws, ws'\}$ with $s' \notin \tau(w')$; for this w', we have

$$[s{*}M(w\,.\,\lambda) : M(w'\,.\,\lambda)] = 1$$

and this $M(w'\,.\,\lambda)$ is the only simple factor of $s{*}M(w\,.\,\lambda)$ having s' not in its τ-set. Vogan in his situation has also a result for $(ss')^4 = 1$, which should be true here too, but his proof cannot yet be carried over. Furthermore, Vogan can prove that

$$\mathrm{Hom}_G(M(w'\,.\,\lambda), s{*}M(w\,.\,\lambda)) \cong \mathrm{Ext}_G^1(M(w'\,.\,\lambda), M(w\,.\,\lambda))$$

for all $w' \in W_a$ with $w'\,.\,\lambda \in X^+$ and $s \in \tau(w')$.

One can show that there is a nondegenerate contravariant form on $s{*}M(w\,.\,\lambda)$. In all cases computed, the multiplicities in $s{*}M(w\,.\,\lambda)$ are at most 1; hence $s{*}M(w\,.\,\lambda)$ is semisimple by the arguments used earlier. Thus we get

$$[s{*}M(w\,.\,\lambda) : M(w'\,.\,\lambda)] = \dim_K \mathrm{Ext}_G^1(M(w'\,.\,\lambda), M(w\,.\,\lambda)) \qquad \text{if } s \in \tau(w'). \tag{3}$$

Vogan conjectures that the analogue of (3) in his situation is true in general; one may ask the same question in our situation. If there were a positive answer, the problem of computing the multiplicities in $V(ws.\lambda)$ would be reduced to determining $\mathrm{Ext}^1_G(M(w'.\lambda), M(w.\lambda))$ for $w'.\lambda < w.\lambda$. As we may suppose the multiplicities in $V(w.\lambda)$ to be known and as we have for $w'.\lambda < w.\lambda$ (cf. [8], 3.11b)

$$\mathrm{Ext}^1_G(M(w'.\lambda), M(w.\lambda)) \cong \mathrm{Hom}_G(\mathrm{rad}\, V(w.\lambda), M(w'.\lambda)),$$

we may hope that this problem is not unapproachable.

References

1. Jantzen, J. C., Darstellungen halbeinfacher algebraischer Gruppen und zugeordnete kontravariante Formen, *Bonner Math. Schr.*, **67**(1973).
2. Jantzen, J. C., Zur Charakterformel gewisser Darstellungen halbeinfacher Gruppen und Lie-Algebren, *Math. Z.*, **140**(1974) 127–149.
3. Jantzen, J. C., Kontravariante Formen auf induzierten Darstellungen halbeinfacher Lie-Algebren, *Math. Ann.*, **226**(1977) 53–65.
4. Jantzen, J. C., Darstellungen halbeinfacher Gruppen und kontravariante Formen, *J. reine angew. Math.*, **290**(1977) 117–141.
5. Jantzen, J. C., Über das Dekompositionsverhalten gewisser modularer Darstellungen halbeinfacher Gruppen und ihrer Lie-Algebren, *J. Algebra*, **49**(1977) 441–469.
6. Vogan, D., Irreducible characters of semisimple Lie groups I, *Duke Math. J.*, **46**(1979), 61–108.
7. Wong, W. J., Irreducible modular representations of finite Chevalley groups, *J. Algebra*, **20**(1972) 355–367.
8. Cline, E., Parshall, B., Scott, L. and van der Kallen, W., Rational and generic cohomology, *Inventiones math.*, **39**(1977), 143–163.

Part C

Geometry and the Sporadic Groups

14.

Four Presentations of Leech's Lattice

J. TITS

1. Introduction

Let $\widetilde{\mathfrak{A}}_r$ denote the "double covering" (nonsplit extension by $\boldsymbol{Z}/2\boldsymbol{Z}$) of the alternating group $\mathfrak{A}_r$. It has been observed by Thompson that Conway's group $\cdot 0$, the automorphism group of Leech's lattice Λ, has a subgroup G_9 isomorphic with $\widetilde{\mathfrak{A}}_9$, and that if $G_9 \supset G_8 \supset \ldots \supset G_2$ is the decreasing sequence of subgroups characterized up to conjugacy by the isomorphism $G_r \cong \widetilde{\mathfrak{A}}_r$, then the centralizers $Z_r = C_{\cdot 0}(G_r)$ form a remarkable sequence of subgroups: $Z_2 = \cdot 0 = \widetilde{\cdot 1}$ (double covering of Conway's simple group $\cdot 1$), $Z_3 = \widetilde{Suz}$ (sextuple covering of Suzuki's sporadic group Suz), $Z_4 = \widetilde{G_2(\boldsymbol{F}_4)}$ (double covering of $G_2(\boldsymbol{F}_4)$), $Z_5 = \widetilde{J}_2$ (double covering of the sporadic group J_2 of Hall–Janko),... (cf. [4], p. 242).

Let R_r be the ring of endomorphisms of the $\boldsymbol{Z}$-module Λ generated by G_r. It is readily guessed, and easily shown, that

$R_2 = \boldsymbol{Z}$;

$R_3 = \boldsymbol{Z}[\zeta]$, where ζ is a primitive cubic root of unity;

R_4 is isomorphic with a maximal order in $\boldsymbol{Q}(i,j,k)$ (they are all conjugate to $\boldsymbol{Z}[i,j,k,\frac{1}{2}(1+i+j+k)]$), and also with the subring of the ring $\boldsymbol{H} = \boldsymbol{R}(i,j,k)$ of real (hamiltonian) quaternions generated by the binary tetrahedral group (embedded in $\boldsymbol{H}$ in the obvious way);

R_5 is isomorphic with a maximal order in $\boldsymbol{Q}(\sqrt{5})(i,j,k)$ (they are all conjugate), and also with the subring of $\boldsymbol{H}$ generated by the binary icosahedral group (embedded in $\boldsymbol{H}$ in the obvious way);

the other rings R_r will not be used here.

We shall be concèrned with Λ viewed as an R_r-module, for $r = 2,3,4,5$. A sufficient motivation for investigating those modules is that, when endowed with a suitable form, their automorphism groups are the interesting groups Z_r. Since one disposes of Conway's handy description of Λ as a quadratic $\boldsymbol{Z}$-module (cf. [3]), it might seem most natural to try to get hold of its

R_r-module structure by explicitly giving the action of generators of R_r on the Z-module in question (cf. [4], p. 244, for the case $r = 3$). However, it turns out to be easier and more enlightening to give a direct construction of the R_r-module itself, a remarkable feature being that the congruences providing that module for the various values of r are almost identical. The main purpose of this expository note is to present a formulary emphasizing that fact.

Over Z, our presentation reduces to that of Conway [3], over $Z[\zeta]$, it is due to Lindsey II [6] (cf. also Tits [9]) and, to the best of my knowledge, the quaternionic approach was investigated first by Conway, Wales, Cohen [2], and the present author [9] [10]. Further details can be found in those references.

2. Formulae

For $r = 2, 3, 4$ or 5, we set:

$p = p_r = 6-r = 4, 3, 2$ or 1;
$d = d_r = 3 \cdot 2^{p-1} = 24, 12, 6$ or 3;
$R =$ the ring R_r of the introduction;
$\zeta = \zeta_r =$ an element of order 3 of the group $R^{\times}$ of units of R, if $r \neq 2$;
$\theta = \theta_r\ (\in R) = 2, 2\zeta+1\ (=\sqrt{-3})$, a square root of -2 in R_4 (e.g. $i+j$ if R_4 is identified with $Z[i,j,k,\frac{1}{2}(1+i+j+k)]$) or 2 according as $r = 2, 3, 4$ or 5;
$\mathfrak{m} = \mathfrak{m}_r =$ the ideal $(p\theta)$ of R (thus, $\mathfrak{m} = (\theta)$ or (θ^3) according as $r =$ or $\neq 5$);

$$\tau = \frac{1+\sqrt{5}}{2} \in R_5;$$

$e = e_r = 4, -3, 2\zeta$ or $\zeta+\tau$ according as $r = 2, 3, 4$ or 5 (essential is only that $e \notin \mathfrak{m}$, $\bar{e}e \in \mathfrak{m}$ and $\bar{e}+e \equiv d \pmod{\mathfrak{m}}$, where $\bar{e}$ denotes the conjugate of e, as usual).

We now consider the right R-module R^d whose elements are denoted by $(x_s)_{1 \leqslant s \leqslant d}$ (with $x_s \in R$)—or simply by (x_s)—and we represent by

$h = h_r$ the standard hermitian form $((x_s), (y_s)) \mapsto \sum_{s=1}^{d} \bar{x}_s y_s$,

$q = q_r$ the quadratic form $x \mapsto h(x,x)$ if $r \neq 5$, and the quadratic form $x \mapsto \lambda(h(x,x))$, where $\lambda : Q(\sqrt{5}) \to Q$ is defined by $\lambda(a+b\tau) = a$ for $a,b \in Q$, if $r = 5$,

and $M = M_r$ the submodule

$$\{(x_s) \in R^d | \textstyle\sum x_s \equiv ex_1 \equiv \ldots \equiv ex_d \pmod{\mathfrak{m}}\}$$

of R^d (note that if $r \neq 5$ one has $\mathfrak{m} = (e\theta)$ and the above congruences imply $x_1 \equiv \ldots \equiv x_d \pmod{(\theta)}$, whereas if $r = 5$ they imply $x_1 \equiv \ldots \equiv x_d \pmod{\bar{e}R}$).

For $r = 2, 3$ or 4, we have $R/(\theta) = \boldsymbol{F}_r$; we then denote by

$\rho = \rho_r : M \to \boldsymbol{F}_r^d$ the homomorphism ("twisted reduction") defined by

$$\rho((x_s)) = \left(\frac{1}{p\theta}(ex_s - \Sigma x_s) \bmod \theta\right)_{1 \leqslant s \leqslant d}$$

and $C = C_r$ a $\frac{1}{2}d$-dimensional subspace of $\boldsymbol{F}_r^d$ containing the vector $(1,1,\ldots,1)$ and intersecting only in 0 the product of any $2p-1$ factors of $\boldsymbol{F}_r^d$.

It is known (cf. e.g. [7], chapter 20 for $r = 2$ or 3) that those properties characterize C up to permutation of the factors of the product $\boldsymbol{F}_{\mathrm{r}}^{\mathrm{d}}$: C_4 is a (binary) Golay code, C_3 a ternary Golay code, and C_2 the subspace of $\boldsymbol{F}_4^6$ defined (up to permutation of the indices) by the equations $x_{2t-1} = \varepsilon^2(x_2 + x_4 + x_6) + \varepsilon x_{2t}$, where ε is a primitive cubic root of unity in $\boldsymbol{F}_4$; note that C_r is self-orthogonal for the canonical hermitian form in $\boldsymbol{F}_r^d$.
Finally, we define:

$$\Lambda = \Lambda_r = \begin{cases} \rho^{-1}(C) & \text{if } r \neq 5, \\ M & \text{if } r = 5. \end{cases}$$

THEOREM. *By restriction of scalars* $R_5 \supset R_4 \supset R_3 \supset R_2$, *the pair* (Λ_5, q_5) *becomes successively isomorphic with* (Λ_r, q_r), *for* $r = 4, 3, 2$. *The quadratic form* q_2 *is integral, even, unimodular (i.e.* Λ_2 *is self-dual with respect to the associated symmetric bilinear form) and does not represent* 2. *The group* $Aut(\Lambda_r, h_r)$ *of all* R_r*-module automorphisms of* Λ_r *preserving* h_r *is the group* Z_r *of the introduction.*

Remark. In R_5^3, the vectors $(2,0,0)$, $(\bar{e},\bar{e},0)$ and $(1,e,1)$ form an R-basis of Λ_5, which provides a simple alternative definition for that module. For $r \neq 5$, bases of Λ_r are also easily found but they are more complicated. In all cases, Λ_r contains the vectors $(p\theta,0,0,\ldots,0)$, $(\bar{e}, -\bar{e},0,\ldots,0)$, $(1-\bar{e},1,1,\ldots,1)$, and every vector deduced from those ones by an arbitrary permutation of the coordinates.

3. On Conway's Uniqueness Proof

In [5], Conway gave a beautiful proof of the uniqueness of Leech's lattice (as an even, unimodular lattice of rank 24 without vector of square norm 2). Here, we sketch the first main step of the proof, adapting it to an arbitrary value of $r = 2, 3$ or 4.

For $z \in N$, let n_z denote the number of elements $x \in \Lambda$ such that $q(x) = 2z$. We know that $n_0 = 1$ and $n_1 = 0$; then, a straightforward application of the theory of modular forms provides a simple explicit formula for n_z (cf. [8], chap. VII, 6.6, and [5]) and one finds, in particular,

$$n_2 = 196\,560, \qquad n_3 = 16\,773\,120, \qquad n_4 = 398\,034\,000, \ldots$$

We denote by

$S = S_r$ the set of all $x \in \Lambda$ with $0 < q(x) \leqslant 2p$ ("short vectors"),
$S' = S'_r$ the set of all $x \in \Lambda$ with $q(x) = 2p$,
$U = U_r$ the group of units $x \in R^\times$ of norm $\bar{x}x = 1$,
$u = u_r$ ($=2$, 6 or 24 according as $r = 2$, 3 or 4) the order of the group U,

and we call *cross* any set of d sub-R-modules of rank one of Λ generated by d elements of S' which are mutually orthogonal with respect to the form h and congruent mod $\Lambda\theta$.

LEMMA. *If two nonproportional elements* x,y *of* S *have the same reduction* mod θ *(i.e. the same image in* $\Lambda/\Lambda\theta = F_r^d$*), then they belong to* S' *and* $h(x,y) = 0$.

Indeed, let V be the group of all elements of U which are congruent to 1 mod θ. It is easily proved that for every $a \in R-\{0\}$, there exists $v \in V$ such that $av + \bar{v}\bar{a} < 0$. Therefore, upon multiplying y by a suitable element of V, we may assume that $h(x,y) = 0$ or $h(x,y) + h(y,x) < 0$. Our assertion readily follows, since

$$4 \geqslant q((x-y)\theta^{-1}) = p^{-1}q(x-y) = p^{-1}(q(x)+q(y)-(h(x,y)+h(y,x)).$$

Putting together the above lemma and the following remarkable identity (valid for $r = 2,3,4$):

$$u^{-1}\left(d^{-1}n_p + \sum_{0<z<p} n_z\right) = (r^d-1)/(r-1),$$

we get the

PROPOSITION. *The reduction homomorphism* $\Lambda \to \Lambda/\Lambda\theta$ *maps* S *onto* $(\Lambda/\Lambda\theta)-\{0\}$ *and the inverse image in* S *of a one-dimensional subspace of* $\Lambda/\Lambda\theta$ *has the form* xU *for some* $x \in S-S'$, *or is the intersection of* S' *with a cross. In particular, every element of* S' *belongs to exactly one cross.*

Note that, as r increases, S_r becomes smaller and the above proof gets "simpler". For $r = 4$, the argument provides a little more: one obtains the following assertion, this time without any use of the theory of modular forms.

Let Λ' *be a free* R_4*-module of rank* 6 *endowed with a positive definite hermitian form* h *such that the associated quadratic form is integral and does not represent* 1, 2 *and* 3. *Then, the number of times the form represents* 4 *does not exceed* 196 560, *and, if it is equal to that number, every* $x \in \Lambda'$ *such that* $h(x,x) = 4$ *belongs to exactly one cross (a cross being defined in the same way as above).*

Having chosen a cross, one is able, following [5] (which can be made slightly more conceptual and less "miraculous", using the ideas of [1]), to completely reconstruct the lattice, which turns out to be unique to within an isomorphism preserving the cross. It follows, in particular, that $\mathrm{Aut}(\Lambda_r,h_r)$ permutes the crosses transitively, which gives a good hold on the group in question. In particular, this provides a very efficient existence proof for Suzuki's sporadic group.

The case $r = 5$, leading to the group of Hall–Janko, can be handled by similar methods but the arguments are somewhat different (cf. [10]).

Acknowledgement

I wish to thank George Glauberman who pointed out several errors in an earlier version of this text.

References

1. Broué, M. and Enguehard, M., Une famille infinie de formes quadratiques entières; leurs groupes d'automorphismes, *Annales Sci. Ec. Norm, Sup.*, 4e *série*, **6** (1973) 17–52.
2. Cohen, A. M., Finite quaternionic reflection groups, Technische Hogeschool Twente, Memorandum no. 229, 1978; *J. Algebra* (to appear).
3. Conway, J. H., A perfect group of order 8,315,553,613,086,720,000, *Bull. Lond. Math. Soc.*, **1** (1969) 79–88.
4. Conway, J. H., Three lectures on exceptional groups. "Finite Simple Groups", Proc. Conf. London Math. Soc., eds. Powell, M. B. and Higman, G., Academic Press, London, 1971, chapt. VII, 215–247.
5. Conway, J. H., A characterisation of Leech's lattice, *Inventiones Math.*, **7**(1969) 137–142.
6. Lindsey, J. H., II, A correlation between $PSU_4(3)$, the Suzuki group and the Conway group, *Trans. Amer. Math. Soc.*, **157**(1971) 189–204.
7. Macwilliams, F. J. and Sloane, N. J. A., "The Theory of Error-Correcting Codes", North-Holland Math. Library, vol. 16, 1977.
8. Serre, J.-P., "Cours d'Arithmétique", Presses Univ. Fr., 2e éd., 1977.
9. Tits, J., Résumé de cours. Annuaire du Collège de France, 1976–1977, 57–67; 1977–1978, 80–81.
10. Tits, J., Quaternions over $Q[\sqrt{5}]$, Leech's lattice and the sporadic group of Hall–Janko, *J. Algebra*, **63**(1980) 56–75.

15.

Buildings and Buekenhout Geometries

J. TITS

The work of Buekenhout on the geometric interpretation of the sporadic groups ([8], [9]) has its origin in an early version of the theory of buildings. My purpose here is to give a brief historical survey of the beginnings of that theory, serving as an introduction to Buekenhout's geometries.

1. Complex Analytic Groups

Originally, the theory of buildings was aimed primarily at understanding the exceptional Lie groups from a geometrical point of view. The starting point was the observation that it is possible to associate with each complex analytic semisimple group a certain well-defined geometry, in such a way that the basic properties of the geometries thus obtained and their mutual relationships can be easily read from the Dynkin diagrams of the corresponding groups.

The definition of those geometries was suggested by the following reconstruction of the n-dimensional projective geometry from the projective group $\boldsymbol{PGL}_{n+1}(\boldsymbol{C})$:

the linear subspaces of the projective space $\boldsymbol{P}_n(\boldsymbol{C})$ can be represented by their stabilizers in $\boldsymbol{PGL}_{n+1}(\boldsymbol{C})$ which, by a well-known theorem of Lie, are nothing else but the maximal connected nonsemisimple subgroups of $\boldsymbol{PGL}_{n+1}(\boldsymbol{C})$;

the conjugacy classes of such subgroups represent the set of all points, the set of all lines, ..., and the set of all hyperplanes of $\boldsymbol{P}_n(\boldsymbol{C})$;

two linear subspaces are *incident*—i.e. one of them contains the other—if and only if the intersection of the corresponding subgroups contains a maximal connected solvable subgroup of $\boldsymbol{PGL}_{n+1}(\boldsymbol{C})$.

For arbitrary complex semisimple groups G, maximal connected nonsemisimple subgroups had been investigated by Morozov [16] and Karpelevič [15]. Their results showed that the conjugacy classes of such subgroups—now called *maximal parabolic subgroups*—are in one-to-one correspondence with the vertices of the Dynkin diagram M of the group G. Generalizing the above example, it was natural to associate with G a geometry consisting of a set (the set of maximal parabolic subgroups) partitioned into classes (the conjugacy classes) parametrized by the vertices of M and endowed with an incidence relation defined as follows: two maximal parabolic subgroups are incident if their intersection contains a maximal connected solvable subgroup‡ (i.e. a Borel subgroup in current terminology).

Two groups having the same Dynkin diagram are locally isomorphic, and it readily follows that the associated geometries are the same; therefore, the above process associates a geometry with every Dynkin diagram. For example, the geometry corresponding to the diagram

(A_n) o——o——o · · · o——o——o (n vertices)

is the n-dimensional (complex) projective geometry whose sets of points, lines, . . . , and hyperplanes correspond to the first, second, . . . , and last vertices of the diagram (from left to right, say). Notice that the duality of $\boldsymbol{P}_n(\boldsymbol{C})$ is visible from the symmetry of the diagram—as is the triality principle for the 6-dimensional hyperquadric from the symmetry of the diagram

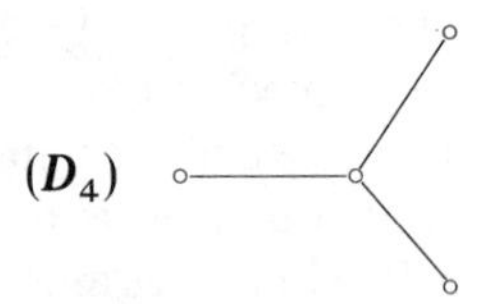

The above definition derives its interest from the following two observations, the first of which is an easy consequence of the explicit description of parabolic subgroups given by Karpelevič.

(i) Let M be a Dynkin diagram, Γ the associated geometry, x an object in Γ, and $v(x)$ the vertex of M corresponding to the class of x; then *the residual of* x *in* Γ—that is, the geometry consisting of all objects of Γ distinct from x but incident with x, with the partition and incidence relation induced by those of Γ—is the geometry associated with the diagram obtained from M by deleting $v(x)$ and all strokes containing it.

For example, in an n-dimensional projective space, the geometry consisting of all linear subspaces which contain or are contained in a given m-dimensional

‡ In this description, I have adopted the "abstract" viewpoint of [19]; in the earlier version [17] which was, in a sense, more geometric, one conjugacy class of maximal parabolic subgroups was distinguished and taken as the "set of points"—like the projective space—and the other parabolic subgroups were represented by subsets of that set—generalizing the linear subspaces of the projective space.

subspace is the "direct sum" of an m-dimensional and an $(n-m-1)$-dimensional projective geometry, corresponding to the diagram

o—o—o . . . o—o—o   o—o—o . . . o—o—o

m vertices $(n-m-1)$ vertices

The second observation is metamathematical.

(ii) When one knows the geometries associated with the Dynkin diagrams of rank 2 (i.e. having two vertices), assertion (i) gives a good hold on the geometry associated with any diagram; indeed, it "almost" characterizes that geometry uniquely (cf. Section 3).

If one excludes type G_2, which can occur only as a direct factor in other groups, the geometries corresponding to rank 2 diagrams are very simple and well-known: associated with the diagrams

$(A_1 \times A_1)$ o o

(A_2) o——o

(B_2) o⇒o

are, respectively, the "direct sum" of two projective lines (where each point of the first is incident with each point of the second), the two-dimensional projective geometry, and the geometry of points and lines of a 3-dimensional hyperquadric (with the obvious incidence relation).

In [18] and [19], examples were given showing how to deduce detailed information on higher rank geometries from a knowledge of the three geometries of rank 2 listed above, using only (i) and a certain "connectivity" property (cf. Section 3, (G3)). In particular, concrete information was obtained about the geometries of the exceptional Lie groups, which was the initial goal of the whole construction, as indicated earlier‡.

2. Chevalley Groups and Algebraic Groups

Since the three geometries of rank 2 described above have obvious analogues over an arbitrary field k, it was natural, taking those as building blocks, to try to associate with every Dynkin diagram a "geometry over k", using (i) as an axiom. This programme was started in [18] with the additional aim of defining, in that geometrical way, analogues over arbitrary fields of the

‡ The properties proved in [19] are part of what I called in [24] the "tableau caractéristique" of the geometries, further investigated by Freudenthal and de Vries in [14]. That "tableau" essentially describes in geometrical terms the double coset decomposition of the group G under consideration with respect to pairs of maximal parabolic subgroups. The method of investigation devised in [19] is rather *artisinale* and now somewhat obsolete: indeed, one can give simple algorithms for deriving the "tableau caractéristique" directly and mechanically from the Dynkin diagram; they are, however, unpublished. Note that, by 3.10 of [27], the problem reduces to a question on Weyl groups.

complex semisimple Lie groups. Shortly afterwards, Chevalley's fundamental paper [11] appeared. As a result, the geometric approach lost much of its value as a way of defining new groups, but a technical basis was provided for a more efficient construction of the geometries in question [20].

As the theory of algebraic semisimple groups over an arbitrary field was developed ([12], [23], [3]), it became apparent that the same geometrical theory would apply to the group $G(K)$ of rational points of an arbitrary isotropic algebraic semisimple group G defined over any field K (for example, cf. [24])‡: the group $G(K)$ has "parabolic subgroups", the conjugacy classes of maximal parabolic subgroups are parameterized by the vertices of the so-called "relative Dynkin diagram" of G over K, and for the geometries (corresponding to various groups G) consisting of the set of all maximal parabolic subgroups partitioned into conjugacy classes and endowed with a suitable incidence relation, the relation between residuals and subdiagrams described by (i) holds. Now, of course, a given diagram—in particular each of the diagrams ($\boldsymbol{A}_1 \times \boldsymbol{A}_1$), ($\boldsymbol{A}_2$) and ($\boldsymbol{B}_2$) considered above—belongs to many different groups and hence to many different geometries. But I soon noticed that all the geometries associated with a diagram consisting of a single stroke of multiplicity $(m-2)$ (here, ($\boldsymbol{G}_2$) must be assigned a quadruple stroke) had the following property:

(GP_m) For any two elements a,b of the geometry, there is a sequence $a = a_0, a_1, \ldots, a_s = b$ of elements such that, for all i,

$$a_i \neq a_{i+2},$$
$$a_i \text{ and } a_{i+1} \text{ are incident,}$$
$$s \leqslant m,$$

and if $s < m$, then the sequence is unique.

(This fact was later explained as a consequence of the BN-pair property: cf. [27], 5.2.) I defined a *generalized m-gon* (because the points and edges of an m-gon are the simplest instance of it) as a geometry consisting of two sets whose disjoint union is endowed with a symmetric binary relation—the incidence—having the following properties: (GP_m) holds, two elements of the same set are incident if and only if they coincide, and every element of the union is incident with at least two other elements (cf. [22], appendix).

3. Geometries of Type *M* and Buildings

At this point it was natural to introduce the following concept of *geometries of type M* (cf. [20], [24], whose terminology we do not quite follow).

We first set a general lexical framework. Let I be a set. By a *geometry over*

‡ Actually, for this author, extending the geometric theory to such a group $G(K)$ was an essential part of the initial motivation behind [23] and [3].

I, we mean a set Γ endowed with a partition $(\Gamma_i)_{i\in I}$ indexed by I and a binary symmetric relation—the *incidence. Par abus de language*, we shall often talk about "the geometry Γ" or "the geometry $(\Gamma,(\Gamma_i)_{i\in I})$". The *rank* of Γ is the cardinality of I. A *flag* is a set of pairwise incident elements of Γ, and two flags are called *incident* if their union is a flag. The *type of a flag F* is the set $\{j\in I|\Gamma_j\cap F\neq\emptyset\}$. If F is a flag of type J, we denote by $\Gamma_{i,F}$ the set of all elements of Γ_i which are incident with F but do not belong to F. The set

$$\Gamma_F=\bigcup_{j\in I}\Gamma_{j,F},$$

partitioned as written and endowed with the restriction of the incidence relation of Γ, is a geometry over

$$I'=\{j\in I|\Gamma_{j,F}\neq\emptyset\},$$

called the *residual* of F in Γ.

Here, we shall only consider geometries which satisfy the following axioms.

(G1) For $j\in I$, two elements of Γ_j are incident if and only if they coincide.
(G2) Every flag which is not of type I is contained in at least two flags of type I. (In particular, the flags of type I are the maximal flags.)
(G3) (Strong connectedness) If i,j are distinct elements of I, the graph of the incidence relation restricted to $\Gamma_i\cup\Gamma_j$ is connected, and the same property holds for the residual of any flag in Γ.

We define a *Coxeter diagram* as a diagram M consisting of a set of vertices I—also denoted by $V(M)$—pairwise joined by "edges with multiplicities", where the multiplicities have values in $N\cup\{\infty\}$‡. (The Dynkin diagrams considered so far were the Coxeter diagrams occurring in the theory of semi-simple Lie groups in which edges of multiplicity 2 or 4 are oriented.) With every Coxeter diagram M, we associate a family $\mathscr{G}(M)$ of geometries over $I=V(M)$, the *geometries of type M*, characterized by the properties (G1), (G2), (G3) and

($G4_M$) If i,j are distinct elements of I joined by an edge of multiplicity m, then the residual of a flag of type $I-\{i,j\}$ is a generalized $(m+2)$-*gon.*

Note the following immediate consequence, generalizing ($G4_M$):

($G4'_M$) If $J\subset I$, the residual of a flag of type J belongs to $\mathscr{G}(M')$ where M' is the subdiagram of M whose set of vertices is $I-J$.

An alternative, "inductive" definition of $\mathscr{G}(M)$ consists of replacing ($G4_M$) by ($G4'_M$) and adding the axiom

(GP) If M consists of two vertices joined by an edge of multiplicity m, then $\mathscr{G}(M)$ is the family of all generalized $(m+2)$-*gons.*

‡ In [24], the possibility of infinite multiplicities was not considered.

If M is the diagram A_n (see Section 1), $\mathscr{G}(M)$ is precisely the class of all "general" projective geometries of dimension n (where lines are only assumed to have at least two points). But for other diagrams M, the axioms (G1)–(G4) fall somewhat short of affording all the properties one would like to have, such as the "linearity" (stated below) which always holds in the geometries associated with algebraic groups (cf. [27], 12.10). If F is a flag and if j is a vertex of M, then by the *shadow* of F in Γ_j we mean the set $\mathscr{S}_j(F)$ of all elements of Γ_j incident with F. For example, the shadows of the elements of a projective geometry in the set of all points are the linear subspaces.

(GL) (Linearity) For $j \in V(M)$, the intersection of two shadows of flags in Γ_j is empty or is the shadow of a flag.

If I is finite, which we assume from now on, a slightly stronger version of that condition, stated here for later purposes, is

(GL′) If $j \in V(M)$, $x \in \Gamma$, and F is a flag, then either $\mathscr{S}_j(x) \cap \mathscr{S}_j(F)$ is empty, or there is a flag F' incident with F and x such that

$$\mathscr{S}_j(F) \cap \mathscr{S}_j(x) = \mathscr{S}_j(F').$$

Examples showing that (GL) is not a consequence of the axioms (G1)–(G4) can be obtained as follows. One considers the geometry Γ associated with an algebraic semisimple group of rank $\geqslant 3$ and a nontrivial group A of automorphisms of Γ such that, for $x \in \Gamma$ and $a \in A - \{1\}$, the distance of x and ax in the graph of Γ is $\geqslant 4$; if Γ/A is the geometry consisting of the set Γ/A partitioned into the subsets Γ_j/A and endowed with incidence relation the image of that of Γ by the canonical projection, then Γ/A will satisfy axioms (G1)–(G4) (for the same diagram as Γ) but, in general, not (GL). A "concrete" example is given by the following: consider a definite quadratic form q on a real vector space of odd dimension $\geqslant 7$, and take for Γ the geometry associated with the complex Lie group $\boldsymbol{SO}(\boldsymbol{C},q)$ (a geometry whose objects are the complex subspaces totally isotropic for q) and for A the group of order 2 generated by the complex conjugation.

In the above example, the "good" geometry Γ may be viewed as a "covering" of the "bad" one Γ/A. This suggests a possible way of defining "good" geometries of type M by adding to the axioms (G1) to (G4) an extra condition expressing some kind of simple connectedness. A remark to that effect was made in [21] (cf. also [24], p. 187, footnote (††)), but the idea was not pushed further because of the discovery of another approach to those geometries based on a completely different set of axioms—the axioms of buildings—which was simpler and more efficient, and which gave a better insight into the structure of the geometries in question ([26], [27]). However, the introduction of Buekenhout's geometries and, more specifically, the work of Buekenhout and Kantor on Lyons' sporadic group has revived interest in an axiomatic approach based on "local properties" together with a condition of simple connectedness. In [30], it is shown that for diagrams M having no subdiagram of type

(C_3) o———o═══o   or (H_3) o———o≡≡≡o

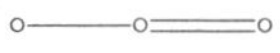

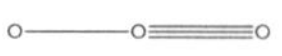

buildings of type M (or, rather, "weak buildings" in the terminology of [27]) are indeed characterized by the axioms (G1), (G2), (G3), ($G4'_M$) and (GP) together with a suitable property of simple connectedness closely related to the corresponding topological notion‡. If one wants to extend this result to arbitrary diagrams, then it is necessary to introduce other—and, in a sense, less natural—axioms of a global nature (for instance, conditions inspired by (GL)), or to include in axiom (GP) a proviso for the geometries of type C_3 and H_3. (If one deals with actual buildings instead of weak ones, the case of H_3 may be handled simply by forbidding diagrams M with a subdiagram of that type: cf. [28].)

4. Buekenhout Geometries

In our previous considerations, the basic axiom (G4)—or (G4′)—appears as a device to define "complicated" geometries by means of simpler ones, like building blocks. It is, of course, not necessary that those building blocks be generalized m-gons or even geometries of rank two. Indeed, the generalities of [20] were excused such assumptions. However, no effective use was made of that freedom until Buekenhout reinvented it as a consequence of his extensive investigations of various geometrical structures ([6], [7], [10]), amongst which geometries associated with sporadic groups played an increasing role. A basic observation of Buekenhout was that interesting geometries could be obtained by allowing families of rank two geometries other than the generalized m-gons, in particular (and quite surprisingly) the following rather trivial family of "circle" geometries, represented diagrammatically by

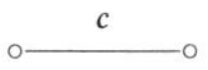

(∗)   o———c———o :

a geometry $(\Gamma,(\Gamma_1,\Gamma_2))$ is said to be of that type if Γ_2 can be identified with the set of pairs of elements of Γ_1 in such a way that the incidence relation becomes the symmetrized inclusion. Here, Γ_1 is understood to correspond to the left vertex in (∗); if the opposite holds, the diagram is

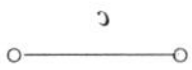

o———ɔ———o .

‡ Those results were obtained during and after the Durham conference. I thank W. Kantor and M. Ronan for stimulating discussions on the subject during the conference.

Buekenhout discovered the remarkable fact that geometrical interpretations of most of the sporadic groups can be obtained (always using $(G4_M)$) by combining those objects with generalized polygons (see Section 5).

Other families used by Buekenhout as building blocks are the following:

the geometries of type $\circ\overset{L}{\text{———}}\circ$ consisting of two sets Γ_1 and Γ_2 such that any two distinct elements of Γ_1 are incident with exactly one element of Γ_2;

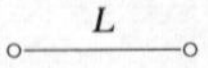

the geometries of type $\circ\overset{Af}{\text{———}}\circ$ consisting of the points and lines of an affine plane.

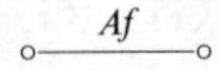

Given any diagram M consisting of vertices which are joined either by a multiple edge or by one of the links described above (or by any other link representing a specified family of rank two geometries), the axioms (G1), (G2), (G3) and $(G4'_M)$ associate a class of geometries which, for the purpose of this exposition, we shall call *geometries of type M*, though it is often desirable to add further restrictions such as the axiom (GL′) of the previous section, usually required by Buekenhout.

5. Sporadic Groups

Let us say that a group G *admits* a diagram M if it acts flag transitively (i.e. transitively on maximal flags) on a geometry of type M. In the following tables, Table 2 summarizes the results of Buekenhout (together with Kantor

TABLE 1. *Residual geometries of rank* 2

Diagram	Geometry
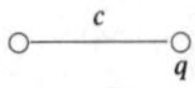 $\circ\overset{c}{\text{———}}\underset{q}{\circ}$	points and pairs of points of a set of cardinality $q+2$
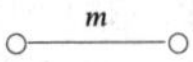 $\circ\overset{m}{\text{———}}\circ$	vertices and edges of an ordinary m-gon
 $\underset{q}{\circ}\text{———}\underset{q}{\circ}$	points and lines of a projective plane over F_q
 $\underset{5}{\circ}=\!=\!=\circ$	points and lines of a quadric of maximal index in $P_3(F_5)$
(resp. 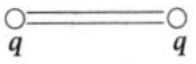 $\underset{q}{\circ}=\!=\!=\underset{q}{\circ}$ with $q = 2$ or 4)	(resp. $P_4(F_q)$)
 $\underset{q}{\circ}=\!=\!=\underset{q^2}{\circ}$	points and lines of a quadric of nonmaximal index—i.e. of Witt index 2—in $P_5(F_q)$ (or, equivalently, lines and points of a hermitian quadric in $P_3(F_{q^2})$)
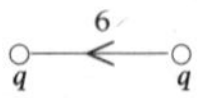 $\underset{q}{\circ}\overset{6}{\Lleftarrow}\underset{q}{\circ}$	points and lines of the generalized hexagon associated with $G_2(q)$ in its usual representation on a 5-dimensional quadric (cf. [22])
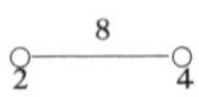 $\underset{2}{\circ}\overset{8}{\text{———}}\underset{4}{\circ}$	points and lines of the generalized octagon associated with the Ree group ${}^2F_4(2)$ (cf. [25])

TABLE 2. *Diagrams admitted by some sporadic groups*

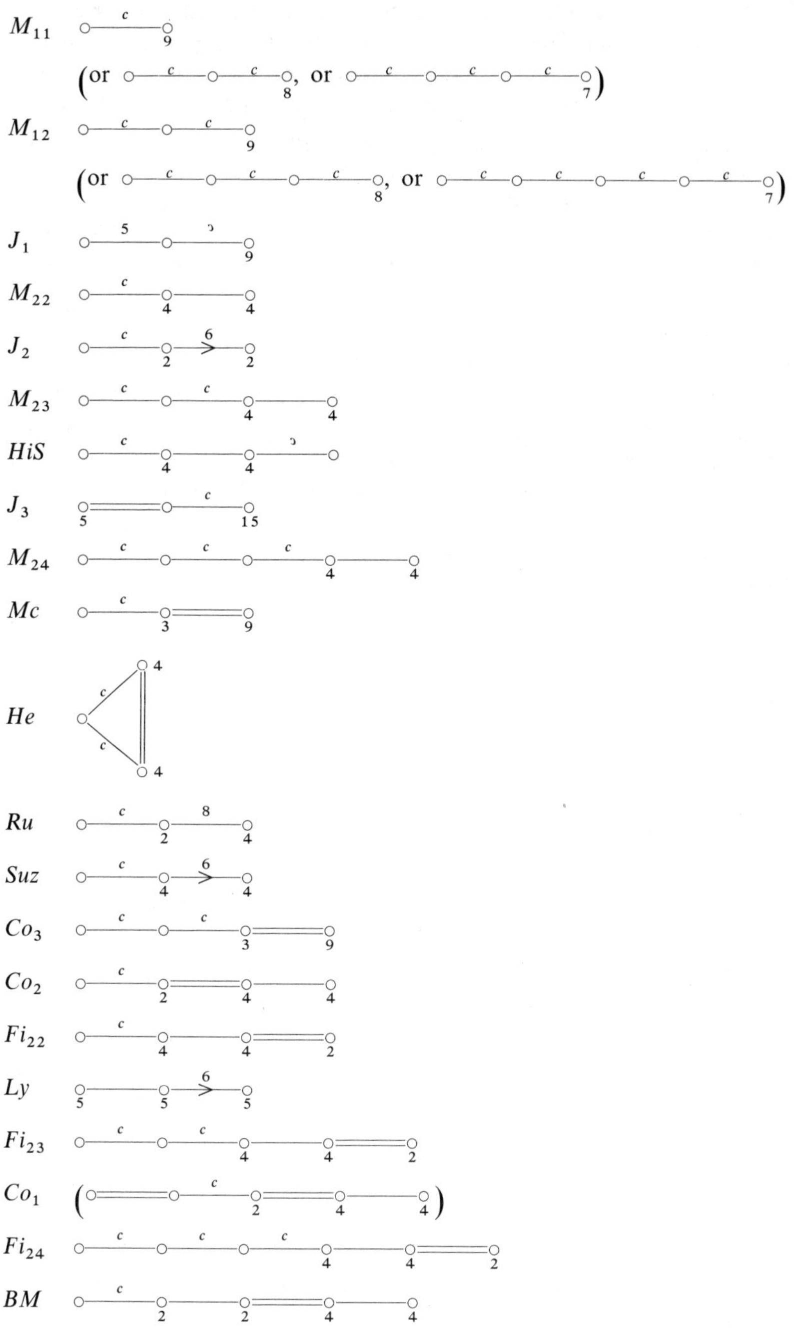

M_{11}

M_{12}

J_1

M_{22}

J_2

M_{23}

HiS

J_3

M_{24}

Mc

He

Ru

Suz

Co_3

Co_2

Fi_{22}

Ly

Fi_{23}

Co_1

Fi_{24}

BM

Notation for groups as in Table 5, pages 22–3, of this book.

in the case of Lyons' group); it lists diagrams which are admitted by 21 of the 26 so far detected sporadic groups. The diagram given for the first Conway group Co_1 is in brackets as it corresponds to a geometry which does not fulfill (GL′); in all other cases in the list, that axiom is satisfied. Edges marked 5, 6 or 8 represent, respectively, triple, quadruple or sextuple edges. The integers attached to the vertices of the diagrams are each one less than the cardinalities of the corresponding residual geometries of rank one, whenever that cardinality is not two; in other words, if M is a diagram in Table 2, I its set of vertices, v an element of I, Γ a geometry of type M on which the group under consideration acts flag transitively, and (q_v+1) the number of flags of type I of Γ containing a flag of type $I-\{v\}$, then the number q_v is written next to the vertex v whenever $q_v \neq 1$. In most cases (and probably all), the data consisting of the diagrams and the numbers q_v characterize the residuals of rank 2 up to isomorphism. However, for the convenience of the reader, we have described these residuals explicitly in Table 1, and used that occasion to explain the meaning of an arrow orienting some of the multiple edges. There we refer to points and lines for the objects corresponding respectively to the left and right vertices of the diagram, and $\boldsymbol{P}_n(q)$ denotes an n-dimensional projective space over $\boldsymbol{F}_q$.

6. Miscellaneous Remarks

6.1. The geometries of type $\circ\overset{c}{—}\circ\overset{Af}{—}\circ$ are those whose objects are the points, pairs of points, and circles of so-called Möbius planes, hence the notation "c".

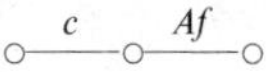

6.2. A finite geometry of rank n and type $\circ\overset{c}{—}\circ\overset{c}{—}\circ\cdots\circ\overset{c}{—}\circ—\circ$ consists of the points, pairs of points, ..., $(n-1)$-tuples of points, and blocks of a Steiner system $S(q^2+q+n-1, q+n-1, n)$ for some integer $q \geqslant 1$. It is known ([31], [13]) that if $n \geqslant 3$, the pair (n,q) must be one of $(n,1)$, $(3,2)$, $(3,4)$, $(4,4)$, $(5,4)$ or $(3,10)$ and that, except for the last one, each of those pairs corresponds to one and, up to isomorphism, only one Steiner system. The systems with $q=4$ are those giving rise to the Mathieu groups M_{22}, M_{23} and M_{24}.

6.3. The geometries of type $\circ\overset{c}{—}\circ—\circ—\circ\cdots\circ—\circ=\circ$, called *locally polar spaces*, have been classified by Buekenhout and Hubaut, assuming the existence of a "sufficiently large" automorphism group [10].

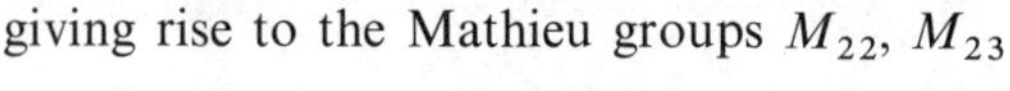

6.4. If M is the Coxeter diagram of an affine Weyl group W of rank $\geqslant 3$, one can construct finite geometries of type M as follows. Let K be a locally compact field and let G be a simple algebraic group over K whose affine Weyl group ([4], [29]) is W. Then a building Δ of type M is associated with G ([5], [29]). Let A be a discrete subgroup of G which is cocompact (i.e. G/A is compact). (N.B. If K has characteristic zero, it follows from [1] that G has "many" cocompact, discrete, "S-arithmetic" subgroups. If char $K \neq 0$, a suitable A might not exist.) Let A' be a subgroup of finite index in A such that, for every vertex v of Δ and every $g \in A' - \{1\}$, the distance between v and vg in the graph of vertices of Δ is at least four: such a subgroup is easily seen to exist. Then it is immediate that the quotient Δ/A'' of Δ by any subgroup A'' of finite index in A' is a finite geometry of type M. Furthermore, if A'' is chosen "sufficiently small", then the geometry Δ/A'' also satisfies the linearity axiom (GL). Note that the automorphism group of a geometry Δ/A'' obtained in this way is usually rather small and has little chance of being flag transitive.

6.5. Consider the geometry Γ of type

$$(\tilde{G}_2) \quad \circ\!\!-\!\!\!-\!\!\!-\!\!\circ\!\!\overset{6}{-\!\!\!-\!\!\!-}\!\!\circ$$

which Buekenhout and Kantor associated with Lyons' group. Let Δ be the two-dimensional simplicial complex whose vertices are the elements of Γ and whose simplices have the flags of Γ as sets of vertices. Let $\tilde{\Delta}$ be the universal covering of Δ. By the results of [30] alluded to in Section 4, $\tilde{\Delta}$ is a building of type $\tilde{G}_2$, but it can be shown that it is not the affine building of an algebraic group over a locally compact local field. Thus it is a "truly exotic" building of type $\tilde{G}_2$. The "points at infinity" of that building (cf. [5], 5.1.13; [2]) provide an interesting "exotic" compact generalized hexagon.

6.6. An essential feature of the theory of buildings of spherical type [27] is that the "local information" provided by the axiom (G4) implies very strong "global" consequences. For instance, the finiteness of all residual geometries of rank one implies the finiteness of the whole geometry. That is due to the finiteness of the Weyl group. As 6.4 and [30] show, the case of arbitrary Coxeter groups is quite different.

6.7. A basic question concerning Buekenhout's geometries is to know precisely how much of the global structure is implied by the local information contained in the diagram. Example 6.2 above and, to a certain extent, also 6.3 (not to mention the case of usual Coxeter diagrams such as $\boldsymbol{A}_n$ or $\boldsymbol{E}_6$; cf. [18], [19], [30]) show that for some diagrams the implication is quite strong. For others the situation may be very different, as is already apparent in 6.4 (cf. also [30]). Buekenhout is currently working on the problem of finding additional "natural" axioms which will "squeeze down" the number of geometries with a given type; the final aim is, if course, a complete classification for certain families of diagrams, leading in particular to geometric characterizations of the sporadic groups.

References

1. Borel, A., Some finiteness properties of adele groups over number fields, *Publ. Math. I.H.E.S.*, **16**(1963) 5–30.
2. Borel, A. and Serre, J.-P., Cohomologie d'immeubles et de groupes S-arithmétiques, *Topology*, **15**(1976) 211–232.
3. Borel, A. and Tits, J., Groupes réductifs, *Publ. Math. I.H.E.S.*, **27**(1955) 55–151.
4. Bruhat, F. and Tits, J., Groupes algébriques simples sur un corps local, *C.R. Acad. Sci. Paris*, **263**(1966) 822–825.

5. Bruhat, F. and Tits, J., Groupes réductifs sur un corps local. I: Données radicielles valuées, *Publ. Math. I.H.E.S.*, **41**(1972) 1–251.
6. Buekenhout, F., Inversions in locally affine circular spaces, I, II, *Math. Z.*, **119**(1971) 189–202; **120**(1971) 165–177.
7. Buekenhout, F., Extensions of polar spaces and the doubly transitive symplectic group, *Geom. Dedicata*, **6**(1977) 13–21.
8. Buekenhout, F., Diagrams for geometries and groups, *J. Comb. Theor. (A)*, **27**(1979) 121–151.
9. Buekenhout, F., The geometry of diagrams, *Geom. Dedicata*, **8**(1979) 253–257.
10. Buekenhout, F. and Hubaut, X., Locally polar spaces and related rank 3 groups, *J. Algebra*, **45**(1977) 391–434.
11. Chevalley, C., Sur certains groupes simples, *Tôhoku Math. J.* (2), **7**(1955) 14–66.
12. Chevalley, C., "Classification des groupes de Lie algébriques, I, II", Séminaire E.N.S., 1956–1958, mimeographed.
13. Dembowski, P., "Finite Geometries", Ergebnisse der Math. 44, Springer-Verlag, Berlin, 1968.
14. Freudenthal, H. and de Vries, H., "Linear Lie Groups", Academic Press, London, 1969.
15. Karpelevič, F. I., O nepoluprostyh maksimalnyh podalgebrah poluprostyh algebr Li, *Dokl. Akad. Nauk S.S.S.R.*, **76**(1951) 775–778.
16. Morozov, V. V., "O nepoluprostyh podgruppah prostyh grupp". Thesis, Kazan, 1943.
17. Tits, J., Sur certaines classes d'espaces homogènes de groupes de Lie, *Mém. Acad. roy. Belg.*, **29**(3) (1955).
18. Tits, J., Sur la géométrie des *R*-espaces, *J. Math. P. et Appl.*, **36**(1957) 17–38.
19. Tits, J., Les groupes de Lie exceptionnels et leur interprétation géométrique, *Bull. Soc. Math. Belg.*, **8**(1956) 48–81.
20. Tits, J., "Sur les analogues algébriques des groupes semi-simples complexes", Colloque d'Algèbre Supérieure du C.B.R.M., Bruxelles, 1956 (1957) 261–289.
21. Tits, J., Short communication, International Congress of Mathematicians, Edinburgh, 1958.
22. Tits, J., Sur la trialité et certains groupes qui s'en déduisent, *Publ. Math. I.H.E.S.*, **2**(1959) 14–60.
23. Tits, J., Sur la classification des groupes algébriques semi-simples, *C.R. Acad. Sci. Paris*, **249**(1959) 1438–1440.
24. Tits, J., "Groupes algébriques semi-simples et géométries associées", Proc. Coll. Algebraical and Topological Foundations of Geometry, Utrecht, 1959 (Pergamon Press, London, 1962), 175–192.
25. Tits, J., "Les groupes simples de Suzuki et de Ree", Séminaire Bourbaki, déc. 1960, exp. 210.
26. Tits, J., "Géométries polyédriques et groupes simples", Deuxième Réunion du Groupement de Mathématiciens d'Expression Latine, Florence, 1961, 66–88.
27. Tits, J., "Buildings of spherical type and finite *BN*-pairs", Lecture Notes in Mathematics, 386, Springer-Verlag, Berlin, 1974.
28. Tits, J., Endliche Spiegelungsgruppen, die als Weylgruppen auftreten, *Inventiones Math.*, **43**(1977) 283–295.
29. Tits, J., "Reductive groups over local fields", Proc. Symp. Pure Math. **33**(1979) (Summer Inst. on Group Representations and Automorphic Forms, Corvallis, 1977), vol. 1, 29–69.
30. Tits, J., "A local characterization of buildings", Collection of papers dedicated to H.S.M. Coxeter (to appear).
31. Witt, E., Über Steinersche Systeme, *Abh. Hamburg*, **12**(1938) 265–275.

16.

Fixed Point Free Involutions and Finite Projective Planes

JOHN G. THOMPSON

This paper is an attempt to use the representation theory of finite groups in the study of finite projective planes.

1. Ovals

LEMMA 1.1. *Suppose P is a projective plane of order n, and $O \subseteq P$ has cardinal $n+2$. The following statements are equivalent:*

(a) $|\ell \cap O|$ *is even for every line* ℓ *of* P.

(b) $|\ell \cap O| \leqslant 2$ *for every line* ℓ *of* P.

Proof. (a) $\Rightarrow$ (b). Let $p \in O$. Each of the $n+1$ lines through p contains at least one point of O distinct from p. Since $n+2 = |O|$, each such line contains precisely two points of O.

(b) $\Rightarrow$ (a). Let $p \in O$, and let $\ell_1, \ldots, \ell_{n+1}$ be all the lines through p. Then the sets $\ell_i \cap O - \{p\}$ are all of cardinal $\leqslant 1$, and they are pairwise disjoint. As $O - \{p\}$ itself has cardinal $n+1$, we have $|\ell_i \cap O - \{p\}| = 1$ for each i.

The subsets of P of cardinal $n+2$ which satisfy (a) are called ovals, and we assume henceforth that O is an oval. Let $G = S(O)$ be the group of all permutations of O.

For each $p \in P - O$, define $t(p) \in G$ as follows:

for each $q \in O$, $q \, . \, t(p)$ is the unique point $r \in O - \{q\}$ such that pqr are collinear.

The existence of the point r is guaranteed by Lemma 1.1(a), and its uniqueness is guaranteed by Lemma 1.1(b). Since pqr are collinear, it follows that

$$q = r \, . \, t(p),$$

and so

$t(p)$ is an involution without fixed points.

In particular,

$$n = 2m \text{ is even.}$$

For each integer k, let C_k be the set of involutions of G which move precisely $2k$ letters. Thus

$$t(p) \in C_{m+1}.$$

Suppose a,b,c,d are distinct points of O. Let $\ell(a,b) \cap \ell(c,d) = p$, where $\ell(x,y)$ denotes the line through x and y. By definition of t, we see that ab and cd are cycles of $t(p)$, so the involution $ab\,.\,cd \in C_2$ and $ab\,.\,cd\,.\,t(p) \in C_{m-1}$. This gives us a map

$$\mu: C_2 \to C_{m+1}$$
$$\mu(ab\,.\,cd) = t(p) \qquad \text{if} \qquad p = \ell(a,b) \cap \ell(c,d).$$

LEMMA 1.2. *If $p,q \in P - O$ and $t(p)\,.\,t(q)$ fixes at least 4 points, then $p = q$. In particular, t is an injection.*

Proof. Let F be the set of fixed points of $t(p)t(q)$. If $f \in F$, and $e = ft(p)$, then ef is a cycle of both $t(p)$ and $t(q)$. If $f' \in F - \{e,f\}$ and $e' = f't(p)$, then

$$p = \ell(e,f) \cap \ell(e',f') = q.$$

There is a module-theoretic interpretation of these combinatorial considerations. Let M_k be the permutation module of G on C_k, so that C_k is a basis for the free abelian group M_k, and G acts by conjugation on C_k. Let

$$v(M_k) = \sum_{x \in C_k} x.$$

If $k \geqslant \ell$, there is a G-homomorphism

$$\lambda_{k,l}: M_k \to M_l$$

given by

$$\lambda_{k,l}(x) = \sum_{\substack{y \in M_l \\ xy \in M_{k-l}}} y \qquad (x \in C_k).$$

And the preceding discussion of O just asserts that

$$(*) \qquad \lambda_{m+1,2}\Big(\sum_{p \in P-O} t(p)\Big) = v(M_2).$$

It is this map $\lambda_{m+1,2}$ which will be studied.

2. Permutation Modules and Cohomology

If M is a ZG-module and G is a finite group, we shall say that M is a transitive permutation module if, and only if, M is a free abelian group which has a basis B which is permuted transitively by G. If M is such a module, we have the augmentation homomorphism

$$\iota : M \to Z$$

which maps each element of B to 1. We set

$$\nu(M) = \sum_{b \in B} b.$$

Thus, $\nu(M)$ is a generator for M^G, the fixed point submodule.

Suppose M,N are both transitive permutation modules, and

$$\lambda : M \to N$$

is a ZG-homomorphism. Then

$$\lambda(\nu(M)) = c \, . \, \nu(N)$$

for some integer c, since $\lambda(\nu(M)) \in N^G = Z \, . \, \nu(N)$. As well as c, there is a second integer associated to λ, namely, $d = \iota(\lambda(b))$ $(b \in B)$. It is obvious, then that

$$(*) \qquad \begin{array}{ccc} M & \xrightarrow{\lambda} & N \\ \downarrow{\scriptstyle \iota} & & \downarrow{\scriptstyle \iota} \\ Z & \xrightarrow{d} & Z \end{array}$$

is commutative, where the lower horizontal map is multiplication by d.

Let $b \in B$, $c \in C$, where C is a basis of N permuted transitively by G, and let H,H' be the stabilisers of b,c respectively. If we use the commutativity of $(*)$ applied to the element $\nu(M)$, we get

$$d|G:H| = c|G:H'|,$$

so

$$c = \frac{d|G:H|}{|G:H'|}$$

is recoverable from d, H and H'. We set

$$c = c(\lambda).$$

LEMMA 2.1. *$\nu(N) \in \mathrm{Im}(\lambda)$ if and only if $c(\lambda) \neq 0$, and $H^1(K)$ is cyclic of order $|c(\lambda)|$, where*

$$K = \ker(\lambda),$$

$c(\lambda)$ is defined as above, and the cohomology group is with respect to ZG.

Proof. For each group G, let $\nu(G)$ be the norm element of G in ZG.

If $c(\lambda) = 0$, then $d = 0$, and so $\lambda(m) \, . \, \nu(G) = 0$ for all $m \in M$, whence $\nu(N) \notin \mathrm{Im}(\lambda)$. From now on, we assume that $c(\lambda) \neq 0$.

Set $L = M/K$, so that

$$0 \to K \to M \to L \to 0 \tag{2.1}$$

is exact. We pick $b \in B$ as a base point, to embed M in ZG; send b to $v(H)$, H being the stabilizer of b in G, and send bg to $v(H)g$ $(g \in G)$. This gives us a second exact sequence

$$0 \to M \xrightarrow{i} ZG \to I \to 0 \tag{2.2}$$

where $I = ZG/i(M)$. Note that as abelian group, I is torsion free, and also $v(G)$ is contained in $i(M)$. This implies that

$$I^G = 0, \tag{2.3}$$

that is, G has no fixed points on I. All the more so,

$$\hat{H}^0(I) = 0,$$

and so the exact sequence for cohomology applied to (2.2) gives

$$H^1(M) = 0. \tag{2.4}$$

Using (2.4) and the exact sequence for cohomology applied to (2.1) gives the surjectivity of

$$\hat{H}^0(L) \xrightarrow{\delta} H^1(K) \tag{2.5}$$

where δ is the connecting homomorphism. This already shows that $H^1(K)$ is cyclic, since $L^G \cong Z$.

Suppose $H^1(K)$ is of order $|c(\lambda)|$. Pick $u \in \hat{H}^0(L)$ such that $\delta(u)$ is a generator for $H^1(K)$. The sequence

$$0 \to L.v(G) \to L^G \to \hat{H}^0(L) \to 0$$

is exact, so we can pick $v \in L^G$ mapping to u. The sequence (2.1) is exact so we can pick $m \in M$ mapping to v. We contend that

$$\lambda(m) = av(N), \qquad (a,c(\lambda)) = 1.$$

In any case, since $v \in L^G$, we get

$$\alpha(g) = m - mg \in K \qquad \text{(for all } g \in G),$$

so that $\lambda(m) = av(N)$ for some integer a. Let

$$e = (a,c(\lambda)), \qquad f = c(\lambda)/e, \qquad b = a/e.$$

Then

$$\lambda(fm) = fav(N) = b\lambda(v(M)),$$

so that

$$fm - bv(M) \in K. \tag{2.6}$$

But the cohomology class of α in $H^1(K)$ is $\delta(u)$, and has order $|c(\lambda)|$ by hypothesis. As (2.6) implies that f annihilates $\delta(u)$, we get $e = 1$.

Now suppose $m \in M$ and $\lambda(m) = v(N)$. Define $\alpha: G \to K$ by $\alpha(g) = m - mg$. Obviously, the image β of α in $H^1(k)$ is killed by $c(\lambda)$, since $c(\lambda)m - v(M) \in K$. Conversely, if c is the order of β in $H^1(K)$, then $c\alpha$ is a principal crossed homomorphism so there is k in K such that

$$c\alpha(g) = k - kg.$$

Hence, $cm - k \in M^G = Z \,.\, v(M)$, so

$$\lambda(cm-k) = \lambda(cm) = cv(N) \in Z \,.\, c(\lambda)v(N),$$

whence $c \equiv 0 \pmod{c(\lambda)}$, and hence β has order $|c(\lambda)|$. The surjectivity of (2.5) and the definition of $c(\lambda)$ show that $c(\lambda)$ kills $H^1(K)$, and so $H^1(K)$ is cyclic of order $|c(\lambda)|$.

3. Unimodular Lattices

LEMMA 3.1. *Suppose Γ, Δ are sublattices of the integral lattice Λ, Λ has rank n, and $(\gamma_1, \ldots, \gamma_n)$, $(\delta_1, \ldots, \delta_n)$ are sequences of length n, with $\gamma_i \in \Gamma$, $\delta_i \in \Delta$. Let $A = (a_{ij})$ be the $n \times n$ inner product matrix defined by*

$$a_{ij} = (\gamma_i, \delta_j).$$

If $\det A = \pm 1$, then $\Gamma = \Delta = \Lambda$ is a unimodular lattice.

Proof. Let $\lambda_1, \ldots, \lambda_n$ be a basis for Λ, and set

$$l_{ij} = (\lambda_i, \lambda_j), \qquad L = (l_{ij}).$$

Since $\Gamma \subseteq \Lambda$, we have

$$\gamma_i = \sum_{j=1}^{n} g_{ij} \lambda_j,$$

where $G = (g_{ij})$ is a square matrix with integral entries. Similarly,

$$\delta_i = \sum_{j=1}^{n} d_{ij} \lambda_j,$$

where $D = (d_{ij})$ is a square matrix with integral entries. Then

$$(\gamma_i, \delta_j) = a_{ij} = \left(\sum_{r=1}^{n} g_{ir} \lambda_r, \sum_{s=1}^{n} d_{js} \lambda_s \right)$$

$$= \sum_{r,s=1}^{n} g_{ir} l_{rs} d_{js},$$

which gives

$$A = GL\,{}^tD.$$

By hypothesis, A is unimodular, so each of $G, L, {}^tD$ is unimodular, whence the lemma.

4. The Character χ of M

From now on, we assume that

$n = 2m$ is even and $\geqslant 4$,

$G = S_n$,

M is the permutation module with C_m as a basis.

Then $\boldsymbol{Q} \otimes_{\boldsymbol{Z}} M = \boldsymbol{Q}M$ is a completely reducible G-module with character χ, and we identify χ.

LEMMA 4.1. *If $\chi^{(\pi)}$ is the irreducible character of S_n corresponding to the partition π, then*

$$(\chi,\chi^{(\pi)}) = \begin{cases} 1 \text{ if every part of } \pi \text{ is even} \\ 0 \text{ otherwise.} \end{cases}$$

Remark. That this lemma might be true was suggested to me by J. Saxl.

Proof. The first step is to show that

(4.1) (χ,χ) is the number of partitions of n, all of whose parts are even.

Since M is a permutation module on C_m, we know that (χ,χ) is the number of orbits of G on $C_m \times C_m$. We need to check that

(4.2) S_n has just one orbit on $\{(a,b) | a,b \in C_m, \langle a,b \rangle$ is transitive on $\{1,\ldots,n\}\}$.

To see this, let $c = ab$. Then $c \in A_n$. Since $\langle c \rangle$ is a normal subgroup of index 2 in $\langle a,b \rangle$, it follows that $c = c_1c_2 = c_2c_1$, where c_i is an m-cycle, and $c_1^a = c_2^{-1}$. There are precisely m elements $a \in C_m$ such that $c_1^a = c_2^{-1}$, and they form a single orbit under $\langle c_1 \rangle$. This gives (4.2).

If we apply (4.2) to the individual orbits of $\langle a,b \rangle$ (where $(a,b) \in C_m \times C_m$), (4.1) follows, the map from orbits to partitions being

$$\mathcal{O} \to (k_1,\ldots,), \{k_i\} \text{ the orbit lengths of } \langle a,b \rangle, (a,b) \in \mathcal{O}.$$

Thus, to complete the proof of Lemma 4.1, it suffices to show that

$$(\chi,\chi^{(\pi)}) \neq 0$$

if every part of π is even.

Let T be a tableau associated to π. For each transposition $t \in S_n$, the two points moved by t are at points (i_1,j_1), (i_2,j_2) of T, and we set

$$h_T(t) = |j_1 - j_2|, \qquad w_T(t) = |i_1 - i_2|.$$

(h = height; w = width).

If $u \in C_m$, we set

$$h_T(u) = \max h_T(t) \qquad \text{and} \qquad w_T(u) = \max w_T(t)$$

as t ranges over the cycles of u. Since every part of π is even, there is a unique $u = u(T) \in C_m$ such that

$$h_T(u) = 0, \qquad w_T(u) = 1.$$

A picture suffices for $n = 6$, $\pi = (4\ 2)$.

$$T: \begin{array}{llll} a_1 & a_2 & a_3 & a_4 \\ a_5 & a_6 \end{array} \qquad u(T) = a_1a_2 \,.\, a_3a_4 \,.\, a_5a_6.$$

Let C,R be the largest subgroups of S_n which fix each column (row) of T, and set

$$e(C)^- = |C|^{-1} \sum_{c \in C} \mathrm{sg}(c)c,$$

$$e(R)^+ = |R|^{-1} \sum_{r \in R} r,$$

$$e(T) = e(C)^- \,.\, e(R)^+.$$

Then $e(T)$ is a primitive idempotent of QS_n, and

$$\dim QM \,.\, e(T) = (\chi, \chi^{(\pi)}).$$

So it suffices to show that

$$M \,.\, e(T) \neq 0.$$

We in fact show that

(4.3) the coefficient of $u(T)$ in $u(T) \,.\, e(T)$ is positive.

Let

$$f(T) = e(T)|C|\,|R| = \sum_{\substack{c \in C \\ r \in R}} \mathrm{sg}(c)cr.$$

The coefficient of $u(T)$ in $u(T)f(T)$ is

$$\sum_{\substack{(c,r) \\ c \in C, r \in R \\ cr \in C(u(T))}} \mathrm{sg}(c).$$

So suppose $c \in C$, $r \in R$ and $cr \in C(u(T))$. Then

$$u(T)^{cr} = u(T), \qquad u(T)^c = U(T)^{r^{-1}}.$$

This gives

$$w_T(u(T^c)) = w_T(u(T)) = 1, \qquad h_T(u(T)^{r^{-1}}) = h_T(u(T)) = 0,$$

(the functions w_T, h_T are constant on C-orbits, R-orbits, respectively), and so

$$u(T)^c = u(T), \qquad u(T)^{r^{-1}} = u(T).$$

Let $c = c_1 . c_2 \ldots c_{2f-1} c_{2f}$, where $2f$ is the number of columns of T, and c_i moves only points in column i. For each odd number i, $u(T)$ maps column i of T bijectively onto column $i+1$ of T, and it follows that

$$c_i^{u(T)} = c_{i+1} \qquad (i = 1,3,\ldots,2f-1)$$

Hence, $\mathrm{sg}(c_i) = \mathrm{sg}(c_{i+1})$, whence

$$\mathrm{sg}(c) = 1.$$

Since $u(T) = u(T)^{1.1}$, (4.3) follows, whence the lemma.

5. The Character of S_n on K

Let N denote the transitive permutation module of S_n on C_2, and let $\lambda = \lambda_{m,2}$ as defined in Section 1 (with $n+2$ as our present n). Let $u \in C_m$, $H = C_{S_n}(u)$. Set

$$K = \ker(\lambda), \qquad I = \mathrm{Im}(\lambda).$$

LEMMA 5.1. *If* $a \in C_2$, $D = C_{S_n}(a)$, *and* $n \geqslant 8$, *then* I^D *has rank* 5.

Proof. We may assume that

$$a = 12.34.$$

Let

$$\begin{aligned} e_1 &= 12.34.56\ldots n-1\,n \\ e_2 &= 13.24.56\ldots n-1\,n \\ e_3 &= 12.35.46.78\ldots n-1\,n \\ e_4 &= 13.25.46.78\ldots n-1\,n \\ e_5 &= 15.26.37.48.910\ldots n-1\,n. \end{aligned}$$

Then

$$\begin{aligned} \lambda(e_1) &= 12.34 + \text{other terms} \\ \lambda(e_2) &= 13.24 + O.T. \\ \lambda(e_3) &= 12.35 + O.T. \\ \lambda(e_4) &= 13.25 + O.T. \\ \lambda(e_5) &= 15.26 + O.T. \end{aligned}$$

The displayed elements $y_1, \ldots, y_5$ lie in different orbits of D, and if $i > j$, then no summand of $\lambda(e_i)$ is in the D-orbit of y_j, whence $\lambda(e_i).v(D)$ are linearly independent, whence the rank of I^D is at least 5. Since

$$(1_D^{S_n}, 1_D^{S_n}) = 5, \tag{5.1}$$

the lemma follows.

LEMMA 5.2. *If $n \geqslant 8$, the character of S_n on QK is $\Sigma_\pi \chi^{(\pi)}$, where π ranges over $P-Q$, P is the set of all partitions of n all of whose parts are even, and*

$$Q = \{(n), (n-2\,2), (n-4\,4), (n-4\,2^2)\}.$$

Proof. Again let $a = 12.34$. Then

$$D \subseteq E = S_4 \times S_{n-4},$$

where the S_4 is the pointwise stabiliser of $\{5,6,\ldots,n\}$. If ϕ is the irreducible character of E of degree 2 which has S_{n-4} in its kernel, then

$$1_D^E = 1_E + \phi,$$
$$1_D^{S_n} = 1_E^{S_n} + \phi^{S_n}.$$

As is well known,

$$1_E^{S_n} = \sum_{j=0}^{4} \chi^{(n-j\,j)}.$$

We need to check that

$$(\phi^{S_n}, \chi^{(n-4\,2^2)}) \neq 0.$$

This we do, for example, by taking the tableau

$$\begin{matrix} \dot{5} & \dot{6} & & \dot{n} \\ \dot{1} & \dot{2} & & \\ \dot{3} & \dot{4} & & \end{matrix}$$

and checking that the primitive idempotent for this tableau does not annihilate 12.34 ($\in N$).

Since D has 4 orbits on the 2-element subsets of $\{1,2,\ldots,n\}$, it follows that

$$(1_D^{S_n}, 1_{S_2 \times S_{n-2}}^{S_n}) = 4,$$

whence

$$(1_D^{S_n}, \chi^{(n-2\,2)}) = 2.$$

Thus, (5.1) implies

$$1_D^{S_n} = \chi^{(n)} + 2\chi^{(n-2\,2)} + \chi^{(n-4\,4)} + \chi^{(n-4\,2^2)} + \Delta,$$

where no irreducible constituent of Δ is of the shape $\chi^{(\pi)}$, $\pi \in P$. This shows that the character χ of S_n on K is of the shape

$$\chi = \sum_{\pi \in P-Q} \chi^{(\pi)} + \chi',$$

where χ' is either 0 or a character. That $\chi' = 0$ follows from Lemma 5.1.

6. $H^1(K)$ is Cyclic of Order $c(\lambda)$

We view M as a lattice with C_m as an orthonormal basis.

If $x \in C_m$, then x is a product of m 2-cycles, so

$$d = \iota(\lambda(x)) = \binom{m}{2} = \frac{n(n-2)}{8},$$

and we get that

$$c(\lambda) = \frac{|C_m|}{(n-1)(n-3)}. \tag{6.1}$$

Also

$$|C_m| = \frac{n!}{|C_{S_n}(x)|} = \frac{n!}{2^m m!}.$$

As before, let P be the set of all partitions of n, all of whose parts are even. For each $\pi \in P$, let $\mathcal{T}_\pi$ be the set of standard tableaux, so that if $T \in \mathcal{T}_\pi$, the underlying partition of T is π, and the integers appearing in T are increasing across rows from left to right, and increasing down columns from top to bottom. Thus

$$|\mathcal{T}_\pi| = \chi^{(\pi)}(1).$$

Set

$$\mathcal{T} = \bigcup_{\pi \in P} \mathcal{T}_\pi.$$

If $T \in \mathcal{T}_\pi$, $\pi \in P$, let $x(T)$ be the unique solution to

(a) $x(T) \in C_m$.

(b) $h_T(x(T)) = 0$.

(c) $w_T(x(T)) = 1$.

And let C_T, R_T be the column and row subgroups of S_n associated to T. Set

$$\gamma(T) = \frac{1}{|C_T \cap C_{S_n}(x(T))|} \sum_{c \in C_T} \mathrm{sg}(c) . x(T)^c \in QM$$

$$\delta(T) = \frac{1}{|R_T \cap C_{S_n}(x(T))|} \sum_{r \in R_T} x(T)^r \in QM.$$

The first observation is that

$$(6.2) \qquad \gamma(T), \delta(T) \in M.$$

This is obvious, since for example, the coefficient of $x(T)$ in

$$\sum_{c \in C} \mathrm{sg}(c) x(T)^c$$

is simply $|C_T \cap C_{S_n}(x(T))|$, the reason being that $C_T \cap C_{S_n}(x(T)) \subseteq A_n$, as already remarked. So in fact, we see that $\gamma(T)$ is a linear combination of members of C_m with coefficients 0,1, or -1. Similarly, $\delta(T)$ is a linear combination of elements of C_m with coefficients 0 or 1.

LEMMA 6.1. *It is possible to order $\mathscr{T}$ in such a way that the matrix of inner products $(\gamma(T),\delta(T'))$ is upper triangular with* 1 *on the diagonal.*

In order to prove Lemma 6.1, we make use of the usual ordering of partitions

$$\pi = (a_1,a_2,\ldots,) > \pi' = (a'_1,a'_2,\ldots,)$$

(with $a_1 \geqslant a_2 \geqslant \ldots, a'_1 \geqslant a'_2 \geqslant \ldots$) provided the first non-vanishing $a_i - a'_i$ is positive.

The first step in the proof of Lemma 6.1 is that

(6.3) if $\pi < \pi'$, $T \in \mathscr{T}_\pi$, $T' \in \mathscr{T}_{\pi'}$, then $(\gamma(T),\delta(T')) = 0$.

Set

$$e_C(T)^- = |C_T|^{-1} \sum_{c \in C_T} \mathrm{sg}(c).c,$$

$$e_R(T)^+ = |R_T|^{-1} \sum_{r \in R_T} r,$$

so that

$$\gamma(T) = u(T)x(T)e_C(T)^-,$$
$$\delta(T') = v(T')x(T')e_R(T')^+,$$

and

$$u(T) = |C_T : C_T \cap C_{S_n}(x(T))|,$$
$$v(T') = |R_{T'} : R_{T'} \cap C_{S_n}(x(T'))|.$$

Since $e_R(T')^+$ is self-adjoint with respect to our inner product, it is enough to show that

$$(6.4) \qquad (x(T).e_C(T)^-.e_R(T')^+, x(T')) = 0.$$

Since $\pi < \pi'$, there is a transposition $\tau = ab$, such that a,b lie in the same row of T' and in the same column of T, whence $e_C(T)^- e_R(T')^+ = 0$ and (6.4) holds. Thus Lemma 6.1 will follow if we can prove the following assertions:

(6.5) If $\pi \in P$ and $T \in \mathscr{T}_\pi$, then $(\gamma(T),\delta(T)) = 1$.

(6.6) There is an ordering of $\mathscr{T}_\pi$ such that if $T,T' \in \mathscr{T}_\pi$ and $T > T'$, then $(\gamma(T),\delta(T')) = 0$.

If $c \in C_T$, $r \in R_T$ and $cr \in C_{S_n}(x(T))$, then we have already used, $c \in C_{S_n}(x(T))$, $r \in C_{S_n}(x(T))$. This gives us

$$(\gamma(T),\delta(T)) = (u(T)v(T)x(T)e_C(T)^- e_R(T)^+, x(T)),$$

and since the coefficient of $x(T)$ in $u(T)v(T)e_C(T)^- e_R(T)^+$ is $+1$, (6.5) follows.

It remains to order $\mathscr{T}_\pi$ so that (6.6) holds. Let the ith row of T be

$$a_i(1)\ a_i(2)\ \ldots$$

Let

$$A_i(T) = \sum_j a_i(j)$$

be the sum of all the integers in the ith row of T. Then we define an equivalence relation of $\mathscr{T}_\pi$ by writing $T \sim T'$ if and only if

$$A_i(T) = A_i(T') \qquad i = 1,2,\ldots.$$

If E,E' are equivalence classes, we write $E > E'$ provided the first non-vanishing $A_i(T) - A_i(T')$ is positive $(T \in E, T' \in E')$. Then (6.6) will follow if we can prove the following assertions:

(6.7) Suppose $\pi \in P$ and $T,T' \in \mathscr{T}_\pi$, $T \neq T'$. If T,T' are in the same equivalence class, then $(\gamma(T),\delta(T')) = 0$. If T,T' are in equivalence classes E,E' respectively, and $E > E'$, then $(\gamma(T),\delta(T')) = 0$.

Let i be the largest natural number such that for $j = 1,\ldots,i-1$, T and T' agree in row j. Let the ith rows of T,T' be

$$\begin{matrix} a_1 & a_2 & \ldots & a_f \\ b_1 & b_2 & \ldots & b_f \end{matrix}$$

respectively. Then

(6.8) $$a_1 < a_2 < \ldots < a_f, \qquad b_1 < b_2 < \ldots < b_f$$

and

(6.9) $$a_1 + a_2 + \ldots + a_f \geqslant b_1 + b_2 + \ldots + b_f$$

either because T,T' are in the same equivalence class, or because $E > E'$. We argue that there are i,j such that $1 \leqslant i < j \leqslant f$, and b_i,b_j are in the same column of T. If this is false, there is a permutation s of $\{1,\ldots,f\}$ such that $b_{s(i)}$ is in column i of T. This means then that $a_i \leqslant b_{s(i)}$, since T,T' are identical in rows $1,2,\ldots,i-1$, and the integers in each column of T are increasing from top to bottom. Coupled with (6.8) we get $a_i = b_{s(i)}$, all i, and since (6.8) holds,

$s =$ identity, against the definition of i. So there is a transposition $t \in C_T \cap R_T$, whence as before, we get (6.7).

Now we apply Lemma 3.1 and conclude that

$$\{\gamma(T) | T \in \mathscr{T}_\pi, \pi \in P\}$$

is a $\mathbf{Z}$-basis for M.

Recall that $Q = \{(n), (n-2\,2), (n-4\,4), (n-4\,2^2)\}$.

Let

$$v(M) = \sum_{T \in \mathscr{T}} a(T)\gamma(T) \qquad (a(T) \in \mathbf{Z}).$$

We argue that

(6.10) if $\pi \in Q$ and $T \in \mathscr{T}_\pi$, then $a(T) \equiv 0 \pmod{c(\lambda)}$.

Let $P = \{\pi_1, \pi_2, \ldots,\}$, $\pi_1 > \pi_2 > \ldots$. Then

$$\pi_1 = (n), \qquad \pi_2 = (n-2\,2), \qquad \pi_3 = (n-4\,4), \qquad \pi_4 = (n-4\,2^2).$$

Let

$$\mathscr{T} = \{T_1, \ldots, T_h\},$$

where

$$\{T_1, \ldots, T_{s_1}\} = \mathscr{T}_{\pi_1}, \qquad (\text{so } s_1 = 1)$$
$$\{T_{s_1+1}, \ldots, T_{s_1+s_2}\} = \mathscr{T}_{\pi_2}, \qquad \text{etc.,}$$

and where the ordering on $\mathscr{T}_\pi$ is such that

$$(\gamma(T), \delta(T')) = 0 \qquad \text{if} \qquad T > T' \quad (T, T' \in \mathscr{T}_\pi).$$

Then the matrix

$$C = ((\gamma(T_i), \delta(T_j))) = (c_{ij})$$

is upper triangular.

In order to prove (6.10), we first prove an easier assertion:

(6.11) $(v(M), \delta(T)) \equiv 0 \pmod{c(\lambda)}$, if $T \in \mathscr{T}_\pi$, $\pi \in Q$.

Namely, by construction of $\delta(T)$, we have

$$(v(M), \delta(T)) = \iota(\delta(T)),$$

and

$$\iota(\delta(T)) = |R_T : R_T \cap C(x(T))|.$$

If $T \in \mathscr{T}_\pi$ and

$$\pi = \begin{cases} (n), & \iota(\delta(T)) = |C_m|, \\ (n-2\,2), & \iota(\delta(T)) = (n-2)!/2^{m-1}(m-1)!, \\ (n-4\,4), & \iota(\delta(T)) = (n-4)!4!/2^{m-2}(m-2)!8, \\ (n-4\,2^2), & \iota(\delta(T)) = (n-4)!/2^{m-2}(m-2)!. \end{cases}$$

Since

$$c(\lambda) = |C_m|/(n-1)(n-3),$$

it is easy to check that $\iota(\delta(T))/c(\lambda) \in \boldsymbol{Z}$ in the indicated cases. For example, if $\pi = (n-4\,2^2)$,

$$\iota(\delta(T)) = \frac{n!m(m-1)2^2}{n(n-1)(n-2)(n-3)m!2^m}$$

$$\frac{\iota(\delta(T))}{c(\lambda)} = 1.$$

Next, let $(g(T_1),\ldots,g(T_h))$ be the dual basis to $(\gamma(T_1),\ldots,\gamma(T_h))$. Then

$$g(T_i) = \sum_{j=1}^{h} g_{ij}\delta(T_j),$$

where $g_{ij} \in \boldsymbol{Z}$. So

$$\begin{aligned}(\gamma(T_i),g(T_j)) &= \delta_{ij}\\ &= (\gamma(T_i), \sum_{r=1}^{h} g_{jr}\delta(T_r))\\ &= \sum_{r=1}^{h} g_{jr}c_{ir}.\end{aligned}$$

Thus, if $G = (g_{ij})$, then

$$G.{}^tC = I, \qquad G = {}^tC^{-1}.$$

Since C is upper triangular, G is lower triangular, which means that if $T_i \in \mathscr{T}_\pi$, $\pi \in Q$, then

$$g(T_i) \in \sum_{T \in \mathscr{T}_\pi, \pi \in Q} \boldsymbol{Z}\delta(T).$$

So

$$(v(M),g(T_i)) \equiv 0 \ (\mathrm{mod}\ c(\lambda)), \qquad (T_i \in \mathscr{T}_\pi, \pi \in Q).$$

But

$$v(M) = \sum_i (v(M),g(T_i))\gamma(T_i),$$

since $\{\gamma(T_i)\}$, $\{g(T_i)\}$ are dual bases, and so (6.10) follows. Now we simply define

$$\tilde{v}(M) = c(\lambda)^{-1}\left\{v(M) - \sum_{\substack{T \in \mathscr{T}_\pi\\ \pi \in P-Q}} a(T)\gamma(T)\right\}$$

and get

$$\tilde{v}(M) \in M, \qquad \lambda(\tilde{v}(M)) = v(N),$$

and our cohomology group is subdued, once we have shown that

(6.13) if $\pi \in P-Q$, and $T \in \mathcal{T}_{\pi}$, then $\gamma(T) \in K$.

We can do better than (6.13) demands, namely we show that

(6.14) $\{\gamma(T) | T \in \mathcal{T}_{\pi}, \pi \in P-Q\} = B$ is a $\boldsymbol{Z}$-basis for K.

Let K_0 denote the abelian group generated by B. Since B is part of a $\boldsymbol{Z}$-basis for M, K_0 is a pure subgroup of M, and it has the same $\boldsymbol{Z}$-rank as K. Thus, it suffices to show that $B \subseteq K$ (in other words, (6.13) and (6.14) are equivalent.)

Since $K = \boldsymbol{Q}K \cap M$, it suffices to show that $B \subseteq \boldsymbol{Q}K$. Now

$$\gamma(T) = u(T)x(T)e_C(T)^-,$$

and $e_R(T')^+ e_C(T)^- = 0$ for all $T' \in \mathcal{T}_{\pi'}, \pi' \in Q$. All the more so, $e_C(T)^-$ annihilates

$$\boldsymbol{Q}M \,.\, e_C(T')^- e_R(T')^+ = M(T'),$$

for all $T' \in \mathcal{T}_{\pi'}, \pi' \in Q$. But

$$\sum_{T' \in \mathcal{T}_{\pi}} M(T') = M(\pi')$$

is a S_n-submodule of $\boldsymbol{Q}M$ with character $\chi^{(\pi')}$. Since $\gamma(T) \in M(\pi')^{\perp}$, we get

$$\gamma(T) \in \bigcap_{\pi' \in Q} M(\pi')^{\perp} = \boldsymbol{Q}K,$$

as asserted. Putting all this together, we have

THEOREM 6.1. *$H^1(K)$ is cyclic of order $c(\lambda)$. There is $m \in M$ such that $\lambda(m) = v(N)$.*

7. Two Special Cases

Let $m \in M$ with $\lambda(m) = v(N)$. Let $M_0 = K + \boldsymbol{Z}m$, and for each natural number k, let

$$a_k(M_0) = |\{x \in M_0 | \,(x,x) = k\}|,$$

so that

$$\theta(q) = 1 + \sum_{k=1}^{\infty} a_k(M_0)q^k$$

is the θ-function of M_0. The problem of ovals in a plane is essentially the problem of finding solutions to

$$(7.1) \quad m \in M, \qquad \lambda(m) = v(N), \qquad m = \sum_{x \in C_m} b_x \,.\, x, \qquad b_x \in \{0,1\}.$$

LEMMA 7.1. *The number of solutions to* (7.1) *is* $\frac{1}{2}a_{(n-1)(n-3)}(M_0)$.

Proof. Suppose $x \in M_0$ and $(x,x) = (n-1)(n-3)$. Let

$$x = c.m + m', \qquad m' \in K, \qquad c \in \boldsymbol{Z}.$$

Since $(y,y) \in 2\boldsymbol{Z}$ for all $y \in K$, while $(n-1)(n-3)$ is odd, it follows that c is odd. Let

$$x = \sum_{s \in C_m} a_s . s.$$

Then

$$\sum a_s^2 = (n-1)(n-3).$$

Since $\lambda(x) = cv(N)$, it follows that $\iota(x) = c.(n-1)(n-3) = \sum a_s$. Thus

$$|\sum a_s| = |c|(n-1)(n-3) \geqslant (n-1)(n-3) = \sum a_s^2.$$

Since $a_s \in \boldsymbol{Z}$, it follows that there is $e \in \{1, -1\}$ such that $a_s \in \{0,e\}$ for all $s \in C_m$, whence the lemma.

If $n = 4$, 6, or 10, then there are solutions to (7.1), since the Desarguesian planes of orders 2,4,8 have ovals. In order to study the cases $n = 8$ or 12, we introduce the following considerations. For each valuation v of $\boldsymbol{Z}$, let K_v denote the v-adic completion of K. We say that K satisfies the v-adic restriction if and only if $K_v + m$ has an element of norm $(n-1)(n-3)$. With this convention, we can show that

(7.2) if $n = 8$, K does not satisfy the 2-adic restriction.

(7.3) if $n = 12$, K satisfies the v-adic restriction for all v.

Suppose $n = 8$. In this case, $\boldsymbol{Q}K$ is irreducible with character $\chi^{(2^4)}$, and of dimension 14. Furthermore, if $T \in \mathscr{T}_{(2^4)}$, then $(\gamma(T),\gamma(T)) = 24$, and if $T' \in \mathscr{T}_{(2^4)}$, $(\gamma(T),\gamma(T'))$ is even, this last assertion being verified by explicit calculation. So K is an even lattice and $(x,x) \in 4\boldsymbol{Z}$ for all $x \in K$.

Now consider the critical coset $m + K_v$ where v is the 2-adic valuation of $\boldsymbol{Q}$, so that $K_v = \boldsymbol{Z}_2 \otimes_{\boldsymbol{Z}} K$, $\boldsymbol{Z}_2$ being the 2-adic integers. Then, $v(M)/3 \in m + K_v$, so we want $b \in K_v$ such that

$$\tfrac{1}{3}v(M) + b = y$$

satisfies $(y,y) = 35 = (n-1)(n-3)$. Since $v(M) \in K_v^{\perp}$, we require that

$$35 = \frac{105}{9} + (b,b)$$

whence

$$(b,b) = \frac{2.35}{3},$$

against $(b,b) \in 4\boldsymbol{Z}_2$.

Of course, it is known that there is no projective plane of order 6, but the above argument seems worth recording nonetheless.

When $n = 12$, we make use of the Mathieu group M_{12} to find a particular solution to $\lambda(m) = v(N)$. On C_6, M_{12} has 5 orbits $\mathcal{O}_1,\ldots,\mathcal{O}_5$, of cardinals $a_i.99$, $a_i = 4, 5, 16, 20, 60$, and if

$$m_i = \sum_{s\in\mathcal{O}_i} s,$$

the multiple transitivity of M_{12} forces $\lambda(m_i) = a_i v(N)$. This implies that if $m = m_2 - m_1$, then $\lambda(m) = v(N)$. In addition, $(m,m) = 9.99$.

Over the field of real numbers, our problem is trivial, since if $y = v(M)/105$, then $\lambda(y) = v(N)$, and $(y,y) < 99$. So we only need deal with p-adic completions. For each prime number, the equation

$$-11 = x^2 + y^2 + z^2$$

has a solution in $\boldsymbol{Z}_p$, the ring of p-adic integers, and if y_1,y_2,y_3 satisfy

$$y_i \in K,\ (y_i,y_i) = 8,\ (y_i,y_j) = 0 \qquad (i \neq j),$$

$$\text{and} \qquad (m,y_i) = 0,$$

then $m + 3xy_1 + 3yy_2 + 3zy_3$ has norm 99 and maps onto $v(N)$.

In order to find the y_i, we take the "standard" M_{12}, namely $M_{12} = \langle\alpha,\beta,\gamma,\delta\rangle$,

$$\begin{aligned}
\alpha &= 0123456789X, && (x \mapsto x+1 \ (\mathrm{mod}\ 11))\\
\beta &= 13954.267X8, && (x \mapsto 3x)\\
\gamma &= \infty 0.1X.25.37.48.69, && (x \mapsto -1/x)\\
\delta &= 18.23.69.4X.\infty 0.57, && (!!)
\end{aligned}$$

Then we take

$$\begin{aligned}
y_1 = \ & \infty 1.08.X6.24.97.35\\
& -\infty 1.08.X6.24.93.57\\
& -\infty 1.08.X4.26.97.35\\
& +\infty 1.08.X4.26.93.57\\
& -\infty 8.01.X6.24.97.35\\
& +\infty 8.01.X6.24.93.57\\
& +\infty 8.01.X4.26.97.35\\
& -\infty 8.01.X4.26.93.57
\end{aligned}$$

$y_2 = y_1^{\alpha}, y_3 = y_1^{\alpha^2}$.

17.

How to Construct a Baby Monster

CHARLES C. SIMS

In this chapter, we shall describe, in general terms, the methods used to construct certain of the recently discovered sporadic simple groups, in particular, the group known as Fischer's Baby Monster. This will not be a detailed report on the construction of the Baby Monster nor will it give enough information to permit the reader to immediately undertake the construction of one of the two sporadic groups whose existence remains in doubt as this is being written. It will hopefully provide sufficient insight into the techniques involved so that the construction of very large groups such as the Baby Monster will be seen to lie within the realm of the possible.

We begin with a few elementary remarks concerning ways that one may describe a particular group. One method is to define an algebraic or combinatorial structure and define the group to be the group of automorphisms of the structure. For example, if we let $\mathscr{G}$ be the graph

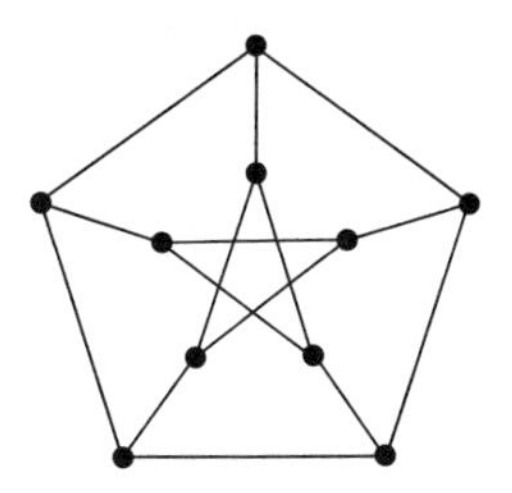

and let $G = \mathrm{Aut}(\mathscr{G})$, then we have defined a group of order 120.

Another approach is to provide explicit generators which are elements of some familiar group, such as a symmetric group. Thus we could define G to be the subgroup of S_7 generated by $(14)(27)$ and $(123)(456)$. In this case G is a group of order 168.

A third method is the use of a presentation by generators and relations. For example, the group generated by two elements x and y subject to the defining relations $x^2 = y^3 = (xy)^5 = 1$ has order 60.

Each of these three ways of defining a group has the drawback that the structure of the group is not immediately apparent from its definition. A random graph has essentially no nontrivial automorphisms. Thus if in defining the graph $\mathcal{G}$ above a mistake was made, then the group Aut($\mathcal{G}$) will almost certainly be smaller than was intended. Therefore, the main problem with groups defined as automorphism groups is to show that they are "big enough".

With groups given by generating elements the problem is just the opposite. Dixon [1] has shown that two or more elements chosen at random in S_n generate either S_n or the alternating group A_n with probability approaching 1 as n goes to infinity. Hence, if a mistake was made in one of the elements of S_7 given above, then the group generated is probably A_7 or S_7. Thus the problem here is to show that the group is "small enough".

With generators and relations the difficulties are much greater than with either of the other two methods. A mistake in the presentation can easily lead to groups which are either too big or too small. Moreover, without any restriction on the type of presentations used, there are no algorithms for deciding such things as whether the group is finite or infinite or whether it is trivial or nontrivial.

At this point I would like to mention the four sporadic groups in whose construction I have been involved. In chronological order they are the Higman–Sims group [2], a group of order about 44×10^6 which was constructed using a graph with 100 vertices, the Lyons group [6], which has order about $5{\cdot}2 \times 10^{16}$ and was constructed by describing generating permutations on just under 9×10^6 points, the O'Nan group [5], which has order about $4{\cdot}5 \times 10^{11}$ and was obtained as a permutation group of degree 122760 and the Baby Monster [4], which has order about $4{\cdot}2 \times 10^{33}$ and was constructed jointly with Leon using permutations of degree about 14×10^9. The entire construction of the Higman–Sims group took only a few hours. However, in each of the other cases evidence for the existence of the groups was obtained many months before the groups were actually constructed and a great deal of machine computation was required to prove their existence. It is the computer techniques used in these calculations which are the main subject of this paper.

Let us start with a simplified view of a modern digital computer. The following diagram shows the essential components of such a computer:

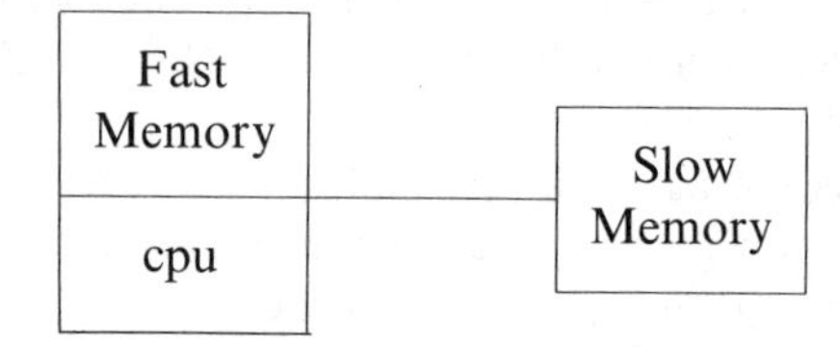

The central processing unit (cpu) performs logical and arithmetic operations. Such operations can be performed in half a microsecond or less. The fast memory contains the program as well as some of the data to be manipulated by the program. The time required to bring a single number from the fast memory to the cpu is less than one microsecond. On a large computer the portion of the fast memory typically available for one program has enough room to store one permutation of the integers $1,2,\ldots,n$, where n is in the range $100\,000 \leqslant n \leqslant 200\,000$. Thus one generator of the O'Nan group can be stored in the fast memory with a little space to spare.

Connected to the cpu and fast memory are additional memory devices such as magnetic disks or tapes. This secondary memory has a much larger capacity than the fast memory but the time necessary to bring one number into the cpu is much greater than for the fast memory. Usually the slow memory is large enough to store several permutations of the integers from 1 to 10^7. It would be possible to store generators for the Lyons group in slow memory but this has not been done. It is out of the question at the present time to consider storing generators for the Baby Monster as explicit permutations.

Let us assume that we wish to study the subgroup G of S_n generated by certain given permutations $x,y,z,\ldots$. From our description of a digital computer it is clear that we need to distinguish three cases:

1. $n \leqslant 10\,000$.
2. $100\,000 \leqslant n \leqslant 200\,000$.
3. $n \geqslant 10^6$.

The boundaries of these cases have been left vague deliberately. In the first case, all the generators of G can be kept in fast memory at one time. In the second case, just one generator can be stored in fast memory. In the third case, only a small portion of one generator can be held in fast memory.

In order to investigate G it is important to be able to perform the following operations:

A. Multiply elements of G.
B. Describe the orbits of G.
C. Determine $|G|$.
D. Given g in S_n, decide whether or not g is in G.

Let us look briefly at the difficulty of carrying out each of these procedures in the three cases mentioned above.

To compute a product such as $w = xyxz^2$ in Case 1 is very easy. We have only to take each point α in turn, compute $\beta = \alpha^x$, $\gamma = \beta^y$ and so on until we have α^w. Since x, y and z are all present in the fast memory, this presents no difficulty. In Case 2 this approach will be too slow. Instead we compute the product of two factors completely, then multiply by the third factor and continue in this manner. The details are a little tricky but not really difficult. The cost of multiplying two permutations of degree 122 760 on the system available at Rutgers using a program written by Andrilli is about \$1.00. Even in Case 3 there is some hope. A program which I recently wrote multiplied two permutations of degree 10^6 for about \$15.00.

Computation of orbits is also easy in Case 1. Given a point α, we can construct the orbit α^G by applying each of the generators of G to α, applying them again to any new points so obtained and continuing in this manner until we find a set closed under the action of the generators. As with multiplying elements, the method used in Case 2 is quite different from that used in Case 1. However, it is not difficult to describe the orbits of a group of degree 100 000. There is, thus far, little experience with general purpose orbit programs for groups acting on millions of points.

Probably the most important question to be answered about our group G is, what is $|G|$? In Case 1 we now have a very efficient algorithm. Based on some of my ideas [7], this algorithm was given its first efficient implementation by Leon [3]. In one test run, the program was given several permutations of degree 1782 which generate the Suzuki sporadic group. It took just 10 seconds to correctly determine the order, which is

$$448\,345\,497\,600 = 2^{13}.3^7.5^2.7.11.13.$$

The algorithm is fairly complicated and I shall not try to describe it in detail here. However, it will be useful to indicate one important idea involved in the algorithm. As was mentioned above, random sets of permutations generate the alternating or symmetric groups. Thus if we want to show G is not A_n or S_n, then we must show that our permutations are in some sense not random. This is done by finding "short" relations which the permutations satisfy.

Procedure D may be thought of as effectively determining G as a subset of S_n. The algorithm for finding $|G|$ in Case 1 yields as a by-product information which makes Procedure D easy to carry out.

Summarizing, we find that all four procedures can be performed efficiently in Case 1 and that Procedures A and B can be carried out relatively easily in Case 2.

Now let us turn to methods for describing permutations on very large sets, methods which were first used in the construction of Lyons' group and later were used in the construction of the Baby Monster. In the first method, we assume that we are given a subgroup H of S_n with $n \leqslant 10^4$ so that H falls under Case 1. More precisely, we assume we are given a generating set for H. We assume also that we have given subgroups $H_1, \ldots, H_r$ of H, also by means of generating sets. Let Ω_i be the set of right cosets of H_i in H and let

$$\Omega = \bigcup_{i=1}^{r} \Omega_i.$$

Elements of H act on Ω by right multiplication and the orbits of the group induced by H on Ω are just the Ω_i.

It is important to observe that the action of H on Ω is effectively defined in the sense that we can perform calculations involving it. Elements of H are normally represented by words in the generators of H. If W is a word in these generators, then $\bar{W}$ will denote the element of H defined by W. An element of Ω is represented by a pair (i,W), where $1 \leqslant i \leqslant r$ and W is a word in the generators of H. The pair (i,W) represents the coset $H_i\bar{W}$. A given point in Ω has many representations. Thus our first problem is to be able to decide when (i,W) and (j,V) represent the same point. Clearly this happens precisely when $i = j$ and $\bar{W}\bar{V}^{-1}$ is in H_i. By Procedure A we can multiply out $\bar{W}$ and $\bar{V}$ and by Procedure D we can decide whether $\bar{W}\bar{V}^{-1}$ is in H_i. Thus we can effectively distinguish points in Ω. We can also apply elements of H to points in Ω. The image of the point represented by (i,W) under the element $\bar{U}$ of H is represented by (i,WU).

The set Ω may be very large. For example, in constructing the Baby Monster we used this method with $n = 3510$, $H = \text{Aut}(M(22))$, $r = 34$ and $|\Omega|$ about 14×10^9.

It would, of course, be possible to describe the action of H on Ω by numbering the points $\alpha_1,\alpha_2,\ldots,\alpha_m$, where $m = |\Omega|$, and for each generator x of H constructing a table whose ith entry is j, where $(\alpha_i)^x = \alpha_j$. This has the advantage that each point has a unique description, namely its number, but it has the disadvantage that the tables require an impossibly large amount of storage space. Our description of points in Ω by pairs (i,W) makes it so easy to apply an element of H that it more than makes up for the added difficulty of having many names for one point.

The group $\bar{H}$ induced on Ω by H is, abstractly, just a homomorphic image of H. If there were no other method of defining permutations on large sets, then there would be no point in considering this approach. However, there is another method. Let X be any subgroup of the symmetric group on Ω. For the time being, X need not have any relation to $\bar{H}$. Let the orbits of X be $\Gamma_1,\ldots,\Gamma_s$ and let γ_j be a point in Γ_j. In addition, let z be a permutation on Ω which normalizes X, that is, $z^{-1}Xz = X$. The map σ of X taking x to $x^\sigma = z^{-1}xz$ is an automorphism of X. The permutation z is uniquely determined by σ together with the points $\delta_j = \gamma_j^z$. To see this, we note that any point in Ω has the form γ_j^x for some j and some x in X. Then

$$(*) \qquad (\gamma_j^x)^z = \gamma_j^{zz^{-1}xz} = (\gamma_j^z)^{x^\sigma} = \delta_j^{x^\sigma}.$$

Therefore $(\gamma_j^x)^z$ is determined by δ_j and σ.

It is not true that for every choice of σ in $\text{Aut}(X)$ and points δ_j, $1 \leqslant j \leqslant s$, there corresponds a permutation z. However, the simple necessary and sufficient condition for z to exist is given by the following lemma:

LEMMA. *If σ is in Aut(X), if $\delta_1,\ldots,\delta_s$ are in distinct orbits of X and if*

$$(X_{\gamma_j})^\sigma = X_{\delta_j}, \qquad 1 \leqslant j \leqslant s,$$

then there is a unique permutation z of Ω normalizing X, inducing the automorphism σ and mapping γ_j to δ_j, $1 \leqslant j \leqslant s$.

The proof of the lemma is very easy and may be found in [8].

In the construction of groups like the Baby Monster we select a subgroup X of the group H and apply the lemma to $\bar{X}$, the group induced on Ω by X. Our first step must be to determine the orbits of $\bar{X}$. The orbits of $\bar{H}$ are just the sets Ω_i. We can find the orbits of $\bar{X}$ if we assume that X is a "large" subgroup of H, say $|H:X| \leqslant 200\,000$. The orbits of $\bar{X}$ on Ω_i correspond naturally to the double coset H_iuX in H. In fact the orbit of $\bar{X}$ containing the element H_iu of Ω_i is the set of right cosets of H_i in H_iuX. If we take the inverses of all the elements in H_iuX, then the result is the double coset $Xu^{-1}H_i$, which by the preceding argument corresponds to an orbit of H_i acting on the set of right cosets of X in H. Since $|H:X|$ has been assumed to be at most 200 000, the orbit computation procedures mentioned above can be used. This leads to a determination of the orbits of $\bar{X}$ and a choice of orbit representatives γ_i.

To apply the lemma we need to make a choice of σ in $\mathrm{Aut}(\bar{X})$ and of points $\delta_1,\ldots,\delta_s$. The way this is done depends heavily on what is known about the structure of the group whose existence it is desired to establish. (Of course, the same is true concerning the choice of the group H and the subgroups H_i and X.) In some cases several choices must be made before the correct one is found. For a given choice we must check the conditions of the lemma. The condition that points $\delta_1,\ldots,\delta_s$ lie in different orbits of $\bar{X}$ is verified using the action of H on the cosets of X and the double coset trick described earlier. The condition

$$(X_{\gamma_j})^\sigma = X_{\delta_j}$$

is verified using the representation of H on $\Delta = \{1,\ldots,n\}$, which we assumed to be of manageable size.

Once the conditions of the lemma have been checked, we can assert the existence of a uniquely defined z. It is important to remark that we can actually compute the image under z of a point in Ω described by a pair (i,W). We do this by writing the point in the form γ_j^x with x in X and using formula (*). This is accomplished with another use of the double coset trick.

Having now constructed the set Ω, the group $\bar{H}$ and the permutation z, we define G to be $\langle \bar{H},z\rangle$. It is now up to us to show that G has the order and the structure of the group we are trying to construct. As mentioned above, we show that G has the right order by finding and verifying short relations satisfied by the generators for G. Relations involving only elements of H can be verified in the representation of H on Δ. The difficult relations are those involving z. Quite often the first relation to be checked is $z^2 = 1$. It is possible to imagine taking each point α in Ω, computing $(\alpha^z)^z$ and showing that the result is equal to α. Unfortunately, the entire annual budget of the National Science Foundation would not be enough to purchase the computer time necessary to perform this computation in the case of the Baby Monster.

There is in fact a much simpler way to verify that $z^2 = 1$. Suppose we show that σ^2 is the identity automorphism of $\bar{X}$, which is easy to do, and also that z^2 fixes each γ_j. Since z^2 normalizes $\bar{X}$ and induces σ^2, z^2 is determined by the points $\gamma_j^{z^2}$. But if $\sigma^2 = 1$ and $\gamma_j^{z^2} = \gamma_j$ for all j, then z^2 must be the identity. Other relations are more difficult to verify but for the Baby Monster no relation had to be applied to more than 200 000 of the 14×10^9 points in Ω.

Once the order of G has been determined, it is still necessary to show that G is simple and that G satisfies whatever additional conditions originally led to the evidence for the existence of the group. This is usually quite straightforward.

References

1. Dixon, J. D., The probability of generating the symmetric group, *Math. Z.*, **110**(1969) 199–205.
2. Higman, D. G. and Sims, C. C., A simple group of order 44,352,000, *Math. Z.*, **105**(1968) 110–113.
3. Leon, J. S., On an algorithm for finding a base and strong generating set for a group given by generating permutations (to appear).
4. Leon, J. S. and Sims, C. C., The existence and uniqueness of a simple group generated by {3,4}-transpositions, *Bull. Amer. Math. Soc.*, **83**(1977) 1039–1040.
5. O'Nan, M. E., Some evidence for the existence of a new simple group, *Proc. London Math. Soc.* (3), **32**(1976) 421–479.
6. Sims, C. C., "The existence and uniqueness of Lyons' group", Finite Groups '72, Proc. Gainesville Conf., 1972, 138–141 (North-Holland Mathematics Studies 7, North-Holland Amsterdam, London; American Elsevier, New York; 1973).
7. Sims, C. C., "Some group-theoretic algorithms", Topics in Algebra, Proc. Canberra, 1978 (Lecture Notes in Mathematics 697, 108–124, Springer-Verlag, Berlin, 1978).
8. Sims, C. C., "A method for constructing a group from a subgroup", Topics in Algebra Proc. Canberra, 1978 (Lecture Notes in Mathematics 697, 125–136, Springer-Verlag, Berlin, 1978).